OIL AND THE ROMANIAN STATE

OIL AND THE ROMANIAN STATE

MAURICE PEARTON

OXFORD
AT THE CLARENDON PRESS
1971

Oxford University Press, Ely House, London W. 1

GLASGOW NEW YORK TORONTO MELBOURNE WELLINGTON
CAPE TOWN SALISBURY IBADAN NAIROBI DAR ES SALAAM LUSAKA ADDIS ABABA
BOMBAY CALCUTTA MADRAS KARACHI LAHORE DACCA
KUALA LUMPUR SINGAPORE HONG KONG TOKYO

PRINTED IN GREAT BRITAIN

Preface

IN the hundred years since the international oil industry began, a controlling factor in its development has been that, except in the United States, its raw material has been drawn from areas which by the standards of the operators were politically and economically under-developed. In these areas—the Caucasus, Romania, Burma, Netherlands East Indies, the Middle East, Mexico, Venezuela, the Sahara—the existing structure of society and way of life did not provide for the concepts and practices of the industry. The problem for the operator therefore was to identify some authority on which to establish a clear title to prospect for and work oil deposits. This was true in varying degrees of the areas named. In some of them the problem was mitigated by the existence of colonial rule, though not necessarily eliminated, since officials, even of the same nationality as the operators, did not necessarily sympathize with their aims and problems. Nevertheless the authority, whether derived from a colonial regime or not, has not remained static in form or in ideology but has tended to become more complex—a process marked by the development of political systems and of nationalist rather than tribal or other group loyalties. This circumstance has been a source of conflict between the emerging state and the oil operators, which originally arose over exploration and production and has latterly extended to manufacturing and distribution. The specific issues are between political and economic necessities as defined by the host government and the right of the external supplier of capital and technique to decide the way in which operations should be carried on and to receive any resultant profits. In this context nationalism is compounded of a number of ingredients: it implies not merely hostility to foreigners or the hatred of colonial rule but also reflects a desire to divert all the profits from oil extraction and processing to domestic purposes, either individual or public, the need to create jobs (and the discovery, as educational programmes are carried through, that foreign labour is not indispensable) and the growing realization of the importance of energy in the development of the economy.

Ironically, all these are stimulated by the success of the original foreign investment.

There is as yet no internationally agreed framework for resolving these conflicts. International law can offer some help in limited circumstances but is completely unenforceable. Diplomatic action is a cumbersome method and from the oil operator's point of view involves the likelihood of losing sight of the main issues in the subtle maze of diplomacy. The only solvent of disputes so far is the entirely *ad hoc* application of self-interest—i.e. a sense of the advantage of obeying the particular conventions about the rights and obligations of lenders and borrowers developed in the course of international investment in order to enjoy the other advantages of membership of an international community of nations. The restrictions written into U.S. foreign-assistance legislation since 1962[1] and the World Bank rules[2] can sharpen this realization, but if a government is sufficiently resolute or unscrupulous in pursuing its own purposes there can be at present no solution. In these circumstances, the point of departure for the investor is the point when the cost of defending the investment exceeds the long-term rate of return. Getting out, short of merely abandoning the investment, is a difficult operation raising problems of compensation—when the host government chooses to recognize them—and valuation.

Romania between 1895 and 1948 offers an interesting case-study in these attitudes and conflicts: it has, however, a number of features distinct from those generally advanced above. At the outset Romania, although underdeveloped, had an established economic structure based on the production and export of cereals. Politically, the kingdom was a recognizable part of the European state system: it therefore already had a series of developed diplomatic and financial relationships with the major European powers. Articulate nationalism already existed in political life and was a determinant of attitudes towards foreigners—being extremely sensitive to the basic political reality that the foreign support which underwrote Romania's independence was necessary to its

[1] See Public Law 87–565 of 1 Aug. 1962, Section 620 (e), amending the Foreign Assistance Act of 1961, and subsequent amendments.

[2] See *Convention on the Settlement of Investment Disputes between States and Nationals of other States* (Cmd. 2745). Neither the Public Law nor the Convention disputes the right to nationalize, but stresses the concomitant requirement for prompt, adequate, and effective compensation.

continued existence. The formulation of a nationalist policy for the country's oil resources followed (1900–11), even though political circumstances did not enable the programme to be carried out until after the First World War. This programme, which necessarily implied denying foreign nationals the same rights as Romanians, was soon hamstrung by technical factors arising from the nature of the industry itself. The success of Romania's oil industry was demonstrated to be related not only to investment—the terms of which the State could control—but also to the operations of the international concerns—which it could not—and in Romania the technically advanced and internationally connected units were foreign-owned. This was the crux of the problem. Romanian nationalist politicians and officials tended to view it essentially as a matter of finance and the creation of jobs, both subsumed under a passionate assertion of sovereignty. Conversely, the foreign operators had to accommodate themselves to the Romanian view of what sovereignty entailed, but were able to demonstrate its practical limits outside Romania both in oil and in finance. The Romanian Government's general need for funds encouraged it eventually and reluctantly to obey the accepted rules of international investment. An equitable and workable arrangement assuring their legitimate rights to both the State and the industry was worked out only on the eve of the Second World War, after which the Romanian Government was no longer dependent on its traditional sources for either technique or finance. When these considerations no longer mattered, the problem of relationships between the industry and the State was solved by violence; shareholders' governments were held off, managements isolated, and workers intimidated. This took place, however, in a particular situation when the whole social and political order which made the original investment possible was destroyed—along with the beneficiaries of nationalist oil policies. At this stage, for the foreign oil interests, economic calculations as to the cost of protecting the investment were irrelevant; the decision whether to stay or to go rested on both sides with government.

It should be remembered that during the period covered by this study the international oil industry, like the Romanian State, was in a particular phase of its own development, which belongs to history. The differences between the industry then and now, in all phases from ownership to retail marketing, are not merely ones of scale but also of complexity and character. Since the Second World War the demand for oil—in a multitude of forms—has grown prodigiously. This expansion has coincided with rapid and virtually universal political change, often of the most fundamental kind. The impact on the international oil business has been too large and complex for summary here, but the key to its analysis, as demonstrated by Mr. J. C. Hartshorn in *Oil Companies and Governments*, is the greater willingness which all governments display to intervene actively in economic life. Before the war politics could never be wholly ignored by the companies: it was one of the factors to be taken into account and could, as in Mexico, where the foreign interests were expropriated in 1938, come to dominate the scene. Nevertheless, for the most part companies could look on their problems as basically both commercial in character and for them—and not government—to solve. Now governments dictate the terms of much of the industry's activities: few problems, even of those which are irrefutably commercial, lack a political aspect, and, in general, the tendency is for the politics rather than the economics to cast the longer shadow. There are signs that appreciation of the costs and risks involved is having a moderating influence, requiring governments to pay more heed to the economics of intervention in the oil business than they have shown hitherto. There is no sign whatever that companies will regain the freedom of manœuvre and decision they once possessed. The shift in initiative between the private operators and government emerges as one of the subsidiary themes of this study: in the attempt to control day-to-day activity, as well as to determine the larger legal and constitutional issues, Romania was a pioneer.

The study which follows is principally concerned with the development of the Romanian oil industry in the particular context outlined above; it is not an economic history of Romanian oil, although economic issues necessarily enter into the argument to indicate both the nature of the industry and particular bones of contention. Similarly, relationships between companies are discussed only to the extent that they illustrate some aspect or other of the over-all relationship between the industry and the State. Technical developments are merely sketched in for the same reasons. The natural gas industry is largely excluded since, from their acquisition by Romania, the sources were under state control and the conflicts resulting from the prior existence of private investment did not therefore arise.

The text was completed in 1967 and takes no account of materials published after that date. It is based on the archives of a number of governments and oil companies and the private records and recollections of individuals who worked in and with Romania during the period covered. Comparable materials from Romanian political life were not available during the time of writing; the archives are now open only till 1918. In consequence one cannot yet establish the inter-play of attitudes in the same way as has been possible in regard to the other countries. In spite of this restricted evidence of motive, enough can be established from available sources without doing essential injustice to the Romanian side of the argument.

Since this book is written for an English-reading public the conventionally known forms of names have been observed: Bessarabia, not Basarabia, Moldavia, rather than Moldova. The same regard for convention prompts the use of Michael instead of Mihai and, contrarywise, Carol in preference to Charles. Where usage does not suggest a particular form, the Romanian has been adopted.

I am deeply indebted to Professor D. C. Coleman of London University and to Mr. Alexander Miller for their detailed advice and assistance throughout this study. I also wish to acknowledge help in various forms received from Mr. F. Bercu, the late H. Bişov, the late Dr. F. A. R. Bruell, the late J. C. Byrnell, Mr. J. N. Cooper, Mr. F. W. D. Deakin, Mr. P. M. Eden, the late M. M. van Kralingen, Mr. C. G. Pearton, Mr. A. F. Peters, the late Sir Albert Stern, Miss Eva Takacs, Mr. E. H. E. Tamchina, Mr. R. E. E. Tamchina, Mr. G. L. Tedder, and the late W. F. W. de Vos.

The following were so kind as to furnish materials or discuss specific issues with me: the late S. W. Barrett, Sir Eric Berthoud, Mr. R. J. Blair, Dr. Helmuth Böhme, Professor R. J. Forbes, Mr. M. Galpin, Mr. E. H. W. Gensch, Mr. E. Guest, Mr. Klaus Jacobs, Mr. J. W. Josephus Jitta, Colonel A. C. Kendall, Mr. H. C. McCarthy, Mr. D. McKellar, Mr. R. K. Middlemass, Miss V. E. G. M. Neagle, Mr. D. J. Peyton-Smith, Dr. M. Pizanty, Mr. W. Schlag, Professor Hugh Seton-Watson, Professor Fritz Stern, Mr. Marton Szabö, Mr. F. W. Simmons, Mr. A. Beeby Thompson, Mr. C. Thompson, Mr. D. C. Watt, Mr. R. R. A. Wheatley, Mr.

G. G. Whitaker, Miss M. D. Williams, and Mr. W. T. Wright. I am also grateful for assistance received from the officers of the Public Record Office, the British Library of Political and Economic Science, the Royal Institute of International Affairs, the Institute of Petroleum, the Weltwirtschafts-Archiv, Hamburg, and the Library of the Peace Palace, The Hague. I am grateful to Mr. R. E. Thompson for preparing the Index.

None of these persons or institutions holds any responsibility for the facts and matters discussed or the opinions advanced in the book.

Contents

List of Figures

List of Tables

List of Maps

Note on Sources

(i) Footnote references to unpublished sources are indicated by a number: if needed, full details should be requested of the author, c/o The Clarendon Press, Oxford.

(ii) References to published sources are shown initially in full, subsequently in contracted form: dates and places of publication are set out in the Bibliography.

(iii) Certain abbreviations have been used in the references, as follows:

AA	Auswärtiges Amt
BFSP	*British and Foreign State Papers*
CER	*Correspondance économique roumaine*
DBFP	*Documents on British Foreign Policy*
DGFP	*Documents on German Foreign Policy*
FRUS	*Foreign Relations of the United States*
MPR	*Moniteur du pétrole roumain*
PITR	*The Petroleum Industrial & Technical Review*
PR	*The Petroleum Review*
PT	*The Petroleum Times*

(iv) All references to "tons" are to metric tons.

(v) Statistics can be found in the publications of the Ministry of Industry and Commerce, notably *Anuarul Statistic al României*, and in the *MPR*. Those given in the Tables and Figures have been selected only to illustrate particular arguments in the text; the source is the *MPR* except where otherwise stated.

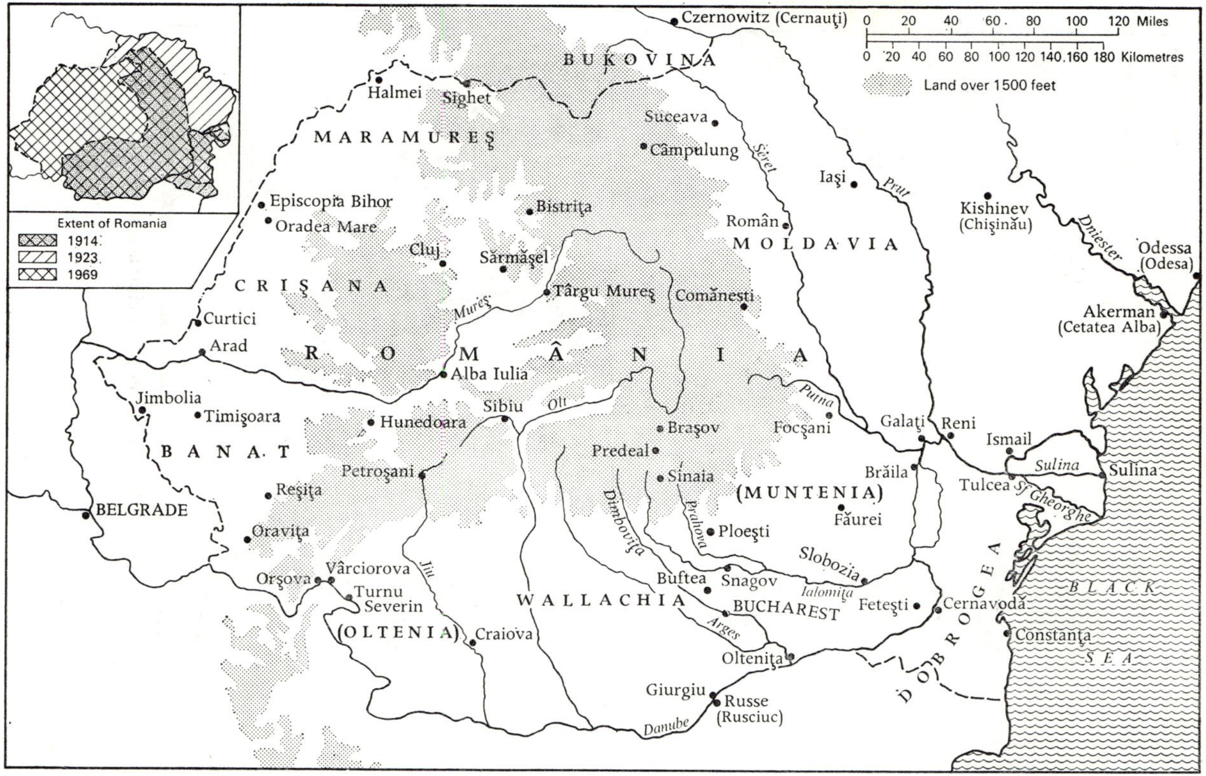

MAP I. Romania

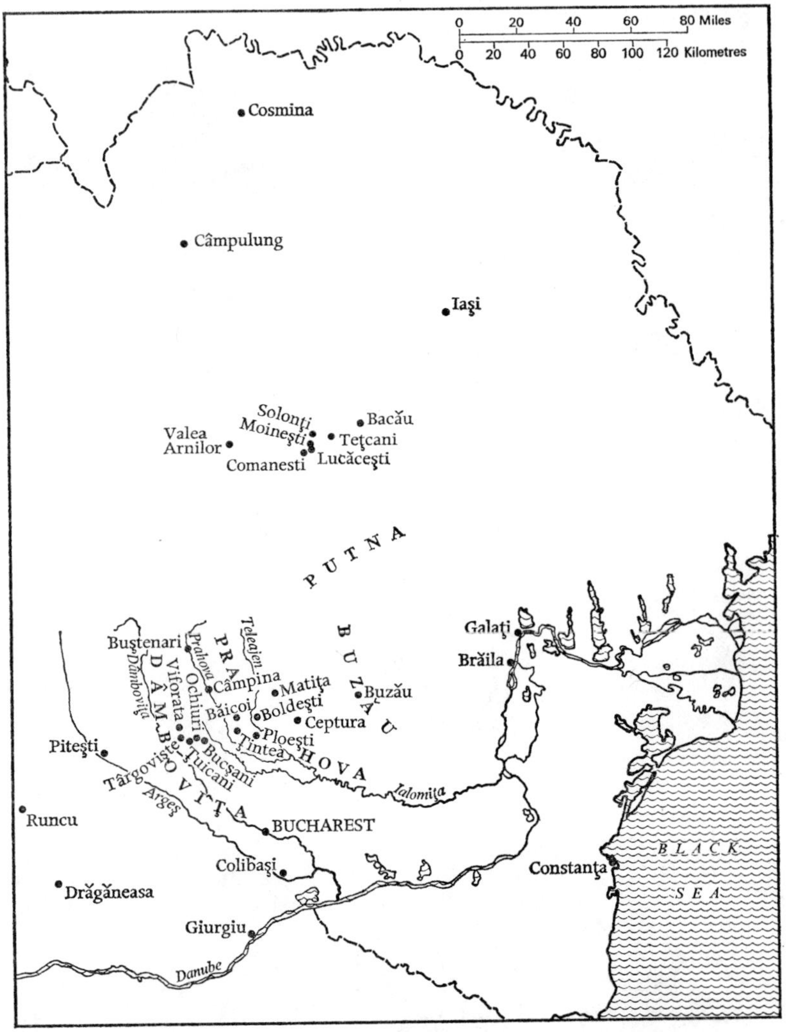

MAP 2. Romania: oil-producing areas

CHAPTER I

The Genesis of the Romanian Oil Industry

The beginnings of the world oil industry

THE manufacturing processes of the oil industry are determined by the physical fact that its raw material, crude oil, is, in its natural state, a mixture, the constituents of which can be separated because they boil at different temperatures. Although crude oil had been known and used since antiquity, the exploitation of the world's resources on a large scale depended on the strict analysis of its composition and developments in engineering and metallurgy. These were achieved only in the nineteenth century, during which research into the molecular structure of organic compounds,[1] and specifically into the nature of crude oil, established that it consisted of compounds of hydrogen and carbon (hydrocarbons)[2] in series. By 1890 all crudes were discovered to contain three major types of hydrocarbons, classified as paraffins, naphthenes, and aromatics. The varying proportions of each in any given crude determined its quality and, to some extent, the treatment necessary to make it marketable. Thus crude oil containing predominantly naphthenes required to be specially refined in order to yield kerosine of an equivalent quality to that obtainable by simple treatment from crude oil consisting mainly of paraffin hydrocarbons.

The processes and techniques by which the results of this progress in organic chemistry were commercially applied were borrowed: the common basic method of treating oil in the laboratory and in the refinery was distillation, a process, taken over from the production of alcohol, by which the component parts of a

[1] For the background of chemical analysis, see R. J. Forbes, "The Chemists and the Composition of Petroleum", in *More Studies in Early Petroleum History 1860–1880*.

[2] Crude oil also commonly contains impurities, mainly sulphur and traces of chlorine, oxygen, and nitrogen. For a discussion of the various series, see A. N. Sachanen, "Hydrocarbons in Petroleum", in *The Science of Petroleum*, vol. v, pt. I.

liquid mixture are separated by evaporation and condensation.[1] Similarly, the means of extraction, drilling, was adopted from quarrying and from boring for water and salt. The first drillings specifically for oil took place in Hanover in 1857, Ontario and Pennsylvania in 1859, Galicia, 1862, Romania, 1863, Alsace, 1870, and Baku, 1871. In all these areas, by reason of the greater depth which could be penetrated, drilling eventually displaced reliance on natural seepages and on the digging of pits or galleries as a means of exploiting resources of petroleum. Various systems were employed: their effectiveness depended largely on the type of strata to be penetrated. In America the drilling tools were generally suspended from a cable of rope or steel, which described a reciprocal movement. The Canadian system was basically the same except that the cable was replaced by poles; it came to be adopted in modified form in Galicia and Romania. Free-fall drilling was practised in Baku:[2] the essential feature of this technique was that the bit was raised and allowed to fall freely, at a rate of 20/40 strokes a minute. Towards the end of the century, rotary drilling made its appearance and eventually superseded all other forms. It incorporated the water-flush method, whereby water was forced through hollow rods to assist the action of the bit and carry upwards the disintegrated material, but, for the lack of suitable steels for bits, its use was at first restricted to soft geological formations.

The search for oil was usually concentrated in areas where seepages had already indicated the presence of oil deposits, or known beds of salt their likelihood.[3] As a result of geological observations in Burma (1855) and the United States (1861) it was

[1] When crude oil is distilled, the most volatile constituents are separated first, and condense as gasoline: they are followed by components with higher boiling points, yielding, in order, kerosine, gas oil, lubricants, waxes. The residue varies with the type of crude oil, and provides fuel oil and bitumen. For the development of the distillation process, see R. J. Forbes, *A Short History of the Art of Distillation*; Sir Boverton Redwood, *A Treatise on Petroleum*, vol. ii; and W. Miller and H. G. Osborn, "History and Development of Some Important Phases of Petroleum Refining in the United States", *The Science of Petroleum*, vol. ii.

[2] On developments in Baku, see Sir Boverton Redwood, "The Russian Petroleum Industry", *Journal of the Society of Chemical Industry*, vol. iv, no. 2 (28 Feb. 1885), and A. Beeby Thompson, *The Oilfields of Russia*.

[3] For a brief account of the development of geological prospecting methods, see E. de Golyer, "Historical Notes on the Development of the Technique of Prospecting for Petroleum", in *The Science of Petroleum*, vol. i.

demonstrated that deposits occurred in anticlinal structures. This correlation enabled prospecting to be carried on in areas where no obvious surface traces of oil were to be seen. The anticlinal theory was first systematically applied to the search for potential oil-bearing strata in 1883 and was gradually extended to embrace other geological formations with which oil was associated, such as domes, faults, monoclines, etc. It became the foundation of prospecting.[1]

Notwithstanding these developments, success in finding oil remained very much a matter of chance, and flush production, followed in many cases by the rapid exhaustion of the particular oil-bearing strata through wastage of underground gas pressure, characterized many early oilfields. This trend was encouraged in the U.S. by the legal arrangements governing extraction; oil deposits were held to belong absolutely to the surface owner, who felt himself compelled rapidly to exploit his subsoil resources which, as fugacious material, under the "rule of capture", would otherwise accrue to adjacent landowners or lessees.[2]

The technical improvements mentioned above in extracting oil from the ground and treating it necessitated improved methods of disposal both of crude oil to refining centres, where these were not situated on the fields, and of products to markets. Wagon-trains over existing roads proved inadequate. Teamsters' charges varied with distance and road conditions, and wide fluctuations in freights encouraged oil-producers to seek alternative means of transport, viz. rail tank-cars and pipelines. The first cars—merely wooden vats mounted on flat-topped trucks—appeared in America in 1865, but were followed three years later by specially constructed vehicles, the use of which extended to Russia and Germany in 1886. Pipeline development was inhibited by the considerable problem of leakage until in 1865 van Syckle successfully transported crude oil through a wrought-iron line over a distance of five miles from the producing field to a railway. This successful

[1] That is, of "surface" geology—the analysis of the substrata by geophysical methods took place only after 1924.

[2] The oil rush in Pennsylvania consequent upon the drilling of Drake's well is the first and classic example: for a graphic account, based on contemporary records and journalism, see P. H. Giddens, *The Birth of the Oil Industry*; Williamson and Daum, *The American Petroleum Industry, 1859–1899: The Age of Illumination*, pp. 375–7, and Appendix E: *The Legal Framework of Crude Oil Production* also discusses this aspect.

demonstration encouraged further investment in long-distance lines and, after the opposition of the railway companies had been overcome, the first trunk-line was constructed in 1874 from the producing fields to Pittsburgh—a distance of sixty miles. By 1900, 18,000 miles of pipeline were being operated in the United States. In Russia gathering-lines from fields to refineries were pioneered by Nobels, but the first long-distance line, completed in 1889, was laid merely to overcome the technical inefficiency of the Baku–Batum railway.

The high cost of barrels in relation to the price of kerosine in Baku occasioned the first tank steamer designed specifically for the transport of oil, which enabled the Russian oil trade to expand beyond the Caucasus and Persia to the main Russian centres—Moscow, Warsaw, and St. Petersburg—and eventually into the European oil market. Transatlantic traffic in oil began in 1861, but the problem of safe storage of the barrels and the difficulty in obtaining wharf-side labour to handle the cargo made this method of transport very expensive. Ultimately, barrels were replaced by bulk cargoes, initially in sailing ships and, after 1886, in specially designed tank-steamers, to serve those markets where sales justified the erection of bulk-storage and handling facilities. These marine developments emphasize the fact that from the first the oil business was primarily an export trade. Production was greatest in the United States, and from 1866 until the late 1880s, in spite of expanding internal demand, the greater part of United States kerosine output went abroad,[1] principally to Europe.

The economic incentive to the exploitation of petroleum resources on an industrial scale was provided by an increase in demand, following the industrialization of North America and north-west Europe, for a cheap illuminant and for lubricants. Traditionally these needs had been met by the use of whale- and rape-oil and tallow, but the depletion of the whale population and the need for longer voyages to new areas caused a rise in price sufficient to encourage the search for a cheaper substitute.[2]

[1] For statistics of exports, see Williamson and Daum, *The American Petroleum Industry, 1859–1899: The Age of Illumination*, Appendix D.

[2] In 1856, in the American market the higher-quality whale-oil was priced at U.S. $2.25 a gallon, rape-seed oil $1.50, kerosine $1; other substitutes varied from 63 to 87¢—see Kendall Beaton, "Dr. Gesner's Kerosene", section iv, *The Business History Review* (Mar. 1955).

Similarly, the mechanization of industry and transport increased the demand for lubricants[1] beyond the existing sources and at the same time created a demand for a lubricant which would not break down in the more exacting operating conditions of machinery.

Crude oil was found to yield products meeting these requirements: kerosine ousted whale-oil and other vegetable illuminants and remained a rival to gas and electricity as a source of light; petroleum-based lubricants virtually eliminated rival forms of lubricants. Refining operations therefore concentrated on the production of the kerosine fraction and the heavier distillate which, treated with sulphuric acid, yielded lubricating oils. All other products ran to waste or were burnt. The stress on kerosine production created a particular problem in Russia, since Baku crude had only a small yield of kerosine and left a large volume of residue. After 1870 the latter was generally pulverized with steam and was found to be suitable for burning in boilers. This new fuel (*mazout*) was used for Caspian steamers and Russian railways, and as an industrial fuel. Largely through difficulties of burner design, the introduction of fuel oil as a combustible into the United States took place only in the 1890s. After that period the invention of the internal combustion and compression-ignition engines created a demand for fractions hitherto wasted—gasoline and gas oil. In consequence, "cracking" entered the commercial manufacturing process for gasoline after 1912 and became an integral part of refining technique. In this process, large molecules are broken up at a high temperature into smaller molecules, often of a different structure. Originally discovered in connection with kerosine, cracking was applied to the production of gasoline when it became clear that reliance on existing methods for meeting the expanding gasoline market for automotive vehicles would involve the creation of an undisposable surplus of heavy residual products.[2]

By this stage, the basic output of the refining industry comprised, in descending order of volatility, gasoline, kerosine, gas oil, and residuals used as fuel oil and as a feedstock for lubricants and greases. The products of distillation (distillates) were then

[1] For the history of research into the problems of lubrication, see R. J. Forbes, "New By-Products for Burners, Bearings and Bitumens", in *More Studies*.

[2] On early cracking techniques, see B. T. Brooks, *et al.*, "The Preparation of Gasoline and Kerosine from Heavier Hydrocarbons", *Journal of Industrial and Engineering Chemistry*, no. 7 (1915), pp. 180 ff. On the history of the process, see J. L. Enos, *Petroleum Progress and Profits: a History of Process Innovation.*

treated with sulphuric acid and caustic soda solutions, to remove various impurities such as sulphides. The refining apparatus, therefore, necessarily included stills and heating boilers, condensers, storage for crude oil and refined products,[1] and agitators and filtration and settling tanks for the chemical treatment of distillates. There was at this time no specific vessel for cracking. A plentiful supply of relatively pure water was essential to refining, both for the production of steam and for washing distillates. From the outset, however, the refinery was a flexible factory in that, within the limits imposed by the type of crude input, the products could be varied by alterations in the manufacturing process. This characteristic enabled refiners to meet varying and fluctuating market requirements.

The first plants operated on batch production but the introduction commercially of continuous distillation processes[2] not only gave higher yields but also reduced unit costs, since the plant was in operation twenty-four hours a day. Continuous production demanded in theory a continuous market: the practical discrepancy between the two was represented technically by investment in tankage and commercially by the search for secure outlets. The former could only be a palliative, since storage was expensive and the costs necessarily burdened that of the current output. The latter, therefore, was a determining factor in the operation of large-scale plants with a continuous process. The production and refining of petroleum necessitated the employment of a small number of drilling and refining specialists, and unskilled labour for such work as digging conduits and preparing sites for drilling rigs—most of the latter being hired *ad hoc*. Once in production, wells needed only inspection and maintenance. Refineries' schedules were equally unexacting. There was no mass labour-force, with consequent management and social problems. The industry was (and has remained) essentially "capital-intensive". The capital needs of the industry therefore had to cover buying or leasing lands, oil extraction, field transport, refinery construction and operation, and distribution and transport. These were not wholly integrated activities, even in the United States, where Rockefeller began in refining and transport, integrated "forward" into the wholesale

[1] See Redwood, *Treatise*, vol. ii, section vi.

[2] Baku, 1883; Romania, 1897; U.S., 1903; see Redwood, *Treatise*, vol. ii, and Williamson and Daum, pp. 263–73.

and retail markets, but still acquired a large percentage of his crude oil on long-term contract from non-integrated producers.[1]

Romania's oil deposits and their exploitation

Romania's oil deposits were located along the southern and eastern slopes of the Carpathian mountains—where, before the development of oil on an industrial scale, the output of various seepages had long been used by the inhabitants for lighting and greasing and for medicinal purposes. These deposits were associated, in Wallachia, with Miocene and Pliocene rock-structures, compressed into anticlines and marked by irregular folds. Deposits in Moldavia were situated largely in the older Tertiary rocks and were, geologically, easier of access. The principal field was in the Prahova district but deposits were also worked in Dâmbovița and Buzău (Wallachia) and Bacău (Moldavia). The crude oil obtained varied in quality. Bușteni (Prahova) yielded a rich oil with an asphaltic base almost devoid of solid paraffins, which subsequently became the standard for marketing purposes:[2] in contrast, oil from Câmpina, also in the Prahova district and only ten miles from Bușteni, had a heavy paraffin base. Generally, however, Romanian crudes[3] were found to have a low sulphur content (seldom more than 0·4 per cent), and a high content of light, volatile aromatics and of naphthenes and paraffins—all of which presented problems in the distillation of the oil for the market. Because of the aromatics in the crude oil, Romanian kerosine could not successfully compete with Russian or American kerosine until 1909, when Edeleanu,[4] the celebrated Romanian petroleum chemist, patented the method of dissolving the aromatics in liquid sulphur dioxide. The naphthenes, with which

[1] See R. W. and M. E. Hidy, *Pioneering in Big Business*, Chap. 7.

[2] See Dr. S. Aisinman, "Bustenari Petroleum", *The Petroleum Review* (26 Oct. 1907) (a translation of a paper read at the World Petroleum Congress, Bucharest, 1907).

[3] On Romanian crude oils, see A. Saligny, "Roumanian Petroleums", *The Petroleum Industrial and Technical Review* (3 Mar. 1900), p. 139; (17 Mar. 1900), pp. 175–6; (14 Apr. 1900), pp. 234–5: also papers presented at the World Petroleum Congress, Paris, 1900, by C. Alimanestenu and Edeleanu, and by N. Concou, published in the special Congress supplement of *PITR* (25 Aug. 1900), pp. 5–9 and 22–6.

[4] See obituary notice, *Moniteur du pétrole romain*, no. 8 (1941), pp. 353–5.

were often combined naphthenic acids, also created difficulties, since the manufacture of lubricants necessarily left a residue with a high ash-content, which was a hindrance to its use as a fuel oil or bitumen.

Petroleum deposits were tapped by means of pits and conduits—a system which could be worked only seasonally and which was in operation long after its supersession in other areas, such as Galicia, and the introduction of machine drilling into Romania itself. The pits were dug by manual labour and measured from 3 ft. to 5 ft. in diameter. In Moldavia, where the oil-beds were not steeply inclined, the pits were round, and were lined with impermeable clay, bound in with planks and wicker-work. In Wallachia, where the shafts were steeper, round holes lined with wicker-work were inadequate to withstand the lateral pressure of the soil; hence holes were square and lined with timber. The "sinker", i.e. the digger, was attached to the surface by a line, and for his job wore leather skins and a tin helmet. Fresh air was supplied to him by a metal pipe, through which it was driven by a rotary fan or a pair of bellows. The latter, although more primitive, had the advantage that the possibility of death through asphyxiation when the fan was turned in the wrong direction was ruled out. Even so the maximum time in the working was only half an hour, and the pit method of extraction was responsible for many industrial accidents and fatalities through excess of gas, an inrush of oil, or sudden flooding by water. A British observer[1] commented (1899) "the workmen who dig [such wells] run a good deal of risk. There are many days on which they consider it most unlucky to work. It is no exaggeration to say that there are about two hundred such days in the year." When the first stage of digging had been completed, the oil and water that collected in the pit were removed by bailing or pumping,[2] which went on regardless of the occult influences that motivated diggers.

Drilling attempts in Romania in the 1860s and early 1870s were unsuccessful. The Pennsylvania system was introduced between 1880 and 1887, and was used on the estates of Prince Cantacuzino at Drăgăneasa in the Prahova valley, but through difficulties of lining the borings all the wells were unproductive,

[1] D. A. Sutherland, see below, p. 21.

[2] On pit extraction, see I. Tănăsescu and V. Tacit, *Exploatarea Petrolului in Rômania*, pp. 89–120.

except one. The output of this well, however, far exceeded any experienced in the country up to this time, and, in default of adequate storage facilities, great quantities of oil ran to waste.[1] Concurrently, similar results were obtained by other pioneers at Sărata (Buzău) and Solonţi and Moineşti (Bacău), where the Canadian system was introduced by Galician drillers. These developments vindicated technically the use of drilling in Romania.

The refining side of the industry was developed on the basis of the output of pits. The first refinery was constructed at Lucăceşti (Bacău) in 1840, and by 1880 total Romanian capacity was about 94,000 tons.[2] With odd exceptions, such as the plant of Hagienof and Câmpeanu at Cernavodă, the Romanian refining industry grew up on the oilfields, with the main centres in Ploeşti and Câmpina (Prahova). The Bacău output was refined at Valea Arinilor. This was in direct contrast to the American practice of having a central refining area to which crudes from different fields were transported. The absence of adequate transport facilities prevented the development of a seaboard refining centre at Constanţa, which became and remained principally concerned with handling refined products for export. The refining sector of the industry comprised a large number of small units. In 1898 a total crude oil production of 134,185 tons was treated in 72 refineries comprising 188 stills.[3] The field location and multiplicity of primitive plants reflected the absence of a Rockefeller or a Nobel in the industry's formative stages in Romania and its essentially domestic preoccupation.

Distribution of products was by ox-cart, until the construction of the railway system offered alternative means of transport. Even during the ox-cart era, Romanian illuminating oil was exported to Brassó, and Pest, and eventually to Marseilles, where it competed successfully for a brief period, with that from the United States and Russia. Exports, however, accounted for only about 15 per cent of the total crude output in 1894.[4]

[1] For a description of the Drăgăneasa operation, see ibid., pp. 121–32.

[2] For refining companies, refinery capacities, and capital investment, see table on p. 75 of Mihai Pizanty's *Le Pétrole en Roumanie.*

[3] See abstracts of "Statistical Report of Ministry of Agriculture, Industry and Commerce for 1898", in *PITR* (1899), pp. 603–4, 631, 659.

[4] See Acting Consul-General Bennett, *Report on the Petroleum Industry of Roumania*, FO, 1896, Miscellaneous Series, no. 411, Dec. 1896, p. 4.

The emergence of the Romanian State

The development of Romania's oil resources took place during a period when the Principalities of Moldavia and Wallachia were achieving, in consequence largely of Great Power rivalry in the Balkans and Near East, first unity and then independence of the suzerainty exercised by the Turks since the fifteenth century.

From the time of Peter the Great onwards, Russia's forward policy south-westwards challenged Turkish rule in the Balkans and resulted in the establishment of certain rights in the Principalities. This advance threatened Austria's trading position along the Danube,[1] particularly after 1812, when Russia[2] became a riparian state controlling the northern part of the delta, and gained rights of free navigation and trade for merchant shipping. Further Russian activity against Turkey involved Great Britain, generally as part of "the Eastern Question", and, more specifically, through Russia's attempt to control the outlet from what increasingly was regarded on the London and Liverpool corn exchanges as an important source of supply. French interest in the Romanian problem was initially ideological. Paris provided the higher education of many emerging leaders of Romanian politics,[3] who successfully engaged the sympathy of Lamartine and, subsequently, Napoleon III[4] on behalf of the independence of their native country. With the Crimean War, the future of the Principalities became a primary question of international politics. By the Treaty of Paris, 1856, Russia's rights were abrogated in favour of a joint guarantee by the signatories of Moldavia and Wallachia, under Turkish

[1] For the part played by considerations of trade and navigation in the achievement of Romanian independence, see J. P. Chamberlin, *The Régime of International Rivers: Danube and Rhine.*

[2] On Russian policy, see C. and B. Jelavich, "The Danubian Principalities and Bulgaria under Russian Protectorship", in *Jahrbücher für Geschichte Osteuropas* (Oct. 1961), pp. 349–66.

[3] In 1839 the Society of Romanian Students was formed in Paris and was supported by donations from both Romanian and French sympathizers, among whom the most influential in their effect on public opinion were Michelet and Quinet: see J. C. Campbell, "The Influence of Western Political Thought in the Roumanian Principalities, 1828–1848", in *Journal of Central European Affairs* (Oct. 1944).

[4] For the question of unification generally and for the crucial role of Napoleon III, see W. G. East, *The Union of Moldavia and Wallachia 1859*; and T. W. Riker, *The Making of Roumania.*

suzerainty. From the post-war settlement there emerged in 1861 a unified regime for both Principalities. Formal independence was achieved seventeen years later in consequence of Romanian participation in the Russo-Turkish war of 1875–8, Romania's claims being recognized by the Treaty of Berlin.

Changes in Romania's international status were accompanied by equally radical changes in its internal situation. Until 1864 the social structure was feudal, with a small class of landowners (boiars) and a large peasantry. Trading and commercial activities were confined to Jews and Greeks in the towns, principally Bucharest, Galați, and Iași. Of a population, according to the 1859 census, of 4,425,000[1] only 10 per cent lived in towns: by 1890 the population had increased to 5,950,000, of which 18·9 per cent was urban. The achievement of unity involved the reorganization of the State: a constitution after the Belgian model was introduced, serfdom abolished, and the State itself acquired 2,500,000 hectares of land as a result of the suppression of the monasteries. The existing assemblies of Moldavia and Wallachia became a Parliament, comprising a Senate and a Chamber of Deputies. Parliament had about 300 members, elected on a restricted franchise expressed through electoral colleges. Changes in government took place under royal management[2] and commonly preceded elections, which were, in the expressive Romanian idiom, "made", and therefore invariably won, by the party newly installed in office. The crumbling of the resultant large majority through internal rivalries prepared the way for the next change to be initiated by the King, whose decision the Opposition attempted to influence by staging street demonstrations. Parliamentary sessions lasted from mid November to the beginning of May, with recesses for religious festivals. The bureaucracy was staffed by political patronage, and the creation of jobs remained one of the objects of administration. The Civil Service reciprocated by assisting its mentors, particularly at election times. Prefects and

[1] Quoted in G. Cioriceanu, *La Roumanie économique 1860–1915*, p. 46. Between 1859 and 1890 Romania had lost three districts of Bessarabia to Russia and had gained the Dobrogea from Bulgaria.

[2] In May 1907, the King complained to the British minister, apropos government changes, that he had been compelled to do what he had never done before, viz. to accept a Cabinet whose members he had not himself selected or at any rate approved: there had been no time. See Sir Conyingham Greene to Sir Edward Grey: Dispatch no. 30, 29 May 1907, *FO Series 371*, vol. 317.

sub-prefects helped to control the vote, and senior officials in departments saw to it that their staff got to the polls.[1]

The two main parties, the Conservative and the National Liberal, were both essentially "connections" built round personalties and both supported by the landowners. The Conservatives generally

> represented the old landed class with their interest and expectations centred mainly on the development of agriculture: and in Roumania they had on their side King Carol, who worked hard to make of the Crown domains model agricultural estates. They wanted industry, but through a gradual and natural growth: . . . they also disliked greatly, as Carol did, the rather "pushing" ways and temperament of the Liberals.[2]

Because they "wanted industry", the Conservatives were in principle prepared to make concessions to foreigners. Their legislative programme on their return to office in 1894 envisaged modifying provisions of the Commercial Code stipulating Romanian nationality qualifications for directors of enterprises, and those of the Constitution barring non-Romanians from owning land.[3] They eventually carried the Mining Law of 1895.[4]

The National Liberals were "liberal" only in the sense that they were the party of 1848, had expropriated monastic lands in 1864, and were interested in Romania's rapid development into a manufacturing and trading state. Their motives in so doing were wholly nationalist, and their liberalism was in the tradition of List rather than Adam Smith or Cobden. There is no formal evidence that List was adopted as a mentor in Romanian liberalism,[5] but his classic work specifically mentioned the Danube area in the discussion of its main theme, and it would be surprising if articulate Romanian politicians remained unaware of the fact, especially in view of the German cultural influence emanating from Carol I and the fact that Romanian economists and engineers tended to

[1] Dispatch of Townley, chargé d'affaires, Bucharest: no. 64, 28 Sept. 1894, *FO Series 104*, vol. 112.

[2] David Mitrany: letter to writer, 9 Mar. 1963.

[3] Dispatch by Townley: no. 94, 28 Nov. 1894 *FO Series, 104*, vol. 112, summarizing the King's Speech at the Opening of Parliament. These restrictions were designedly anti-Semitic but applied to all foreigners.

[4] See below, p. 18.

[5] Although Manoilescu, Liberal Minister of Industry and Commerce in 1931, echoed List's arguments on the shortcomings of what both took to be Adam Smith's ideas: see *CER*, no. 2 (1931), pp. 1–15.

study in Germany. Further, List's notions of the nation-state as the basic economic entity and the *manufacturing* nation-state as the only tolerable form were clearly congenial to the mind of Romanian Liberals. It was axiomatic in their political vocabulary that the interests of the State were "higher" than those of individuals. The programme of November 1882 defined the party's economic objectives as the "protection of our economic interests, not of those of others, and the economic emancipation of the country as a natural and logical consequence of its political emancipation".[1] Economic emancipation entailed national effort *prin noi înșine* ("through ourselves alone"), in the slogan coined by Dimitrie Sturdza. The Liberals explicitly rejected the international effort implied by the western Liberal theory of the division of labour. Foreign loans were necessary and constituted recognition of state sovereignty; their application was exclusively an internal matter: other investment was "colonization". Romania's resources should be developed by Romanians—particularly if they were Liberals.[2]

As, during the period of the Old Kingdom, 1866–1920, they were in office for a total of forty years, the Liberals had every opportunity to implement their ideas: they were responsible, *inter alia*, for the recovery of the railways for the State, the foundation of the National Bank,[3] the tobacco monopoly, the denunciation of the ten-year-old trade agreement with Austria-Hungary in 1886 and its replacement by a stiff protectionist tariff, and for the first law for the encouragement of national industry in 1887. In this way the State became the biggest entrepreneur.[4] In their public utterances the Liberals tended to be overtly hostile to foreign investment and enterprise on the terms then understood in the main creditor nations. The British Consul-General in Galați

[1] See F. O. Manoliu, *La Reconstruction économique et financière de la Roumaine et les partis politiques*, p. 148.

[2] For the results of this policy after the First World War, see below, pp. 103 et seq.

[3] This was managed in such a way that the shares were reserved to adherents of the Liberal Party even when they could only pay for them out of future dividends. Wealthy Conservatives were financially ostracized. See dispatches by W. A. Winter (Bucharest) to Granville: no. 72, 7 June 1880, *FO Series 104*, no. 15; and Townley to Grey: 3 Jan. 1912, *FO Series 371*.

[4] On this topic see N. Spulber, "The Role of the State in Economic Growth in Eastern Europe since 1860", in *The State and Economic Growth*, ed. H. G. J. Aitkin, pp. 255–86.

noted in his report for the year 1890[1] that "the tendency towards protection has become more marked every day and great exception is taken to the most-favoured-nation clause as hampering measures of retaliation when a foreign country imposes an import duty on corn", and that it had been officially decided that foreign chambers of commerce among foreign merchants in Romania were to be "neither recognised nor encouraged".

There also emerged in the domestic politics of the Old Kingdom various dissident groups, of which the most important, from the viewpoint of its permanent impact on Romanian politics, was the "Junimists", whose leaders included Petre Carp, Titu Maiorescu, and Alexander Marghiloman.[1] The party advocated an explicit programme of financial and agrarian reform and a close connection with Germany and Austria-Hungary. In the parliaments of the eighties the Junimists supported the Liberals on foreign policy but opposed them on their handling of domestic issues.

There were no wide differences between the parties on foreign policy: fear of Russian-sponsored Pan-Slav expansion in the Balkans united Romania's political leaders. The only question at issue was that of the external guarantor necessary to the maintenance of the kingdom's new sovereignty. British policy was already anti-Russian but was also committed to Ottoman Turkey, from which Romania was only lately emancipated. Romania itself was not central to British policy in the Levant. Ideological sympathies with France could not overcome the fact that France, by reason of the defeat of 1871, was militarily no longer of the first importance. From that period onwards, and certainly after the Congress of Berlin, Germany appeared the most efficacious guarantor on both dynastic and military grounds, and in 1883 an alliance was concluded with Germany and her partners in the Triple Alliance. The treaty was kept secret, being disclosed by the King successively to a few Romanian political leaders, as their arrival in office demanded. This is the most striking instance of a general practice, viz. that the King was his own foreign minister.[2] Carol I was also the creator and Commander-in-Chief of an army, based on compulsory military service, with a peacetime establishment of 45,600 men and a normal wartime force of 207,000: its artillery was supplied by Krupps.

[1] *F.D. 1891 Annual Series No. 837*, pp. 5 and 8.

[2] See R. W. Seton-Watson, *A History of the Roumanians*, Chap. XII, pp. 346–70.

The gradual abridgement of Turkish control also permitted Romania to enter into independent commercial relationships. Before the Treaty of Adrianople, 1829, Moldavia and Wallachia functioned as a granary for Constantinople. The opening, under the treaty, of Romanian harbours to foreign shipping and the foundation in the same year of Die erste Donau-Dampfschiffahrts-gesellschaft (DDSG) operating along the Danube as far as Cernavodă,[1] linked Romania with Western and Central Europe and assisted in the reorientation of her foreign trade which took place in the 1860s. From the unification onwards, Turkey declined to third place (after Austria-Hungary and Great Britain) as a market for the agricultural products which were, and remained, Romania's largest single export item until after the First World War.

At this period any significant extension of foreign trade was inhibited by a number of internal factors—poor communications, the rudimentary nature of the credit system, lack of indigenous capital, and the diversity of currency in circulation. In consequence, Romanian governments sought to remedy these deficiencies by reforming the currency[2] and by contracting loans[3] for expenditure on public works. Between 1864 and 1875 450 million lei was raised—350 million in Germany, 72 million in France, and 33 million in Great Britain; 100 million derived from Romanian sources. Two-thirds of the total sum was expended on communications, the rest on covering budgetary deficits. It remained characteristic of Romanian finance that the State's receipts from the exports of

[1] Russian hostility to Austrian commerce was aimed at protecting Odessa, as against Galaţi and Brăila, and successfully prevented the DDSG from operating in the Danube delta. Passengers and freight were transhipped at Cernavodă to go overland to Constanţa. The setting up of the Danube Commission after the Crimean War put an end to Russian obstruction in the Delta until 1946. Nevertheless, a policy of low freight rates on Russian railways enabled Odessa to handle timber from Galicia and Bukovina which otherwise would have gone to Galaţi. (See *Trade and Commerce of Roumania 1909*, *FO*, May 1910, p. 28.)

[2] Since the right to mint money was regarded as a fundamental attribute of statehood, Turkish opposition considerably delayed the establishment of a new Romanian currency: this was not achieved until 1890, when the gold leu became the standard unit, supplanting the silver leu and various intermediate currencies, and being valued, until the First World War, at par with the Swiss and French francs.

[3] The first loan was authorized by the Law of 15 Jan. 1864; it was for 23 million francs at 7 per cent over 24 years and was negotiated on the London and Paris markets by the Ottoman Bank and Stern Brothers: on Romanian financial policy, see G. Cioriceanu, *La Dette publique de la Roumanie*.

grain had to meet current expenses and interests on loans; their scheduled repayment was usually made out of further loans. Thus the State's financial position rested on the vagaries of the harvest and the goodwill of foreign lenders.

The public works programme, however, strengthened the new kingdom internally. By 1875 5,420 km. of metalled roads had been completed and about 1,300 km. of railway were in operation. Romania had been linked with the main European system, westwards to Hungary through Vârciorova and northwards to Galicia through Român: Berlin and Vienna had been brought within two and a half days' journey from Bucharest. There was also a branch-line from Iași to link with the Russian railway to Kishinev (Chișinău) and Odessa. The Romanian Government bought out the shareholders, both native and foreign, in the various companies, and created a united state administration, the Căilor Ferate Române (C.F.R.) in 1880.[1]

The organization of credit was determined by the needs of agriculture, and the first public banks, founded either by the State or by private funds, served a limited clientele. Banking facilities for foreign trade were provided either by those private banks which developed out of money-lending, such as those of Marmarosch and Chrissoveloni, or by foreign capital, e.g. Anglo-Austrian in the Banque de Roumanie 1865,[2] or German, as in the case of Banca Generală Română, founded by the Disconto-Gesellschaft-Bleichroeder group in 1895.[3]

This banking investment was part of a general move by German and Austro-Hungarian capital into Romania, particularly in those sectors requiring technical expertise, e.g. foundries, manufacture of agricultural machinery, brewing, and oil. Investments in the latter were necessarily made on the basis of concessions, since

[1] The railway question became entangled with the attempts by the Powers to compel Romania to grant civic rights to Jews and exacerbated German–Romanian relations from 1867 to 1879. See reports of Degré, Romanian agent at Berlin, of 27 and 29 May 1877 in *Correspondance diplomatique roumaine sous le roi Charles Ier (1866–1880)*, ed. Nicolas Iorga; and N. M. Gelber, "The Intervention of German Jews at the Berlin Congress 1878", *Year Book V*, Leo Baeck Institute.

[2] Subsequently the Austrian participation was withdrawn and the Bank reorganized as the Bank of Roumania Ltd., with British capital. On the question of foreign-owned banks in Romania, see D. Kastris, *Les Capitaux étrangers dans la finance roumaine.*

[3] See dispatch by H. E. Browne (Second Secretary, British Legation Bucharest): no. 29, 1 Aug. 1897, *FO Series 104*, vol. 134.

Article 7 of the Romanian Constitution barred foreigners from holding landed property. The first attempts were uniformly unsuccessful. From 1879 onwards the Viennese banking house of Suchard and Co. financed drillings at Colibași (Dâmbovița), but the output was insufficient to meet the royalties involved, and the bank retired. Similarly, in 1887, the German company Hildebrand failed with drillings at Matița and Ploeşti. Austro-Hungarian capital met with its first unqualified success in 1895.[1] British and French experience of Romania's oil resources dated from 1868, but had been equally unfruitful, partly for technical reasons and partly through incompetent management.[2]

In monetary terms, the investment in Romanian oil during the period 1865–95 totalled only 25 million lei (£1·05 m.). Progress was retarded by the absence of an adequate land survey and the complexity of the system of land-holding, particularly in the peasant sector. In 1864 about a quarter of the land had been distributed to the peasants, whose individual holdings were subsequently subdivided by a law of succession providing for the equal partition of estates between all heirs. In the Wallachian oil area this process had reduced the average size of plots to about one acre. Further, in the peasant sector, there was also the *Moșneni*, or Elder's land, which passed down without any apparent title. The boundaries of such holdings were marked often by accidental features such as trees and stones. It was thus difficult in many cases to ascertain the real owner of petroliferous land and, in consequence, for an intending concessionnaire to get a good title, there being no documents to prove ownership or to define the extent of the holding. Furthermore, since for competitive reasons concessions were often taken out in a hurry, the complexity of the system of landownership meant that negotiations could be carried on with the wrong parties. When this was discovered, concessions

[1] See below, p. 23.

[2] e.g. the failure of Jackson Brown & Co., who in 1868 appeared at Ploeşti as the "Wallachian Petroleum Company" and invested 8 million francs in the acquisition of lands, subsequently discovered to be highly productive, in Băicoi, Țintea, and Buștenari. The company sank pits and, in order to enter the kerosine market, constructed a refinery at Brăila and organized wagon transport on a scale hitherto unprecedented. All these activities resulted in the expenditure of the entire capital of the company in four years, at which juncture the loss of oxen through disease and the destruction of the refinery by fire led to its dissolution. For the beginnings of investment of British capital in Romanian oil, see "Roumania: a review and retrospect", *PITR* (3 Mar. 1900).

taken out in good faith had to be cancelled. Connivance between peasants and intermediaries increased the hazards of investment.[1] The effect, as noted by C. M. Pleyte on his exploratory visit to the country for the Royal Dutch, even as late as 1906, was the proliferation of large numbers of small concessions with an insecure legal basis, making the country "an el Dorado for lawyers".[2]

The Mining Law of 1895

This chaotic situation was aggravated by the fact that the legal status of petroleum deposits and the right to exploit them varied considerably throughout the different provinces of Romania, and there was no over-all state policy with regard to intending concessionaires. A beginning was made in 1890, concerning the oil resources on state lands, i.e. those that had reverted to direct state ownership after the expropriation of 1864 and the dissolution of the monasteries. A single regime for the exploration and exploitation of all mineral deposits on both state and private lands was instituted by the Mining Law of 20 April 1895.[3] It was introduced by Carp, as Minister of Domains, in accordance with the party policy, announced at the opening of Parliament in November 1894, to extract other resources from the soil besides those provided by agriculture. The Law incorporated two features designed to remedy the existing difficulties: it established a distinction between surface and subsoil rights, and further provided (Article 4) that "throughout the whole territory of Romania, any person without distinction of nationality and without any other formality than authorization of the Minister of Domains, is free to devote himself to the discovery of mines, to the investigation and prospecting of the surface without causing its deterioration. This authorization is not necessary for the owner of the surface".

At the same time, the Law declared the interest of the State in the exploitation of mineral deposits by reserving (Article 5) the right of the State to demand of the owner of the land on which deposits were discovered whether or not, within a given time, he

[1] See memorandum of Consul Liddell on "The Petroleum Industry and its Abuses", enclosed in Dispatch no. 26, 7 Sept. 1897, *FO Series 104*, vol. 135.

[2] I.

[3] Text in *Lois, réglements, décrets concernant les mines et le pétrole*, ed. D. R. Rosetti, pp. 4–139. Regulations on the application of its provisions, ibid., pp. 143–207.

intended and was able to exploit them, and, in the case of a negative answer, to proceed to exploration and subsequent exploitation itself or to concede the mining rights to a third party. An exploitation right thus conceded lasted for seventy-five years, and was (Article 7) regarded under the category of real property, transferable, mortgageable, and distinct from the ownership of the surface. Similarly, concessions renounced by the concessionaire reverted to the State either for direct exploitation or for concession to a third party for the remainder of the original seventy-five-year term (Article 12).

These considerations established the relationship between the State and the individual owner of subsoil rights with regard to the exploitation of mineral deposits in general. Petroleum deposits were considered as a special type of mineral,[1] and by Article 65 their exploitation rested at the "full and entire disposal of the surface owner, provided that the exploitation does not imperil the safety of his neighbours".

In practice, this clause meant that the exploitation of petroleum was subjected to no prior or residual state direction, since Article 4 specifically ruled out the necessity to obtain the authorization of the Ministry of Domains in the case of a surface owner exploiting resources on his land. According to the Law, the State's interest in petroleum resources was confined to the dues and taxes stipulated and to the supervision or the relationship between a person or company engaged in the petroleum industry and the owner of neighbouring property and with employees (Articles 67–69). Furthermore, under Article 70, the exploitation of those lands the State held in its own right was submitted to the general terms prescribed by the Law for other surface owners, except that royalties, due to the State as landowner, were to be established by subsequent regulation.[2] Thus, apart from these minor rights retained in respect of its own property and its general concern for industrial welfare, the Romanian State played only a limited role in the exploitation of petroleum;[3] it levied taxes, and determined the

[1] They were not the subject of a specific *petroleum* law till 1942, see below, p. 239.

[2] Promulgated 10 May 1895: text in Rosetti, pp. 143–207. See also commentary by A. de Richard in *Loi sur les mines*.

[3] The mining of solid substances was attended by a greater degree of state supervision or control, and, in respect of deposits of salt, Articles 77–80 of the Mining Law ascribed to the State a complete monopoly; the State also exercised

size of exploration and exploitation areas and the scale of royalties. It did not, however, attempt to discriminate in favour of any particular class of owners or exploiters: it did not attempt to determine the organization of any exploiting company, and did not single out its own lands for any special regime.

Since rights to petroleum deposits were vested in the owners of the surface of the land in which those deposits occurred, much of the confusion resulting from the different systems of land-tenure was eliminated, although for reasons explained above (p. 17) the "surface owner" had to be carefully investigated by intending exploiters. Accordingly, a consolidation law was necessary to complement the Mining Law in this respect. Consolidation as established by the law of 8 May 1904[1] was a means of creating a legal basis for exploration and exploitation by the registration of the applicants' right to do so by a tribunal in the locality concerned, after an investigation of land-holdings by public inquiry. Once registered, the right was established against all comers.

Notwithstanding this lacuna and the delay in filling it, the terms of the Mining Law of 1895 were themselves sufficient to throw open Romania's petroleum resources for rapid development.

certain rights in the Dobrogea, where all mines were state property, but no petroleum development took place in that area during the period of the 1895 Law.

[1] See Rosetti: text, pp. 210–33; regulations, pp. 236–67. For *Exposé des motifs* see *MPR* no. 5 (1904), pp. 181–5. See also A. de Richard.

CHAPTER II

The Foreigners Arrive

The beginnings of corporate investment

THE technical and legal changes already described made possible the development of Romania's oil industry from primitive reliance on pits and ox-carts to a highly developed industrial undertaking whose operations became an important factor in Romanian economic life. But the introduction of the new techniques in all phases of the industry from exploitation to marketing demanded greater resources of capital than had hitherto been necessary.

These changes in cost were discussed by D. A. Sutherland in a report published in 1899 which is the first appraisal of the industry published in English by a technically qualified observer.[1] Sutherland noted that the cost (based on labour costs and the provision of well-lining materials, bailing equipment, and tanks) of a 150-metre pit was 6,000 francs[2], and of a 300-metre pit 15,000–18,000 francs. These costs were doubled for difficult strata. Maintenance costs of such wells he averaged at sixty francs a month, "including the wages of a driver and the keep of the unfortunate horse which is doomed to gyrate like the turnspit of a mediaeval kitchen".

Mechanical drilling could penetrate strata as far as 1,000 metres beneath the surface and open up wells which were more productive. It was, however, much more expensive—75,000 francs to a depth of 600 metres—on account of greatly increased primary and drilling costs, and the much longer time taken to complete a well. Drilling also demanded specialist labour, which in the absence of skilled Romanians was imported from Galicia.[3] Sutherland's figures are generally supported by Brackel and Leis in their discussion of Romania in the context of U.S.–Russian oil rivalry,[4] though no

[1] See *PITR* (1 Apr. 1899), pp. 95–101 and 104; (15 Apr. 1899), pp. 131–7 and 146.

[2] At this period francs 24 = £1.

[3] A government drilling school was opened at Câmpina in 1904.

[4] O. von Brackel and J. Leis, *Der dreissigjährige Petroleumkrieg*, pp. 257–60.

exact comparisons can be established. Nevertheless, from their study of Romanian sources (not available to the present writer) they arrive at an average combined extraction and refining cost of thirty francs a ton—"which in no other production area has been achieved as an average and which in itself secured competitiveness on the world market for Roumanian producers".

Capital was needed not only for the extraction itself but also for expenditure on marketing facilities—tankage, pipelines, railways in refineries, and oil docks, and so on—the cost of which related basically to the cost of iron and steel. Sutherland reported on pipelines which ran from fields to refineries or to railway loading-points, that "the cost in steel tubing of 3″ diameter is about 11·5 francs per metre, so that a line of 12 miles (the largest then in existence in Romania) inclusive of conical threads, freight and laying would cost about £8,000".

The advent of foreign capital and the controversy over state-owned oil lands

Capital of the order required for large-scale operation with the new techniques was not forthcoming from Romanian sources. Romania was not a money economy: agriculture was the predominant industry and the wealth of the boiars consisted not in cash or securities but in latifundia,[1] farmed to provide a wheat crop which earned a premium at Liverpool. Oil was regarded by the generality of landowners as a secondary 'crop', the importance of which was entirely overlooked. The outstanding exception in this respect was Prince Cantacuzino, who financed and supervised the successful venture on his estates at Drăgăneasa.[2] Public corporate enterprise was restricted by the relatively small number of companies and their limited capital basis.

Of the indigenous companies operating before the first Mining Law, one of the most important was the Societatea Română pentru Industria și Comertul Petrolului (Romanian Company for the Manufacture and Marketing of Oil Products) founded in 1889 by Romanian capitalists, with assistance from the Vienna

[1] Even when estates were sublet, metayage was widespread, particularly in Wallachia. On the prevalence of latifundia, see D. Mitrany, *The Land and the Peasant in Rumania*, pp. 255 et seq.: on wheat production, ibid., pp. 288–9.

[2] See above, p. 8.

banking house of Offenheim and Singer. The company rapidly became prominent in the inland petroleum trade—having refineries at Bucharest, Câmpina, and elsewhere, and possessing land in its own right in Moinești and Solinti. However, after five years' operations it ran into financial difficulties, which induced its directors to seek the help of the Hungarian Bank for Commerce and Industry of Pest, a creation of the Wiener Bankverein. The Hungarian bank financed a new Romanian company registered on 17 September 1895 as the Steaua Română S.A. The Bank also sought additional support on the London market, where the Romanian Oil Trust Limited was set up to bring in British finance as a guarantee of the Bank's advances to Steaua. Under the reorganization of the Societatea Română, Steaua Română acquired all its commercial and industrial assets and took over Societate Română's landed property at Câmpina on mortgage. Initially, Steaua had a capital of 2,400,000 lei: the problem for its management was "to develop a business originally adapted to the requirements of an inland trade into one capable of competing successfully in outside markets of Europe and Asia".[1] Accordingly it set up an integrated exploration, refining, and marketing business—the first in Romania. The organization's immediate progress was spectacular: increased drilling activity met with success and the company's production between 1896 and 1899 rose from 23,655 tons to 152,500 tons or from 29 per cent, to 61 per cent of the total for Romania. Dividends for these years were from 5 to 7½ per cent. These results demonstrated beyond question that Romania was a suitable field for oil development provided that capital and skills were forthcoming on a sufficient scale.

Capital arrived from the Netherlands from 1897 onwards, but the first group to draw the correct conclusions from the Steaua experience was that led by Frits Olie, who in 1899 invested the fortune derived from the East Indian tobacco trade in the Internationale Rumeensche Pet.-Mij. of Amsterdam. The company, for an outlay of 5 million francs, purchased properties at Buștenari and Gura Ocniței and an existing refining company, the Aurora, whose Băicoi plant was modernized and extended. Henceforward,

[1] Dr. Lagerwall, Managing Director, in his report to the second general meeting of shareholders of the Romanian Oil Trust 29 Dec. 1899, summarized in *PITR* (30 Dec. 1899), p. 718; (6 Jan. 1900), pp. 4 and 19.

Internationale concentrated on production and Aurora on refining and marketing.

During the same period, the United States and Germany first showed interest in Romanian oil. From 1895 onwards a series of representatives of Standard Oil Company of New Jersey appeared in the Prahova to inspect the oil workings and assess their possibilities, while in 1898 Richard Sorge, a noted German geologist, made a similar survey on behalf of the Disconto-Gesellschaft, the main supplier of credit to the Romanian government. Rockefeller was interested in Romania as a source of supply to offset the current American deficit against Russian output, and to strengthen his position in face of increased Russian competition in India and Central Europe. The motives of the Disconto-Gesellschaft differed, inasmuch as they had not hitherto made an investment in oil, but were concerned to establish a German-owned source of supply for the German market.

The opportunity to transform these investigations into investments was provided, not by the operational frailty of existing concerns, but by a financial crisis of the Romanian State itself. Since independence the financial basis of Romania had been almost wholly provided by the receipts from grain exports and by foreign loans. The disastrous harvests in the seasons 1895–9 curtailed exports and therefore the return to Romania, contributing to a deficit in the financial year 1898–9 of 74 million lei, or about one-third of the total annual budget.[1] In consequence the German banks, including the Disconto-Gesellschaft, refused further advances. At this juncture the Carp Government resorted to a policy of realizing the country's most valuable commercial assets—the petroleum-lands belonging to the State, and a concession for a pipeline from the oilfields to Constanța—to create a new sphere of activity and make the economy less dependent on the vagaries of the harvest.[2] The short-term financial difficulties were overcome by the direct personal intervention of the King, which

[1] On 24 Oct. 1899 Sir John Kennedy, the British minister, reported the King as stating, over lunch at Sinaia, that Romania needed £1 million for current expenses, even at 7–8 per cent, and that the forthcoming renewal of the Bank of Romania's concession might be used to bring pressure to bear on its directors to help in floating a loan in the U.K.: see dispatch in *FO Series 104*, vol. 139.

[2] This idea had been in the Junimist programme since 1894. See dispatch by Townley: no. 94, 28 Nov. 1894, *FO Series 104*, vol. 112.

induced the banks to grant the equivalent of £7,000,000 for five years[1] as a preliminary to the negotiation of a long-term loan.

In the developing financial crisis, the Carp Government commissioned Angel Saligny, Director-General of the C.F.R., to report on the possibilities of pipeline construction. In September 1899 he reported in favour of a line from Băicoi via Ploești-Buzău and Fetești to Cernavodă, where there was to be a Danube terminal, and then to Constanța—a total distance of 310 kilometres. The route was to run via Buzău in order to provide the greatest incentive to exploitation in Moldavia, hitherto impeded by lack of transport facilities. Alternatively, Saligny suggested, a line should be constructed direct from Băicoi to Giurgiu on the Danube—156 kilometres. The adoption of either would depend on the level of output and exports. Saligny's plans were based on the transport of kerosine, the loss to the C.F.R. in this respect being offset by increasing rail-transport of residuals. The cost of the Băicoi–Constanța line, including ancillaries, he estimated at 13 million francs, which should be met by concessioning the construction and operation to a private undertaking for twenty five years on a "common-carrier" basis, with a fixed uniform tariff and level of profit. On the expiry of the concession, the line would revert to the State.[2]

In March 1900 the Government announced the general conditions on which it was prepared to consider offers for the construction of a Constanța line: these included, apart from technical provisions, fixed transport and storage charges for users, and state supervision of the concessionaires' accounts. The attraction to the operator of a concession on state lands was twofold: the lands themselves were geologically favourable, and a concession offered an easily defined legal basis for operations—both necessary inducements to the profitable application of new techniques. The pipeline was complementary in that discoveries on state lands would create serious transport problems for the already strained resources of the C.F.R. and their solution was necessary to realizing the full advantages of any concession.

The first approach was made by the Disconto-Gesellschaft for

[1] See dispatches by Kennedy: 8 Nov., 9 Nov., and 1 Dec. 1899, *FO Series 104*, vol. 139. On the uses and its outcome see also report by C. S. Francis (U.S. minister), "Financial Conditions in Roumania", in *Foreign Relations of the United States 1902*, pp. 905–9.

[2] For text of the report (16 Sept. 1899), see *MPR*, no. 1 (1940), pp. 49–51.

a concession of 10,000 hectares on state petroliferous lands, comprising those areas favourably reported on by Richard Sorge, and for the construction of a pipeline to Constanţa alongside the existing railway. In return the Disconto offered a 10 per cent royalty on the crude oil produced and a rent of 30 lei per hectare per annum. At about the same time Standard Oil asked for a concession to construct a crude-oil pipeline from the main producing centres to Constanţa. The Romanian Government requested both parties to make a joint approach: negotiations took place during November and December 1900.

In the proposed agreement[1] the Romanian Government undertook to grant the Standard Oil Company for fifty years an exploitation-concession of about 15,000 hectares on state lands recognized as oil-bearing. The question of whether the work of exploitation was to be carried out by Standard Oil under its own name or as a Romanian joint-stock company was left open, but during the first eight years of the concession the exploiting company undertook to expend 7 million francs on drilling wells of 500 metres in depth in such a manner as to prospect the whole area (Article 1). Royalties for the crude oil extracted were estimated at 8 per cent rising to 10 per cent, and, in addition, certain specific taxes were agreed upon (Articles 2 and 3). Standard Oil was also to be granted an exclusive right for eight years to search for oil on state lands outside the recognized petroliferous area, against payment of a royalty (Article 4.)

Articles 5 to 18 of the draft convention concerned the concession of a pipeline. The line, which was to be used for the transport of crude oil from the state lands and from those belonging to private owners, was to run alongside the railway. It was expressly stated that every producer had the right to make use of the line for the transport of his oil. Standard undertook to construct all the necessary storage and pumping-stations, to which all producers had the right to establish branch lines. The company was to be permitted to import free of duty all machinery and accessories necessary to construction, including the pipe itself (Article 11).[2] The route was established (Article 12) as being from the region of Câmpina to Băicoi and hence to Constanţa, and it was further

[1] Text subsequently published in *MPR*, no. 9 (1904), pp. 273–6.

[2] This provision was in accordance with the Law for Encouragement of Industry of 1887.

provided that, if production in Moldavia increased so that the combined output of Moldavia and Wallachia reached 600,000 tons per annum, then Standard Oil would be bound to lay a second pipeline. The transit rates which the company would be permitted to charge other producers for transporting their oil through the pipeline were to be fixed by law. Standard undertook to pay a predetermined royalty to the State on the quantity of oil going through the line, amounting to 95 centimes a ton per 100 kilometres for the Câmpina-Constanța line and 70 centimes for the Moldavia line. Article 16 set out the conditions of government control of the working of the line and of the installations belonging to it, and provided for arbitration in case of dispute, with appeal to Romanian courts. Furthermore, it explicitly reserved to the State the right to inspect all the accounts and operational records of the company. The concession was to run for thirty years, after which the Government reserved the right to purchase the line outright. By Article 19 Standard Oil bound itself "to participate in the advance of the sum of 10 million francs which the Disconto-Gesellschaft makes to the Roumanian Government on the security of the revenues of the lands conceded".

The Standard Oil proposals presented the Liberals with an opportunity to stage that political crisis on which their chance of returning to office depended. Sturdza accused the Government of wishing to sell out Romania's vital interests: the scheme provided for the exploitation of Romanian oil to serve the interests of a rival producing state; this would be achieved by price manipulation by Standard Oil to ruin Romanian producers; the resultant monopoly would then deter other foreign investors. Against the Government, the Liberals urged that the State should not part with such important assets merely in order to satisfy temporary needs: even if it were decided to realize the state lands and pipeline concessions, this should be done only on the basis of tenders and not by bargaining between the State and particular interests. Opposition policy was not only publicly anti-Standard but also privately pro-German. In October 1900 Alimăneștianu, the Liberal responsible for industry within the Ministry of Agriculture, Industry, Commerce, and Domains, wrote to a German correspondent[1] stressing the identity of German and Romanian interests in oil matters and urging large-scale German help in realizing

[1] Professor Oebbeke of Munich, see *Petroleum* (15 June 1918), p. 725.

Romania's oil assets, which offered a convenient source of supply to combat Standard Oil in Germany. In the political clamour the attempts of the Standard representatives to state their case and to rebut the charges went unregarded,[1] and on 8 December 1900 the proposals were withdrawn.

Two days later the Disconto-Gesellschaft came forward with an alternative proposition,[2] undertaking to form a company with sufficient capital to exploit petroliferous land on a large scale—for which right they offered in return various tax and royalty payments and a loan of RM 6,480,000. At this stage, however, the political agitation was now against any concession of whatever kind and the negotiations lapsed. In February 1901 the Carp Government resigned, on failing to get a vote of confidence over its budget proposals, and was succeeded by the Liberals, Sturdza combining the offices of Prime Minister and Minister of Finance. His remedy for the financial crisis relied on retrenchment, good harvests, and the patience of Romania's creditors.[3] The new government was also assured of further immediate German financial support in addition to that negotiated by the King, in that at the beginning of 1902 the Disconto-Gesellschaft opened an unlimited account in favour of the Romanian Government at 5 per cent, pending the conclusion of a long-term loan.

It became an article of faith with Liberals, thereafter, that "the powerful and rapacious Standard Oil" had attempted to despoil Romania of her oil resources. The terms of the draft agreement and the declared intentions of the Government make it clear that, however that company might have behaved either then or previously elsewhere, it was prepared to agree to official supervision of its activity in Romania. Nevertheless, Liberal opinion then and thereafter consistently ignored this evidence of the company's willingness to accommodate itself to Romanian conditions, and official party policy continued to insist on the attempt at monopoly by Standard Oil.[4] The state-lands crisis became therefore a determinant

[1] See interview with *Moniteur des intérêts pétrolifères roumaines*, translated in *PITR* (24 Nov. 1900), p. 476.

[2] Text subsequently published in *MPR*, no. 10 (1904), pp. 303–5.

[3] On Sturdza's financial reforms see Kennedy, dispatch to F.O. 8 Mar. 1902: *FO Series 104*, vol. 151; also O. von Brackel, *Rumäniens Staats-Kredit in deutscher Beleuchtung*, Chap. IV.

[4] The Romanian verb *acaparere* implies 'grab' rather than 'monopolize', although customarily translated as such.

of political attitudes and was a major factor in the formulation of a specific nationalist programme for Romania's oil resources.[1]

The weakness of Standard's[2] position lay in its misunderstanding of the strength of nationalist feeling in a country which had within living memory been liberated from a long period of foreign rule, and its apparent disregard of the techniques of political change in Romania. The Liberals, having been out of office since April 1899, needed some big issue on which to base their agitation in the Parliament and in the streets to demonstrate to the King that the time had come for a change of government. Standard Oil presented them with one. No attempt appears to have been made to sound Liberal opinion—with or without appropriate *douceurs*—or to obtain some bipartisan support in Romania on such a fundamental issue. The maladroit nature of this approach was, presumably a result of its being Standard's first oil-producing (as distinct from refining or marketing) venture outside the United States and therefore set in unfamiliar territory. It was evidently supposed that a deal with a state involved only the same sort of commercial nexus as a deal with a private partner. The company enjoyed no diplomatic support. At the time, U.S. representations to Romania were made by an envoy based in Athens, whose bailiwick also included Greece and Bulgaria.[3] Standard Oil was put in a disadvantageous position simply by being American: among all Romanian political groups there was hostility to the United States on account of the U.S. Government's persistent representations about the treatment of Jews in Romania.

Both rebuffed undertakings returned on a private basis in 1903. In June of that year Standard Oil representatives toured the oil areas and inspected the properties of Telega and Steaua Română with a view to buying either or both of these companies. Such ideas, however, were abandoned in favour of a directly owned Standard enterprise, and in December 1903 G. F. Southard registered himself as a firm "for trading in all kinds of petroleum business".

[1] See below, p. 66.

[2] There is only an implied mention and no discussion of the state-lands controversy and Standard Oil's role in *Pioneering in Big Business* (see p. 516), which R. W. Hidy confirmed to the writer was due to the absence of company records.

[3] See J. B. Jackson's account of his reception, as reported to Hay, U.S. Secretary of State, on 9 Feb. 1903, in *Foreign Relations of the U.S. 1903*, pp. 699, 700, and 701.

The Liberal Government remained unwilling to see Standard in Romania but could not, without prejudice to the legal basis of the economy, deprive it of the right to commence operations on privately owned lands.[1] This did not apparently preclude Sturdza from using his position to hamper the new company. His attitude was summed up in a speech at Prahova made in June 1904: "Taking as a model the motto of Americans, 'America for Americans', I have as my motto 'Europe for Europeans' and especially 'Romania for Romanians'. . . . Guard yourself against the Standard Oil Trust and all who are in league with it."[2]

In June 1904 the Southard concern was taken over by Româno-Americană S.A., which was formally incorporated as a Romanian company with a capital of 2 million francs, entirely owned by the Standard Oil Company of New Jersey.[3] Registration was protracted since the competent tribunal rejected the application of the new company, which had to obtain a High Court decision in its favour before it could properly be constituted and operate. Româno-Americană announced its intention to build a refinery and concentrate its activities on the export market,[4] to explore deeper strata than hitherto—using for this purpose the Pennsylvanian drilling system—and to train and employ Romanian labour, wherever possible. Finally, Româno-Americană reiterated that it had no intention of applying for a concession on state lands. The basis of Româno-Americană's operations was land held at Cosmina (Prahova) by the European Petroleum Co.—a British concern whose primary oil interests were in Baku, but which had in Romania concessions of 400 hectares together with other assets, including a short pipeline and 10,000 tons of tankage. All these assets were taken on long lease by Româno-Americană against a royalty of 10 per cent on the crude output and a stipulated drilling programme. Româno-Americană also acquired from private owners lands at Comănești (Bacău) and Boldești (Prahova) totalling 4,600 hectares. The early history of the company was punctuated by lawsuits over failure to fulfil contracted drilling obligations. In April 1905 Româno-Americană increased its capital to 6 million francs, acquired tankage at Constanța, and began work on a refinery at Ploești, which went on stream in December of that year.

[1] See *MPR* for Mar. and Apr. 1904. [2] See *MPR*, no. 11 (1904), p. 333.

[3] Statutes in *MPR*, no. 12 (1904), pp. 425–30.

[4] The export market intended was Germany (see below, pp. 40–43).

The Disconto-Gesellschaft/Bleichroeder group also reappeared privately in Romania in 1903 by buying itself into existing concerns. In co-operation with the Internationale and Aurora, it assumed control of the Telega Oil Company,[1] a British enterprise with about 10 per cent of Romanian production, but whose operations were curtailed by insufficient capital and lack of a refinery. Later in the same year, the Disconto-Gesellschaft took a leading part in re-forming the Buştenari Company—a combination of small producers in the Buştenari district. The capital was increased to 10 million francs and the company was amalgamated with the existing National Pipeline Company, which had a capital of 500,000 francs and owned a line from Buştenari to Băicoi. The new resources enabled Buştenari[1] to initiate a policy of drilling for its output rather than of relying on hand-dug wells. Apart from reorganizing existing concerns, the Disconto-Gesellschaft/Bleichroeder group also extended its investment by founding, in December 1904, two new companies, the Vega, for refining, and Creditul Petrolifer, for transport and storage. A French concern, the Compagnie Internationale des Pétroles, contributed about one-third of Vega's capital. In both companies a minor shareholding was held by Romanian nationals.

During 1902 operational difficulties led to the control of Steaua Română passing from Austro-Hungarian into German hands. Steaua's policy of achieving the quickest possible increase in output tied up an excessive amount of capital in the oilfields. In particular, Steaua had on concession about 7,500 hectares, the technical advantages of which were to some extent offset on the financial side, since the land in question was held on leases, which in many cases stipulated an annual amount of development work. Such clauses imposed wasteful expenditure. Furthermore a decline in prices for crude and products, and the temporary elimination in 1902 of its market in the Balkans and Asia Minor[2] deprived the company of income which it had expected to use as working capital. Expensive technical innovations not only failed to yield the expected economies but suffered from teething troubles. In July 1901 Steaua had introduced electric power[3] into its opera-

[1] In Jan. 1907 these two producing interests were fused into the company Concordia: see F. Haase, *Die Erdöl-Interessen der Deutschen Bank und der Direction der Disconto-Gesellschaft in Rumänien*, p. 75.

[2] See below, p. 39.

[3] On the advantages of electricity, see article 'Electricity as a Motive Power

tions by an overhead-wire system radiating from a generating station at Sinaia but frequent breakdowns in the supply of current entailed continuous reliance on reserves of steam power. Similarly, the water-flush boring system was widely introduced on the company's lands in place of the Canadian pole system, but again technical difficulties caused unexpected expense. As a result Steaua Română's liabilities stood at 9 million francs on 30 April 1902. In November of that year the Hungarian Bank failed and the Steaua was necessarily involved. The Wiener Bankverein intervened to take control of the company during the liquidators' investigation of the Hungarian Bank. It soon appeared that, even if Steaua repaid advances made by the bank and provided for its own floating debt in Romania and for outstanding debentures, no assets would remain to represent the major part of the share capital. Hence, to satisfy the bank's liquidators and to keep the company going, new capital was necessary. Negotiations were opened with the Deutsche Bank, which agreed to salvage Steaua Română on terms involving the liquidation of the Romanian Oil Trust in London. In return for sufficient funds to enable the bank's liquidators to settle outstanding claims and also to keep Steaua Română in being, the Deutsche Bank assumed effective direction and control and raised its share capital to 17 million francs (£680,000).

The Deutsche Bank's intervention in Romania was prompted by developments in the German oil market, particularly with regard to kerosine,[1] but the reconstruction of Steaua Română led to the appointment (as its President) of the Bank's Director, Artur von Gwinner, who was an ardent supporter of the Baghdad Railway project, and who saw in the German-controlled company a source of fuel for the operation of the Balkan and Anatolian stretches of the line.[2]

The initiatives of the Disconto-Gesellschaft and the Deutsche Bank were followed by other German institutions. In July 1904 another powerful combination, the Dresdner Bank and the

in the Petroleum Fields', *PITR* (20 Apr. 1899), pp. 170–1: on its application to Romania see "The Electrical Installation at Câmpina, Roumania", ibid. (9 Dec. 1899), pp. 663–4.

[1] See below, p. 41. The Deutsche Bank also made similar investments in Galicia and Russia. On both the Disconto-Gesellschaft and Deutsche Bank ventures see F. Haase, *Die Erdoel-Interessen.*

[2] See below, p. 50.

Schaffhausen'scher Bankverein, entered the Romanian industry by founding the Câmpina-Moreni company, which, with a capital of 3 million francs, quickly acquired large areas in Moreni, hitherto largely unexplored, but soon discovered to be very productive. Involved in Câmpina Moreni were also the Henkel von Donnersmarck interests together with those of Anton Raky, founder of the Internationale Bohrgesellschaft, Erkelenz. The latter strengthened his position in Romanian oil in 1906, by founding the S.A. Română Raky as a drilling concern, in order to find use for his assets when the working of German mines was closed to private enterprise.[1]

The investment of the Royal Dutch in Romania effectively resulted from its competition with Standard Oil in the European gasoline trade.[2] The company originally featured (1903) as a purchaser of the gasoline output of another Netherlands company, Internationale. Subsequently, the possibility of a Standard monopoly through Româno-Americană of the Romanian gasoline output determined the Royal Dutch on direct investment. They adopted the same procedure as Standard; a pilot firm explored the possibilities and was then transformed into a Romanian limited company, Astra Română S.A.[3] The reasons, however, differed: Royal Dutch was concerned not to jeopardize its impending amalgamation with the Shell Transport & Trading Co.

By 1906, therefore, the foreign interests which were to assume such importance in the development of Romanian oil were all represented by subsidiary companies established for that purpose. The total capital investment was computed at 185 million francs, of which 74 million were of German origin.[4] This accession of

[1] On these, and other smaller German ventures, see F. Wirth, *Deutsche Arbeit und deutsches Kapital in der rumaenischen Erdoelindustrie*, 1927, pp. 45–8.

[2] The company had shown some interest in acquiring state lands in 1900, but this was dispelled by the controversy over the Standard–Disconto project: see F. C. Gerretson, *History of the Royal Dutch*, vol. ii, Chap. 5.

[3] For details of the development of the private firm of C. M. Pleyte into Astra Română, see Gerretson, vol. iii, Chap. 2.

[4]

German	74m.	Italian	15m.
French	31m.	U.S.	12·5m.
Dutch	22m.	Belgian	5m.
Romanian	16m.	U.K.	3m.

Miscellaneous 6·5m.

MPR, no. P. 261 (1907); see also Paul Schwarz, "Die Kapitalsinvestitionen in der rumänischen Petroleum-Industrie in ihrer geschichtlichen Entwicklung" (a paper read before the International Petroleum Congress, Bucharest, 1907).

German capital and the consequent reorganization of the various managements increased German influence in Romanian oil to an extent greater than the aggregate of German participations, when compared to the number belonging to other nationalities, would suggest. All the United States capital was in one company (Romåno-Americanǎ), while much of the rest of the capital was in minority holdings in various undertakings (e.g. British and Belgian in Telega, and French in Vega). The Romanian stake, although totalling some 16 millions, was spread over a large number of small undertakings or confined to minority interests in large ones. Furthermore, German capital was invested in those concerns with the exception of Romåno-Americanǎ, which were the market leaders in the industry.[1]

The advent of this foreign capital enabled Romanian oil to expand rapidly in all phases of activity. It implied easy access to foreign markets, both as a source of capital and as an outlet for products. The investment took place in response to market considerations, and specifically the preponderance of Standard Oil in the trade of Central Europe. Even the German Government, which by reason of its standing alliance with Romania had political reasons independent of Germany's oil needs for urging the Deutsche Bank to participate in Romanian oil, subscribed to these considerations. Nevertheless, the investment's foreign character left open the possibility of political complications with the governments of the shareholders concerned, particularly in view of the endemic nationalism of Romanian political life.[2]

[1] After the First World War this preponderance of German capital was widely regarded as a sinister manifestation of German imperialism: at the time even the Liberals thought of Berlin as their primary source of finance and technique.

[2] See below, p. 62.

CHAPTER III

Production, Trade, and Policy

Refining and marketing

THE application of new capital, both Romanian and foreign, resulted in increased exploration activity through all the known oil-bearing areas. The digging of pits diminished in favour of drilling wells (see Table 1), and the annual output of crude oil

TABLE 1

PITS AND WELLS DRILLED (AS AT 31 DECEMBER)

1902	577	136
1903	635	195
1904	745	220
1905	680	340
1906	701	425
1907	643	595
1908	583	681
1909	568	748
1910	547	819
1911	560	847
1912	488	850
1913	537	944
1914	562	987

increased rapidly, exceeding 1 million tons for the first time in 1907 (see Figure 1). Similarly, refinery construction by the new companies or combinations of companies extended the range and the capacity for manufacture. The principal plants were:

Company	Date of construction	Place	Initial distilling capacity (tons p.a.)
Steaua Română	1897	Câmpina	516,890
Vega	1905	Ploeşti	283,260
Aurora	1898	Băicoi	275,520
Romåno-Americană	1905	Teleajen	151,680
Colombia	1906	Ploeşti	143,720
Aquila	1904	Ploeşti	141,909
Aurora		Târgovişte	79,960
Astra Română	1908	Ploeşti	44,890

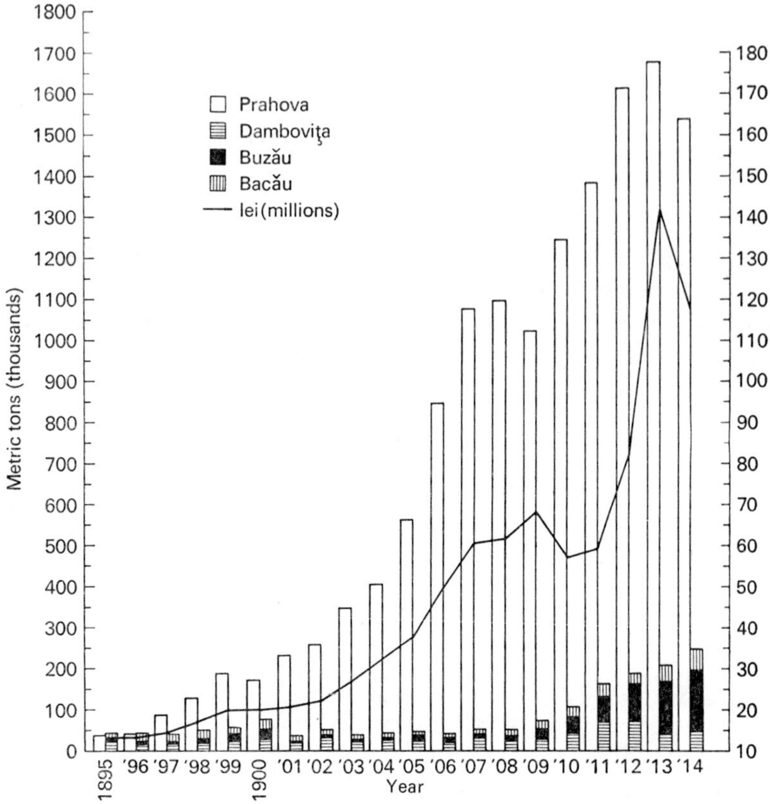

FIG. 1. Annual crude-oil production by areas.

The new plants, in addition to the then conventional distillation processes, provided for the manufacture of the higher-grade products, motor gasoline and kerosine, demanded by the Central and Western European oil markets. All lubricants were manufactured in refineries owned by foreign-capital companies. The growth in refining was superimposed on the existing pattern of manufacturing in the petroleum industry; the small refineries remained in being, working for the domestic market. The increase in over-all capacity put an end to the practice whereby small producers, by reason of their limited storage facilities, were forced to sell their crude oil virtually as it was produced, and enabled them better to withstand fluctuations in the price of their output by forward selling on contract to the large refiners. The latter used such supplies to make up any shortfall in their own crude output so that they were able to employ their capacity to the maximum extent and, if ever united, could enforce market leadership on their own terms.

For the purposes of a report[1] on the working capacity of Romanian refineries, in June–July 1908, the Government Inspector of Finances, Constantin Hălăceanu, distinguished three categories of refining, viz.:

Group I, comprising nine refineries of over 40,000 tons each.

Group II, comprising five refineries of between 10,000 and 40,000 tons each.

Group III, comprising forty refineries of under 10,000 tons each.

Subsequently, the numbers varied but the pattern remained. Until the end of the First World War, Group I was entirely "foreign"-owned and Groups II and III entirely Romanian. There were many different techniques used within this sector of the industry, but the broad difference between Group I and the others was that the former could carry out continuous distillation, and had better storage facilities and a more varied crude supply. These refineries were therefore better placed to manufacture to more exacting specifications, with the concomitant possibility of bigger unit returns.

The main products were kerosine (produced both for domestic consumption and for export as an illuminant), gasoline (also known as benzine, produced virtually for export for use in automotive vehicles and chemical manufacture), and residual fuel oil. The latter was a low-value product, and the possibilities for export were regulated by rail freights: residuals were, therefore, used within Romania as a fuel in manufacturing industry and for locomotives and ships. In response to the demand for higher-grade products, refining methods were developed to overcome the disadvantages in quality of the yields from the various Romanian crude oils. The first Edeleanu[2] plant in Romania to wash out the aromatic fractions in kerosine, which caused soot in lamps, went into operation at Astra's refinery in Ploești in 1913. Further, the demand for motor gasoline increased the use of crude oil from Băicoi and Câmpina, which were rich in gasoline fractions, e.g. Casin or Tețcani crudes, distilling 38 per cent–47 per cent

[1] See below, p. 53. [2] See above, p. 7.

gasoline at 150 °C. in contrast to Gura Ocniței or Sărata crudes, which yielded only 20 per cent.[1]

The market structure of the industry also reflected the division between "Romanian" and "foreign" capital. The domestic retail market provided the basis for the continued existence of small-scale concerns, as producers, refiners, or distributors (the latter comprising at least two sets of middlemen). They were and remained, however, particularly vulnerable to changes through their limited scale of operations, particularly as regards shortage of tankage. Thus, for example, as specifications became more stringent, they found themselves saddled with a type of crude oil commanding lower prices or with insufficient refining capacity to produce high-value products. In 1900, however, both these disadvantages were latent, and in spite of them the small-scale producers and refiners survived, largely through political pressure in general, and government price and transport control in particular, as beneficiaries of Romanian nationalist policies.

(*a*) *The export trade*

The export market provided the *raison d'être* of the foreign capital concerns, but prices could not be similarly controlled by the Romanian State, since they depended ultimately on the predominant U.S. contribution to the world oil trade—a position which was strengthened by the decline of Russian output in the years 1901–5 and of the Galician after 1908. In the United States the beginnings of production in the Gulf Coast area created a rival centre to Pennsylvania and Indiana[2] which was more favourably situated for the export trade, with the result that New York and Philadelphia were replaced as bases for shipment by the Gulf ports. In consequence of the predominance of the United States in the world export trade, quotations "f.o.b. U.S. Gulf" became the accepted international yardstick until about 1950. This was a matter not only of the proportion of export output in relation to other sources, but also of the structure of refining in the Gulf area.[3]

[1] As at this time gasoline engines had low compression ratios of 2½–3, the quality requirements, and hence the refining process, were not exacting: anti-knock aromatics, such as toluene, were *extracted*, the residual product being still adequate for automotive purposes.

[2] On the shift in American production, see H. F. Williamson *et al.*, *The American Petroleum Industry: The Age of Energy 1899–1959*, Chap. 2.

[3] See D. C. Hamilton, *Competition in Oil: the Gulf Refinery Market 1925–1950*.

During a period when oil movements in the international market outside the Western Hemisphere originated from a few sources only, viz. Romania, the Netherlands East Indies, Iran, and later Iraq, their export prices were necessarily conditioned by the price in the major alternative and preponderant market, and the cost of freight. In the European market the standard Romanian export price, f.o.b. Constanța, customarily enjoyed a premium over the Gulf price in virtue of the possibility of supply at shorter notice and in smaller quantities, to suit the general storage and handling conditions of the trade. The high cost of transport in Romania restricted the area in which this premium could be earned. Nevertheless the Constanța quotation did reflect state financial policy in that payment in lei by the exporter, before the products left the refinery, of the costs of rail (later pipeline) transport, customs and excise duties, and manipulation expenses in the state-owned installations was a prerequisite of delivery to Constanța. These costs, which accrued to the State through either the C.F.R. or the fiscal authorities, became known as the "f.o.b. charges", and approximated to between 35 per cent and 50 per cent of the total f.o.b. quotation. This was also subjected to various *ad hoc* charges, depending on government financial policy. The Romanian State thus exercised an indirect command over the oil exporter through the possibility of varying the f.o.b. charges, which entered into the total f.o.b. price quotation.

The development of Romania's oil resources by foreign capital was based primarily on the demand for kerosine and lubricants in Western and Central Europe. This required the production of higher-grade kerosine than was necessary for the original export market, Bulgaria and Turkey, and of gasoline, for which there was no demand in these countries. Returns from the Balkan market were inadequate to support continued development of the industry and were too sensitive to influences from Baku.[1] Until the mid eighties, virtually all supplies to Western and Central

[1] In Russia, during the 1890s, the metallurgical industry's work on government contracts for the railway system had caused the Baku industry to concentrate on the output of fuels, to the neglect of kerosine. The reduction in kerosine output in turn caused a rise in prices which encouraged additional investment in drilling. The completion of the government contracts, in 1901–2, entailed a reversion to the distillation of kerosine, which now, took place at a rate greater than the domestic market could absorb. Russian producers therefore turned to exports and undersold their rivals in the Balkans and the Levant.

Europe were made by Standard Oil from the United States: after that time its position was challenged by Russian sources in consequence of the reduction of railway freights in 1885, and the organization of the Nobel distribution system.[1] Between 1883, when U.S. kerosine ceased to be competitive in St. Petersburg, and 1899, when Russian kerosine began to be sold in Portugal, the competition between Russian and American oil interests intensified their search for other sources of supply and necessarily involved Romania.[2]

Romania was geographically well placed to participate in the Western and Central European oil trade both by sea from Constanţa and by the Danube from Giurgiu to Regensburg. The rail connection with Constanţa was improved after the opening of the Cernavodă bridge in 1895. The sea route from Constanţa to Hamburg was only fifty nautical miles shorter than that from New York, but freight costs before shipment could be expected to be cheaper, since Ploeşti was nearer Constanţa than the main U.S. sources of supply were to New York. The improvement of storage facilities in Constanţa[3] and the pipeline project inhibited the development of the Danube waterway, which appeared better on

[1] The advantages enjoyed by Baku producers were to some extent offset by difficulties, owing to the quality of the product and to retail practice. Russian kerosine commonly contained more gasoline fractions than the best Pennsylvanian product and therefore "flashed" more easily: lubricants from the two sources tended to be complementary—the Russian having a high viscosity at low temperatures which made them suitable as machine oils, while the American, having a high viscosity at high temperatures, were more suitable as cylinder or spindle oils. In the main illuminant market, Germany, kerosine was sold wholesale by weight, and retail by volume: since the Russian product was 4 per cent heavier, German dealers had an initial preference in favour of the Americans. See "The Importation of Russian Petroleum into Germany", *PITR* (1899), p. 519.

[2] The competition also changed the character of marketing inasmuch as both Standard Oil and Nobel discontinued selling through independent importers and organized their own marketing subsidiaries, viz:

Nobel: Oesterreichische-Russische Nafta Import G.m.b.H.
Deutsche-Russische Nafta Import G.m.b.H.

Standard: Anglo-American Oil Company, 1888.
Deutsche-Amerikanische Petroleum Gesellschaft, 1890.
Soc. Italo-Americana del Petrolio, 1891.

On the organization of the Nobel interests see A. Beeby Thomson, *The Oil Fields of Russia*; for the rivalry between the two companies see Hidy, *Pioneering*, and Williamson and Daum, Chaps. 19 and 24.

[3] On Constanţa, see reports by Vice-Consul Pogson, 1905 (*FO* June 1906, pp. 3–5), 1906 (*FO* Oct. 1907, p. 13).

a map than in actual practice. Although it was considerably improved after the opening of the Iron Gates Canal in 1896, operations were still hampered by the winter freezing and the much reduced draught in summer. The first was offset by the construction of winter harbours at Deggendorf (Bavaria) and Passau, but the second factor compelled the loading of vessels to less than full capacity. Even in normal conditions, Regensburg could only be reached by vessels with a maximum draught of one metre (i.e. in terms of capacity, 300–400 tons). Larger vessels (600–1,000 tons) operated as far as Budapest, where their cargo was transhipped. The trip from Giurgiu to Regensburg took four weeks. Because of these operational restrictions, Romanian kerosine, using the Danube route, was effectively competitive only in south Germany and Switzerland: raw gasoline enjoyed a special rail tariff enabling it to reach central Germany. From 1906 onwards, i.e. after the large-scale foreign investments in Romanian oil, only 10–13 per cent per annum of Romania's exports went via the Danube.[1] Through the technical shortcomings of the C.F.R. overland rail communications[2] were of less importance until the Dardanelles were closed.[3]

There were also difficulties over the quality of the product. Until it came to be treated by the Edeleanu process, Romanian kerosine tended to burn unevenly and leave deposits of soot in the lamps then in use, so that despite propaganda by Romanian officials and publicists, it was less acceptable than the established brands from American and Russian sources.

As the first undertaking founded for the export trade, Steaua Română set the pace by shipping cargoes of kerosine to Regensburg in 1898. At the same time, this company erected storage facilities and a case factory at Constanța for use in an export trade by steamer, mainly with Bulgaria, Turkey, and Italy. Under the aegis of the Romanian Oil Trust, Romanian oil began to appear in quantity in the U.K. This ceased in March 1904 when, after the reorganization of Steaua Română by the Deutsche Bank, the company's exports, other than those to the local Balkan market, were directed to Germany as part of the Deutsche Bank's challenge

[1] On Danube trade, see K. Obbeke, "Bayern und die Rumänische Petroleumindustrie", in *Festschrift für den III Internationalen Petroleum Kongress*, Berlin, 1907; and Brackel and Leis, pp. 325–45.

[2] See below, p. 45.

[3] See below, pp. 72–73.

to Standard Oil. Other Romanian companies lacking similar export organizations sold f.o.b. Constanţa.

The development of Romanian oil in the export markets,

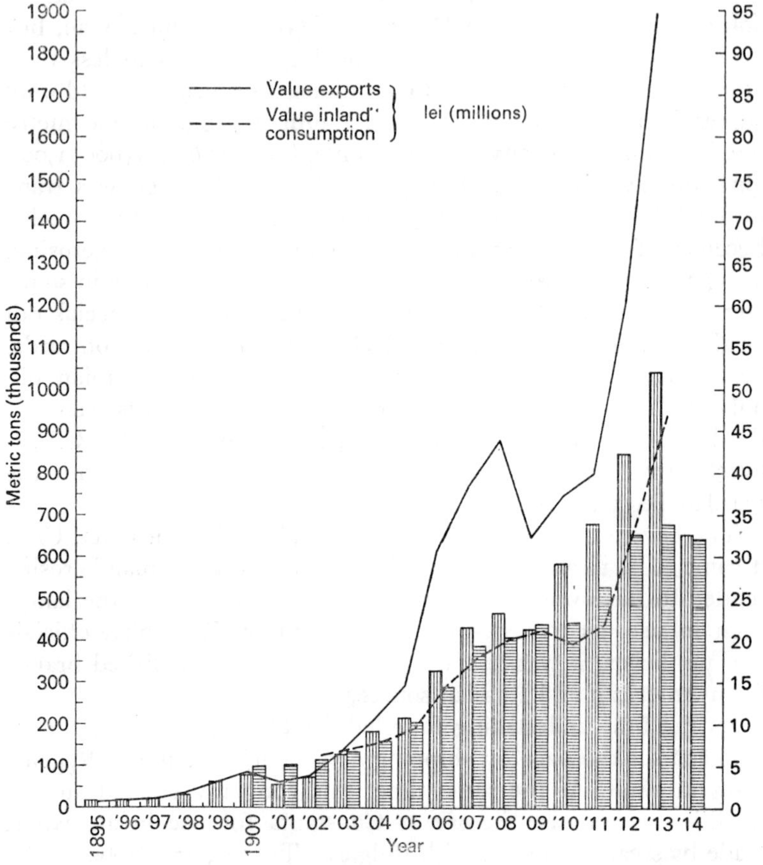

FIG. 2. Exports and inland consumption, 1895–1914.

particularly Germany and Austria-Hungary, disappointed the expectations of those who had invested in its production. Romanian kerosine and gasoline were to be used in a battle with Standard Oil, and for this purpose were sold through organizations set up by the Deutsche Bank—first the Petroleum Produkte A.G.,[1]

[1] The shareholding companies in the Petroleum Produkte A.G. were the Deutsche Bank, the Deutsche Petroleum A.G. (the organization which managed the Deutsche Bank's oil investments in Romania and elsewhere), the Berliner

and then the Europäische Petroleum Union[1] (E.P.U.). Both were defeated in the market—an episode which had not only commercial repercussions but nationalist implications.[2] Notwithstanding this

TABLE 2

EXPORTS OF PETROLEUM PRODUCTS

Year	Crude oil and residuals	Kerosine	Gasoline	Total tons	Value in francs/lei
1895	15,718	2	..	15,720	686,000
1896	17,269	529	..	17,798	910,000
1897	21,339	48	..	21,387	1,069,350
1898	27,480	4,253	..	31,733	1,745,315
1899	48,108	14,283	..	62,391	3,119,550
1900	48,899	24,612	4,145	77,656	4,001,775
1901	30,753	16,819	8,300	55,872	3,057,025
1902	23,073	32,291	16,344	71,708	3,916,434
1903	58,723	45,897	21,601	126,226	6,923,220
1904	45,204	78,270	36,970	160,444	10,696,620
1905	49,515	118,134	46,696	214,345	14,466,970
1906	54,799	190,914	74,493	325,206	30,572,310
1907	78,423	262,480	89,522	430,434	38,384,230
1908	78,765	263,633	122,860	465,258	43,622,780
1909	54,265	262,587	108,736	425,588	32,406,500
1910	122,492	337,036	126,334	586,151	37,371,810
1911	231,327	323,012	125,060	679,887	39,986,480
1912	318,441	353,564	173,816	846,420	60,323,380
1913	380,077	418,622	237,168	1,036,446	94,773,850
1914	191,545	297,800	164,143	654,024	..

particular set-back, Romanian oil, under the foreign capital regime, found new and regular outlets, based on the two routes to Central and north-west Europe discussed above. The progress in the export trade is shown in Tables 2 and 3.

(*b*) *The problem of transport—the Constanța pipeline project*

The export bias to the industry's activity focused attention on problems of transport—at which point the State became involved, in that the railways were already under its control and the disputes

Bank, Steaua Română S.A., and the Shell Transport and Trading Company. After the Steaua agreement the Disconto-Gesellschaft was admitted to P.P.A.G., which henceforward sold *all* Romanian oil in Germany on behalf of its principals. Supplies were drawn from Steaua Română (Deutsche Bank share), Telega (Disconto share), and Aurora (under contract to Shell Transport and Trading). On the origins and operations of P.P.A.G. see Gerretson, vol. iii, pp. 83–8; and R. Henriques, *Marcus Samuel* (references, see index).

[1] See Gerretson, vol. iii, pp. 88–106.

[2] See below, p. 63.

TABLE 3

ANNUAL EXPORTS BY COUNTRIES OF DESTINATION

(In tons)

Country	1903	1904	1905	1906	1907	1908	1909	1910	1911	1912	1913	1914
Austria-Hungary	20,712	22,107	19,240	16,479	22,959	3,623	5,409	33,881	46,311	86,013	77,184	84,253
Belgium	164	22	87	7,944	33,390	18,726	28,857	40,280	27,673	11,907	25,136	18,076
Bulgaria	3,909	4,995	4,643	3,448	3,243	4,062	6,285	7,364	8,550	10,358	13,149	18,689
Egypt	..	25	..	..	20,910	79,945	66,317	73,286	115,170	142,418	121,642	62,714
France and French North Africa	13,148	40,631	118,925	91,618	116,076	123,052	134,564	142,601	127,396	170,732	159,606	58,384
Germany	24,271	25,757	37,091	38,040	48,381	34,109	24,972	53,294	54,059	84,041	126,295	99,165
Greece	298	166	..	..	6	19	17	18	100	1,502	6,958	12,406
India	..	..	..	5,466	58,371	15,715	..	..	..	..	..	..
Italy	10,140	8,026	77	34,206	21,466	21,549	31,104	28,479	18,048	25,921	118,643	108,144
Japan	..	..	..	..	..	..	..	18,631	1,549	..	..	..
Netherlands	298	166	..	..	6	19	17	18	100	1,502	6,958	12,406
Russia	..	..	575	..	..	3	26	33	47	1,210	9,114	7,016
Scandinavia	..	3,348	2,751	3,435	5,183	6,406	10,417	12,724	20,531	15,843	22,039	19,412
Serbia	7	34	704	25	403	121	291	622	1,044	2,614	3,675	24,369
Spain	..	..	..	..	71	..	63	14	648	64	111	..
Switzerland	250	119	25	26	5	127	..	10	..	..	1,155	836
Turkey	1,241	2,100	7,660	22,524	24,587	32,568	37,838	29,951	49,655	53,549	73,790	31,463
U.K.	37,085	36,869	22,570	91,262	64,575	109,372	70,979	125,687	180,614	214,195	232,880	77,971
U.S.A.	..	..	..	..	..	..	..	2,353	3,827	..	..	..

in 1900 over the pipeline scheme had made it clear that any pipeline construction and operation would take place only on its terms.

After the withdrawal of the Standard Oil–Disconto-Gesellschaft scheme, the Minister of Finance in the Sturdza Government, Emil Constinescu, suggested in 1902 that the Constanța line should be constructed, owned, and run by the State—the estimated cost being 12 million francs. In the following year he introduced a law whereby 500,000 francs from the current budget surplus was set aside for the necessary preliminary studies of routes to the Black Sea and also to the Danube. Furthermore, an engineer was sent to the United States to study problems of pipeline construction. Various estimates of cost were made, and the Government, through Alexandru Beldiman, Romanian minister in Berlin, sought the advice of Richard Sorge. Sorge made comparisons with the Caucasus, where oil exports along the single-track railway from Batum to Baku were made at a million tons a year: he argued that even when Romanian exports reached that figure there were two other export routes by water available, and, provided that the railways could move the larger quantities of oil, there was no reason why a pipeline should be necessary. A pipeline would be helpful but was not either then or in the foreseeable future essential; Romanian exports would be better stimulated by a cut in rail freights. Beldiman reported in these terms to Sturdza and the project was shelved.[1] The proposals concerning the line advanced by the Deutsche Bank and Banque de Paris et des Pays Bas in 1905 (see below, p. 57) merely emphasized the fact that the State now had the whole pipeline project firmly in its own hands.

The premiss on which the construction of a pipeline was put into abeyance, viz. that the railways could adequately move the increasing oil exports and at rates which did not nullify Romania's competitive advantages in European markets over current Russian and American production, was not long valid. By 1906 the performance of the railway caused the convening of a conference, attended by representatives of the Ministry of Industry, the C.F.R., and the oil companies. The latter complained particularly about congestion and delays in transit, resulting in restriction on exports—the current programme of three oil-trains a day moved

[1] In consequence, Steaua Română and Aquila, who had been induced by the scheme to plan refineries at Constanța and even buy sites for their construction, reverted to extending their up-country plants.

only 32,000 tons a month, whereas the net export availability was 40,000 to 45,000 tons a month. It was generally admitted that this was due to a shortage of locomotives, which private concerns had not been allowed to provide for themselves.[1] Further, in the view of the oil industry, the existing routes[2] suffered from two major drawbacks—they all converged on Fetești and then followed the same route over the Cernavodă bridge to Constanța, and—since they passed through the grain areas—oil-trains had to give way during the harvest to grain exports.[3]

Various remedies[4] were canvassed to increase both the number of trains and their running speeds, and to cut down unloading time at Constanța. The conference agreed that the oil companies should draw up a memorandum of shortcomings and suggest "means for satisfying the present and future needs" for both inland and export trades. The problem was complicated by the fact that the types as well as the volume of exports increased. Residuals, hitherto a burdensome waste product, were now in some demand as fuel,[5] but competition in export markets was so severe that Romanian residuals could not bear the high freights levied by the railway authority.[6] The kerosine trade also increased rapidly and Asia Minor was no longer supplied in cases, but in bulk.

[1] Official permission was granted in 1906, but the type of locomotive had to be the same as those used by the C.F.R.—which ruled out any rapid increase in numbers available.

[2] There were two main routes by which oil was transported and to this a third was being added.

Câmpina—Ploeşti—Mogaskaia—Feteşti and Cernavodă—Constanţa: 282 km.

Ploeşti—Buzău—Făurei—Feteşti and Cernavodă—Constanţa: 279 km.

Ploeşti—Slobozia—Feteşti and Cernavodă—Constanţa: 256 km.

(Completed September 1909).

[3] The industry also suffered by reason of the army manœuvres for 1907, which, taking place in the Dobrogea, occupied the Constanţa railway with military traffic during preparations in September and closed the port for a fortnight in October.

[4] In his report for 1906 the British Vice-Consul in Constanţa wrote, "the critical question of the moment is the improvement and extension of transport facilities. It has been estimated that the State railway alterations would need at least £4 million, whilst a pipeline from the chief oil fields at Câmpina . . . would be made at an expense only of £400,000. It is therefore probable that the cheaper alternative will be adopted in the near future." The Vice-Consul was over-sanguine: see *Trade and Shipping of Constanţa for the year 1906*, no. 3939 in the series *Roumania: Diplomatic and Consular Reports*, Foreign Office, Oct. 1907.

[5] See below, p. 49.

[6] Approximately one-third of the f.o.b. Constanţa price of residual fuel oil represented the freight.

The problem of the disposal of the residuals that necessarily resulted from refining operations enforced continuous official consideration of improving transport facilities, but the immediate cause of the decision to build the pipeline was the sudden increase in output from Prahova in 1910.[1] In consequence, at the end of 1911 the Ministry of Industry decided on the construction of three lines,[2] one for residuals and two for kerosine.

The necessary law was introduced on 25 March 1912. Permission to finance (to the extent of 18 million lei) a pipeline along a route to Bâicoi—Constanța via Ploești, Buzău, Făurei, Fetești, and Cernavodă—was given to the Minister of Finance (Article 1). Construction was to be undertaken by the Ministry of Public Works through the management of the railways: competence in commercial and tariff questions was ascribed to the Ministry of Industry and Commerce (Article 3). It was foreseen that other lines from the producing areas to the main trunk-line could also be constructed, though not in competition with existing private lines (Article 4). Existing refineries, as specified in the Law, were to secure a partial amortization for a maximum period of twenty years by means of rebates in the transport tariffs for crude and kerosine to a level of 15 million lei.[3] The necessary land for lines, pumping stations, and all ancillaries was declared expropriated on grounds of public utility by Article 8.

Although the law as a whole reflected an initiative and the ultimate continuing competence of the State, concessions were made to the industry in Articles 4 and 5 whereby existing refiners whose plants were inland and therefore dependent on the line were safeguarded against loss of export markets (in consequence of the raising of transport tariffs) to any refineries operating from the seaboard. This was done, as explained by the Prime Minister, P. P. Carp, when introducing the Bill, not as a right but as a simple matter of equity.

As defined in the Law, the route to be followed was 287 kilometres in length: the head of the line was well situated to gather

[1] See Figure 1 above.

[2] Originally two, but, in view of the different qualities of kerosine exported to the European and the Balkan and Near Eastern countries respectively, the number was increased to three.

[3] The levels were based on the transport of a minimum quantity of 500,000 tons of crude and kerosine through the line p.a. at 1·5 centimes per ton per kilometre.

crude oil from the existing producing areas in Prahova—all within a radius of 30 kilometres—and to take kerosine from the concentration of refineries in Ploeşti. The Directorate of the Railways sent two missions—to Russia and the United States—to study the organization and running of pipeline systems. It was decided that the crude lines were to be 9 in. in diameter from Băicoi to Ploeşti, and 11 in. from Ploeşti to Constanţa, the kerosine lines 6 in. and 5 in. In this way it was hoped to transport 1·8 m. tons crude oil and 500,000 tons kerosine a year. Storage problems were discussed by an *ad hoc* commission,[1] which recommended which types of crude oil should be transported and into which type of tankage they had to be segregated: a total capacity of 140,000 tons would be required.

Tenders were put out for 600,655 metres of 5 in. pipe, 90,300 metres of 9 in., and 210,027 of 14 in. The successful bid was for 11,680,247·80 francs put in by the firm of Stokois and Zonn, Rotterdam, on behalf of the United States Steel Products Company. The contract was approved by the Council of Ministers on 2 December. It was estimated that the work would be finished and the line in operation by 1915. Construction was interrupted by the Balkan Wars of 1913, by the end of which year the 18 million lei originally appropriated was exhausted, and a further subvention became necessary.

(*c*) *The inland market—the Proration Law of 1908*

Romanian oil-production went through the same phases of flush production described above[2] as characteristic of early developments in Pennsylvania. The multiplicity of new wells and the inadequacy of storage forced producers to sell almost immediately and kept prices low.[3] They attempted to offset this disadvantage by combining to stabilize and maintain prices: the effects were, however, local and temporary. The setting up of Steaua Română introduced into the Romanian industry an organization sufficiently large in relation to other producers and refiners to exercise leadership in both these phases. In consequence, the Steaua management initiated

[1] Composed of representatives of Steaua Română, Astra Română, Creditul Petrolifer, Româno-Americană, and of small firms (Aisinmann).

[2] p. 3.

[3] See A. P. Bennett, *Report on the Petroleum Industry of Roumania*, (FO, 1896, Misc. Series, no. 411); and article in *PITR* (9 Dec. 1899).

the control of production, through agreements, and of surplus refining capacity, through subsidies. Again, the effect was short-lived, since unorganized production increased and depressed the price of crude oil. There was nothing in liberal theory or practice against the operation of cartels. In 1901 the Minister of Domains (responsible for oil affairs) sponsored the formation of a 'Petroleum Producers' Association', under the Presidency of Constantin Alimăneştianu, of the Ministry. This Association contrived to maintain prices at between 3 and 4 francs per 100 kilogrammes. The problem remained, however, of putting the Romanian oil industry on a solid and profitable foundation. In a speech[1] to members at Ploeşti, on 4 February 1902, Alimăneştianu suggested a programme of measures comprising the organization of a selling syndicate disposing of at least 70 per cent of the yearly output of the home retail trade, including arrangements for storage and transport, of a technical and commercial bureau for advice on problems of manufacturing and materials used in petroleum engineering, and of a syndicate of home and foreign bankers to furnish capital and provide credit.

By this means Alimăneştianu hoped to raise prices to producers and reduce them to consumers through economies of scale. As an immediate step a central sales bureau for crude oil was organized from within the Association in February 1902: it was unsuccessful in maintaining prices—during 1903, the latter dropped from 3 francs in January to 1·7 francs in December. Alimăneştianu's programme was never implemented, and the problem of disposing of crude oil output disappeared through the enlargement of manufacturing opportunities represented by the investment in refining[2]—to be replaced by that of disposing of the resultant products.

Refining operations were designed to yield kerosine, gasoline, and, where possible, lubricants, in that order. They necessarily gave rise to a large yield of residuals, the price of which affected the earning capacity of the developing industry. The disposal of these residuals remained a persistent problem, though new outlets were found in the conversion of C.F.R. locomotives from coal- to oil-burning and the adoption of fuel oil by the Romanian Department of Navigation in 1905 and the Romanian Navy in 1906.

[1] Text of speech, in translation, in *PITR* (1 Mar. 1902), pp. 240 f.

[2] See above, p. 35.

Steaua Română also began to interest itself in the marine bunker business, setting up tanks at Alexandria for this purpose. In 1909 the company began supplying residual fuel oil to the Anatolian railway, thus fulfilling, in part, the aim of the original investment by the Deutsche Bank. Further, the naval authorities of the Great Powers began to make contracts for supplies of residuals for use as fuel oil. Nevertheless, such outlets only enabled refiners to keep abreast of their yields, not to overcome the difficulties they created. Steaua Română, after various attempts, managed to form a residuals cartel in May 1906, with the approval of the Government, the latter imposing a limit on any upward movement of the price, viz. that it should not exceed the price of a quantity of Cardiff coal c.i.f. Brăila with an equivalent thermal output.

The main product for the inland market was kerosine, for lighting and heating. On this market, specifications were not so exacting as for exports, and therefore Steaua Română's technical superiority could not be used to compel smaller refiners to fall into line. Nevertheless, the company was able to control a proportion of the output of kerosine through a joint office established in August 1901 with Aurora and four smaller refiners, through which all sales of kerosine meeting a guaranteed specification were made, at prices decided upon monthly. Steaua and its partners were able to hold prices in this way, and the arrangements were renewed in May of each year. The reorganization of Steaua Română by the Deutsche Bank, and the company's consequent participation in the Europäische Petroleum Union, strengthened the position within Romania of the cartel, which sought to persuade new entrants to the industry to join, in order to maintain the price structure. The most formidable newcomer was Romåno-Americană, which, from the completion of its refinery in 1905 until the spring of 1907, exported all its kerosine output: after that time it began to take a strong interest in the domestic kerosine market. For this purpose, the company erected storage and distribution depots and introduced improved methods of selling direct to retailers and consumers, a practice inimical to the existence of the various middlemen traditionally engaged in distribution.[1] Moreover it introduced a higher-quality product, made from Moreni crude oil.

The competition thus provoked reduced the home market price

[1] This technique of bypassing traditional intermediaries caused great antipathy to Standard Oil in Europe and subsequently touched off the agitation

for kerosine during 1907 from 12 lei per 100 kilogrammes to 6 to 7 lei, and brought about an agitation in both political and commercial circles, during which all the charges of monopoly made in 1900 were revived. It was alleged that Românо-Americană was deliberately forcing down prices in order to eliminate its competitors: having arrived at a *de facto* monopoly of the market it would try through increased prices to recoup itself for the losses necessarily sustained. The suggestion derived its plausibility from the hostility to Standard Oil caused by the state-lands controversy and from what were construed as that organization's malpractices elsewhere.[1] It was, however, not simply an argument of small business against big: the small manufacturers and distributors whose interests were damaged by Româno-Americană's marketing methods were all *Romanian*: therefore they could expect official sympathy, if not support.

Furthermore, the competition had corroded the existing cartel, whose members, other than Steaua, saw themselves as likely to be forced out of business after the impending defeat of their organization: the inland trade would then be left to the only companies possessing distributing assets, viz. Steaua and the Americans. These cartel members therefore suggested the foundation of a neutral sales bureau, for a period of ten years, which Steaua should be invited to join.[2] The Steaua countered by proposing its own scheme for such a bureau,[3] in March 1908: the scheme was not immediately acceptable and the cartel agreement was due to be renegotiated in May. While these proposals were being debated within the cartel, Româno-Americană extended its operations and prices continued to fall, reaching the lowest point of 4 lei per 100 kilogrammes ex-refinery in April 1908.[4]

In May 1908 the Romanian Government put an end to this situation by announcing that it would in future allow no competition in the home retail trade, which instead should be regulated by agreement among refiners: in default of such agreement, the

for a kerosine monopoly in Germany and Austria-Hungary; see Gibb and Knowlton, *The Resurgent Years*, p. 203; and Gerretson, vol. iv, p. 170.

[1] As to which, see R. W. and M. E. Hidy, *Pioneering*, Chap. 8; and A. Nevins, *Rockefeller*, vol. i, Chaps. V–VIII.

[2] 2.

[3] 3.

[4] On the competition between Româno-Americană and the cartel, see article "Le Commerce du pétrole en Roumanie", *MPR* no. 7 (1908), pp. 245–50.

Government proposed to set up a state monopoly. During the last three days of the parliamentary session, the Government rushed through a bill giving effect to these ideas in order, as the Minister of Finance, Emil Constantinescu, explained, "to extirpate the evil transplanted from elsewhere before it takes root in Romanian soil".

The Law for the Proration of the Inland Consumption of Illumination Oil empowered the Government (Article 1) "to arrange the distribution of the total consumption of illuminating oil [kerosine] among existing refineries and those which may be established", and provided that (Article 2) quantities for distribution were to be fixed among refiners on the basis of the producing capacity of each refinery. Control of output was to be complemented by control over prices. Article 7 declared that the maximum selling price[1] would be determined by the Government and would be binding on all refiners. There were heavy penalties for non-compliance, either by over-production or by manufacturing an inferior quality or by selling at prices higher than those permitted (Articles 10–14). Only refineries capable of meeting the State's prerequisites would be permitted to manufacture kerosine for the home market. Further, the Law empowered the Government to construct storage facilities, in the case when refiners would not themselves construct them, at railway stations and ports (Articles 15 and 16). The Law was to be applied at the latest on 1 October 1908 (Article 17).[2]

As published, the terms of the Law required the definition of refining capacity and also of the average price of crude oil. Further, it had merely established the basis for the maximum price of kerosine ex-refinery, not the retail price to the consumer: it therefore left open the vexed question of the retail system to be adopted, i.e. whether it was to operate with or without intermediaries. The question of refining capacity—whether constructed capacity or actual throughputs were to be taken into account in arriving at the ex-refinery price—was considered at Constantinescu's request by

[1] This was to be, ex-refinery per 100 kg., "between frs. 3·50 and frs. 4·50, representing the cost of refining and the refiner's profit"—to which figure was to be added "the average price of the crude oil delivered to the refinery. The total of these two figures will form the maximum selling price."

[2] It was in fact applied from 1 July 1908: see "Report of the Minister of Finance to the Council of Ministers, 19 June 1908", *MPR*, no. 18 (1908), p. 638.

an *ad hoc* commission of refiners[1] and by Constantin Hălăceanu,[2] an official of his ministry. On the basis of Hălăceanu's report, the Government announced the first proration on 24 June 1908, and established the price for the next quarter at 8·70 francs per 100 kilogrammes.[3] The Proration Law induced that combination among refiners which the Government wished: in this it emphasized the position of Steaua Română and Româno-Americană, the only concerns possessing storage and distribution facilities for the internal market.

Steaua Română and Aurora announced the formation of a new distributing company, the "Societatea anonimă pentru distribuirea produselar petrolului".[4] Other companies were invited to participate. Româno-Americană did so, after its point about selling direct to consumers instead of through intermediaries had been conceded. On 28 July 1908 the three companies registered with the tribunal of Ilfov (Bucharest) contracts by which their production would be sold through the new organization (which became known as Distribuția). Kerosine quotas would be those assigned to each of the participants by the Law: other products, *pro rata* to the distillation capacity, as established, of their respective refineries. At the same time, the three companies agreed to surrender their distributing organizations and their government contracts to Distribuția, and on this basis the new company began to consolidate and extend its sales network throughout the country and tried to secure the adhesion of other refiners. In this way Distribuția put an end to the other product cartels, which had been organized and regulated the supply to the domestic markets in a way which

[1] Consisting of Bergher (Steaua Română), Edeleanu (Vega), Aisinmann, Mony (Colombia), Bossal (Aquila), Parascheva and Frischoff (small refiners). Hălăceanu took part in the discussions. The commissioners advocated that capacity be determined by reference not only to the crude intake capacity of a refinery, but also to the redistilling capacity and storage and loading facilities.

[2] Hălăceanu broke down the refining process into its component parts—according to which he then examined the plants in the existing refining industry, in order to assess the seven different manufacturing systems in use on a common basis (see translation of this report in *PR* (18 July 1908), pp. 15, 45, 46, and 71). Hălăceanu was also commissioned to examine "the average price of crude oil"—a concept difficult to establish, since 70 per cent of crude output belonged to refinery companies and therefore was never quoted in the market, while 20–25 per cent was sold under long-term contract: the basis for an open-market price was thus slender.

[3] The small refiners protested at the basis of the award, and at the price, which they argued should have been 10·70 francs.

[4] *MPR*, no. 13 (1908).

ensured the continued existence of the small producers. Even so, the latter were dissatisfied with the results of proration and advocated a complete state monopoly.[1]

The Proration Law was explicitly anti-American. In the debate in the Chamber, Româno-Americană was criticized, for example, for bringing down prices to the point where two kilogrammes of kerosine were sold for the price that obtained for one before it

TABLE 4

INLAND CONSUMPTION

Years	Gasoline	Kerosine	Gas oil	Fuel oil	Others	Total (tons)	Value francs/lei
1900	1,500	30,000	500	65,000	..	97,000	
1901	1,555	30,904	675	66,753	169	100,056	
1902	1,365	32,210	2,685	77,876	671	114,807	6,318,706
1903	2,120	30,273	2,976	97,098	744	133,211	7,195,261
1904	2,369	32,440	2,739	119,735	685	157,968	7,843,512
1905	2,696	31,558	5,046	162,213	1,261	202,804	9,657,470
1906	4,059	35,243	7,878	237,477	1,970	286,627	14,766,048
1907	5,689	38,467	7,238	332,999	1,809	386,202	18,126,879
1908	9,055	38,422	9,564	347,329	2,391	406,761	20,604,106
1909	14,041	39,451	12,558	366,703	3,815	436,568	21,211,849
1910	20,314	41,849	14,035	360,351	4,279	440,828	19,680,814
1911	24,450	43,941	17,921	434,094	5,624	526,030	21,888,700
1912	30,656	49,941	23,198	540,385	6,937	651,117	34,154,420
1913	30,131	51,396	26,980	560,492	8,170	677,169	46,852,153
1914	31,672	51,710	32,653	524,254	9,579	642,868	

came on the scene. This sort of practice, it was alleged, would cause other foreign organizations to abandon Romania. The language of the Minister in the official *exposé des motifs* was similarly highly charged.[2] Româno-Americană's attempts to defend itself by arguing from the benefit to the consumer in both quality and price, and by showing that its quotations had not been the lowest in a number of public adjudications, were ignored.[3] The agitation was not concerned to protect the Romanian consumer but the Romanian producer, and, as Gerretson notes,[4] the Standard was regarded as the evil genius of the industry. The Law was also timely for Steaua Română, whose position in the inland market was assured by the Government's action. The Bill was introduced unexpectedly and passed unanimously, after four hours' debate

[1] See Grigorescu in *MPR*, no. 13 (1908), pp. 465–6; no. 14, pp. 508–10.

[2] See *MPR*, no. 11 (1908), pp. 385–8.

[3] See letter to the Bucharest journal *Curventul Financiară* reprinted in *PR* (25 Apr. 1908); see also table in *MPR*, no. 10 (1908), p. 256.

[4] Gerretson, vol. 3, p. 123.

in the Chamber of Deputies and none in the Senate. It was the first of four measures fundamentally altering the relationship between the industry and the State that were so enacted—the others being the Mining Laws of 1924[1] and 1937[2] and the Nationalization Law of 1948.[3] The progress of the internal market can be seen in Table 4.

State oil lands and state policy

The Romanian State impinged on the development of the oil industry not only through its capacity as legislator and customer, but also as owner of the most likely source of raw material. The disputes of 1900 did nothing to further the exploitation of the State's resources but merely established the conditions on which such exploitation would be carried out. Given that the Government was not willing to wait until it commanded sufficient capital for a company of its own, some form of concession was necessary, and the questions at issue were concerned with who was to receive it and the arrangements under which it was to be granted. Applicants had a continuing interest in gaining concessions on state-owned lands in order to strengthen their position in Romanian oil in relation to that of their competitors in other markets.

In 1903 the Disconto-Gesellschaft made another approach to the Romanian Government with proposals similar to those made in the name of the syndicate in 1900. The Romanian Government's terms were stiffer and included a demand for state participation in profits and some control over the exploitation of the areas conceded. These clauses were unacceptable to the Disconto-Gesellschaft, and the negotiations, which took place in Berlin, were broken off. The next year Rothschilds, Nobel, and Mantascheff,[4] all producers in Baku, sought to use Romanian resources against Standard Oil. Negotiations were opened by a Viennese lawyer, Dr. Herman Fialla, who revealed the identity of his principals only in July 1904. On 10 September he concluded with Sturdza a secret agreement[5] whereby they, the Baku group, would found a company to exploit the state oil lands on concession

[1] See below, p. 115. [2] See below, p. 205. [3] See below, p. 318.

[4] On this episode see Haase, pp. 37–42.

[5] See *MPR*, no. 22 (1905), pp. 500–3, and editorial comment in no. 23 (1905), p. 645.

for fifty years on terms explicitly allowing comprehensive supervision by the State. Its purpose was stated in Article 1, according to which the group was to set up in Bucharest a Romanian State Petroleum Company, under Romanian law, to exploit petroliferous lands belonging to the State. The company was to have no connection in any way with Standard Oil. It was to receive the support of the Romanian State in the exercise of its rights, and in return undertook "to establish its enterprises on a basis which best corresponds with the interests of the Romanian State, particularly in regard to employment as far as possible of Romanian nationals as technicians and workers". The company was to obtain the exclusive rights (concessions) to explore and drill state lands for fifty years (Article 2), the relevant areas were to be subsequently enumerated by the Government but the concession was not to exceed 30,000 hectares in total (Article 4). The choice of actual plots was at the company's discretion for a period of ten years, but a mixed commission of state and company representatives was to determine their extent in each case.

In return for these rights the company undertook (Articles 5–7) to make certain annual payments together with a proportion of dividends above a certain level. The Government was to have the right to inspect and annually audit the company's books.

Articles 8 to 11 set out the reciprocal obligations of the State and the company, concerning the actual working of the areas under concession. Article 12 reserved to the State the exclusive right to erect and operate public, i.e. trunk, pipelines. By Article 13 the State undertook to draw, so far as possible, its own supplies of crude and products from the company. Finally, the Romanian Government reserved the right (Articles 14–15) to take part in the drafting of the statutes of the company which would, when completed, have to receive official approval. The validity of the contract was to be sanctioned in a constitutional manner and the Government undertook to use its good offices for this approval to be given as soon as possible.

The agreement was signed by the Prime Minister, Sturdza, and Dr. Fialla at Munich on 10 September 1904 and was covered by a letter[1] from the Prime Minister in which was set out an undertaking by both parties to observe complete secrecy on the terms for a

[1] Given by Sturdza in the Senate, 13 May 1905; *MPR*, no. 15 (1905), pp. 432–3.

period of seven months, during which the Romanian Government bound itself not to discuss with any other party the questions involved, and during which the company was to be set up. The final working agreement would be reached on the basis of the agreement under reference. Sturdza also stipulated that the board of management would include members appointed by the Government and envisaged an increase in the suggested royalty to 8 per cent.

The terms specified in Sturdza's letter expired on 10 April 1905, by which time his government had given way to that of the Conservatives under Cantacuzino. On 11 January 1905 Fialla inquired of Sturdza as to the effect of the change of government on their Munich agreement, to be informed that it still remained in force. On 24 February 1905 Fialla informed Sturdza that he had, as foreseen by that agreement, succeeded in forming a company, but that the principals in the case were now the Deutsche Bank and the Banque de Paris et des Pays Bas.[1] At the beginning of March he visited Cantacuzino, accompanied by Spies, the General Manager of Steaua Română, and entered into negotiations on the basis of the Munich agreement, whose terms were at that juncture not known to the general public, on behalf of his new principals. The terms[2] varied slightly from those originally negotiated by Fialla on behalf of the Russian interests, and the text began significantly: "New project for a concession, worked out in conformity with the desires, exigencies, and conditions expressed by the Romanian Government." As before, a company was to be set up, at the latest by 1 August 1905, and was to be conceded the subsoil rights to an area of 30,000 hectares (later reduced to 10,000) on state lands for a period of fifty years, in the first ten of which the company undertook to sink 150 wells in stipulated districts. In return the State was to receive a royalty of 10 per cent on the output per annum, either in value, determined at well-head prices, or in kind, at the State's option, together with taxes and a share in profits.

These proposals, dated 17 May 1905, evoked a great political

[1] Haase, (p. 39) suggests that the substitution may have had to do with the 'threads between the Deutsche Bank and the Russians'—as seen later in the establishment of the Europäische Petroleum Union. The archives of the Deutsche Bank disclosed no evidence on this point.

[2] See *MPR*, no. 23 (1905), pp. 645–9; also letter by Ion Lahovari (Conservative Minister of Domains in 1905) published in *MPR*, no. 5 (1908), p. 168.

uproar,[1] not so much on the grounds of the foreign nature of the capital, but of the extent of the area involved, which was considered too much for any commercial purpose.[2] At this stage also the terms of the Fialla–Sturdza agreement were made public in government journals as a test of public reaction. From the resultant political controversy there emerged a proposal put forward by Lascar Catargiu and Alexandre Marghiloman, for a company financed by exclusively Romanian capital to explore and exploit state lands, the rights to which would be guaranteed by the Government. The company, called "România", was hurriedly brought into legal existence from 29 May to 2 June.[3]

The proposals of the Deutsche Bank and the Banque du Paris and des Pays-Bas and the counter-proposals of the putative "România" company led to the final announcement of the policy on the future development of state lands. On 6 June 1905 the Prime Minister, Cantacuzino,[4] responding to an interpellation of Sturdza, laid down three criteria:

1. Part of the state lands was to be kept as a reserve for the future (i.e. could not be conceded at all).
2. There was to be no monopoly of any lands conceded.
3. The State was to retain in its own hands all means of transport.

Sturdza welcomed the terms of this declaration, which resolved the differences between the parties on this issue, and was designed as the basis for a comprehensive programme of development: the declaration did not in itself exclude foreign interests. Sturdza's volte-face, on the technical grounds of the change of principals, put him in the position of denying to the Germans and French what he had been willing to concede to the Russians nine months previously. His defence[5] was that in the meantime, and as a natural consequence of economic progress, Romanian nationals were in a position to undertake the financing of exploitation of the

[1] The *MPR* editorial (no. 1 (1906), p. 7) also reported "irritation" in Romanian Government circles at the forceful intervention of the German minister (Kiderlen Waechter) on behalf of the banks.

[2] See H. E. Browne to FO, 12 June 1905: *F.O. Series 104*, vol. 165; and *MPR*, no. 15 (1905), pp. 421–4.

[3] For a prototype of the "România" scheme, dating from 1867, see *MPR*, no. 36 (1906), p. 32.

[4] See *MPR*, no. 16 (1905), pp. 458–61.

[5] See Sturdza's speeches in the Senate of 6 June 1905: *MPR*, no. 16 (1905), pp. 461–5; and his pamphlet *La Question du pétrole en Roumanie*.

state lands: in these circumstances the concession of state lands to foreigners was detrimental to the interests of the State. The British minister, Sir John Kennedy, offered another explanation:[1] that Sturdza foresaw the accession to office of a Conservative government which would include as Foreign Minister Take Ionescu, who was legal adviser to the Standard Oil interest in Romania. In order to forestall possible concessions to the Americans, Sturdza offered the state lands (through Fialla) to the Russians and then to the Germans.

In consequence of the declaration, Dr. Fialla withdrew the propositions of his principals and the formation of the "România" company was proceeded with. There was no public issue of shares but 60 per cent of the capital was taken up privately by Romanian citizens. The remainder of the capital was to be issued subsequently, in the form of bearer shares, which were to be freely negotiable. The company was to be managed by Costinescu, a former Liberal Finance Minister. Immediately after Cantacuzino's statement an official inspection of state lands was made, during which all applications for concessions were blocked.[2] This examination resulted in the Law of 30 December 1905[3] intended to secure the State's interest in the exploitation of Romania's petroleum resources and to remedy the concentration of capital in areas known to be oil-bearing to the detriment of those which were not.

The Law divided state lands into lots with a maximum of 100 hectares of proven oil-bearing lands and a maximum of 1,000 hectares of non-proven lands. The Council of Ministers was given final authority on the approval or dismissal of any demand for concessions (Article 1). No one concessionaire could have more than three lots altogether, and before acquiring them the concessionaire had to prove that he disposed of sufficient capital for development, determined (Article 2) as at least two million francs for each lot of 100 hectares of proven land and 1,000 hectares of non-proven. Articles 3 to 13 discussed restrictions on concession-holding drilling obligations, transport, and royalties. Article 15 introduced a novel feature by stipulating that only Romanian courts were competent to try disputes between the State and the concessionaire.

[1] Dispatch to FO, no. 31, 5 June 1905: *FO Series 104*, vol. 164.
[2] *MPR*, no. 16 (1905), pp. 473–4.
[3] *MPR*, no. 38 (1905), pp. 1064–6.

The Law did not define "proven" and "non-proven" lands, which were by subsequent Ministerial explanation specified respectively as those on which there were borings or hand-wells in exploitation and those which were extensions of "proven" lands but were hitherto undeveloped.[1] The Law signified the end of ideas of exploiting state lands *en bloc* through an individual concessionaire who would be responsible for all matters of detail: this in itself deprived the "România" company of its *raison d'être*, and the project collapsed. It is noteworthy, however, that it did not discriminate between Romanian and foreign nationals as applicants for concessions.

In introducing the Law,[2] the Minister of Domains, Lahovary, had expressed his conviction that under its aegis the petroleum operations of the State would multiply and prosper and that, as the rights of the State and those of concessionaires were equally protected, the national economy would find in the oil industry an abundant source of riches for the future. These hopes were not realized: the State had arrived at a policy for the development of its own lands, but could not compel others to adopt it. No concessions were taken out in the next three years. During this period, production increased from 641,070 tons (1905) to 1,142,448 (1907), all from private lands. Moreover, although in 1908 the output increased further to 1,180,000 tons, the rate of increase fell off from an average of 34 per cent to one of under 4 per cent. In this situation the only prospects appeared to be offered by the state lands. This was the incentive to revision of the Law, which was proposed by Take Ionescu in the debate on the Royal Message at the opening of the parliamentary session 1908–9.[3] He advocated different regimes for proven and non-proven territories, with the former split up into small plots of 15 to 20 hectares and concessioned by public adjudication against royalty payments, and the latter reserved for Romanian capital and technicians: in this way the state lands would attract concessionnaires and the stagnation would be broken. Vintila Brătianu, then Finance Minister, gave the orthodox Liberal rejoinder, stressing that the State should explore and then provide openings for "Romanian capital, Romanian intelligence, and Romanian labour".

[1] At the time the State domain included 5,750 hectares of proven lands and 82,000 hectares of non-proven lands.

[2] *MPR*, no. 35 (1905), pp. 985–90.

[3] *MPR*, no. 36 (1908), pp. 1337–8.

The implications of this brief statement were formally developed in the Law for the Exploitation of State Lands passed on 5–6 April 1909.[1] This authorized the Ministry of Industry and Commerce[2] to determine which territory on state lands was to be declared oil-bearing: such territories could be leased for thirty years by public adjudication

> according to the following rules: a quarter of the chosen area will be reserved to the State and cannot be conceded except on the basis of a special law: the second quarter will be leased only to companies with Romanian capital, with nominative shares reserved exclusively for Romanian nationals: the third quarter to companies in legal existence at the time of the present law disposing of a paid-up capital of at least 500,000 million; the fourth quarter . . . without distinction among all comers to whoever offers the highest royalties. (Article 1.)

Fusions of companies for exploitation in the same region were forbidden (Article 2). Articles 3 to 8 set out the details of the administration of the concessioning procedure, royalty payments, and drilling obligations.

In the Chamber of Deputies Djuvara claimed that the solution given by the Law was "just, honest, and Romanian" (*soluție dreaptă, cinstită și românească*). In the long term, the Law's significance lay in the last ascription, since it introduced a distinction between Romanian and non-Romanian participants in the oil industry which became fundamental to the country's mining legislation, and which in consequence had important effects on the relationship between the industry and the State. This policy was underpinned by the stipulation on nominative shares. It was a novel feature in that shares in Romania were traditionally "to bearer": *nominative* shares, registered with the company in the name of the owner, made it possible to ensure that the capital stayed in Romanian hands. Citizenship thus became a criterion of ownership. It was maintained through subsequent formal changes—when, for example, as early as 1911, Liberal policy abandoned concessions to individuals as a technique of exploration, in favour of joint ventures between the State and private individuals, with

[1] *MPR*, no. 9 (1909), pp. 313–18 (Djuvara's statement); *MPR*, no. 10 (1909), pp. 355–60: details summarized *MPR*, no. 11 (1909), pp. 391–5.

[2] Industry and Commerce was separated from its former departmental association with Domains and Agriculture in 1908.

the former contributing lands and some capital, and the latter, the remainder of the capital and day-to-day management.[1]

The Romanian State's right to decide the terms on which its own property was to be exploited was uncontestable, however restrictive they might be. At this time, however, the State's domain, though extensive and comprising areas most favourably situated for working, was by no means all the land available for oil exploration. It was nearly twenty years before this trend was worked out to the point where all lands not already in private ownership became state lands.[2]

The demand for a "national" industry

The foregoing analysis of the industry demonstrates that, even in the heyday of international private capital transactions, foreign capital enjoyed no privileged position or even the possibility of unhindered operation: there was in Romania no counterpart of the ex-territorial rights enjoyed by foreign nationals investing in other countries lacking the means to develop their own resources, as, for example, Turkey. The State was able to impose its own legal and technical requirements in matters both of principle—as over the exploitation of its own territories—and of detail, as over, for example, the setting up of companies.[3]

The Conservatives were in general satisfied with the state of affairs: they took the view that under proper safeguards on state-owned land, and subject to Romanian law, exploration and exploitation should be open to all comers: Romanian capital should be encouraged to participate, but foreign capital should be invited on equal terms, from no matter what source. This reflected not so much enlightened internationalist opinion as a deep commitment to Romanian agriculture. The general assumptions on which the foreign investments were made, however, remained anathema to

[1] See exchanges in the Senate between Carp and Ionel Brătianu, in *MPR*, no. 35 (1911), pp. 1356–7.

[2] See below, pp. 112 et seq.

[3] Pleyte wrote to his principals, Geconsolideerde Hollandsche Petroleum Compagnie, on 22 Jan. 1907: "Permission must be requested from the Ministry of Domains to set up a second registered office of the Company in Bucharest. The application must be accompanied by: (1) a sworn translation of the Memorandum of Association and the Articles of Association . . . (2) A power of attorney for the Representative . . . In the Application with which these documents are sent the Representative also states that he submits to Romanian law." 4.

important sections of Romanian opinion, made politically articulate by the Liberal party. The advocates of "Romania for the Romanians"[1] regarded the "foreign capital" companies as, by definition, inimical to Romanian interests, but until the retirement of Sturdza from public life at the end of 1909 the cry remained a slogan without a programme, and its ultimately xenophobic implications were not worked out. This task was taken up by Sturdza's successors, the younger members of the Brătianu dynasty, Ionel and Vintila. The latter, when Mayor of Bucharest, first announced his ideas for the "nationalization" of the oil industry to an international audience at the World Oil Congress held in Bucharest in 1907.[2] These were designed to rectify "the weak not to say utterly insufficient participation of the national element in all its forms in an industry we call national" and rested on two general arguments, namely that Romania was sufficiently advanced technically to be able to stand on its own feet, and that its oil resources should not be monopolized by foreigners (Standard Oil) or, as exports, suffer through market policies not in accordance with Romanian interests (the failure of E.P.U. against Standard Oil in Germany).[3] These arguments were embedded in the received Liberal version of Romanian petroleum developments since the state lands crisis of 1900. The remedy lay in state action "reserving the greatest possible part in oil exploitation to the national element", whether in capital or labour, by, for example, the creation of an industrial credit bank and protection of small producers through state-owned means of transport and storage, and the work of the State Geological Institute. The basis of royalty payments to small landowners should be changed to allow them progressively to participate in the wealth created by the output of wells on their land. The State should act powerfully before it was too late: otherwise "it will be difficult to struggle with the unequal arms we possess".

Vintila Brătianu's analysis was inadequate: in 1907 the training of Romanian technologists and technicians had not advanced

[1] Sturdza in a speech at Prahova, June 1904 (see above, p. 30). "Romania for Romanians" was also anti-Semitic and anti-Hungarian, i.e. it implied a definition of nationality as a test of citizenship, excluding some who lived under the jurisdiction of the Romanian State and including some who did not. The grievances of the Romanians in Transylvania became an official part of Liberal policy after a speech by Sturdza in the Senate on 9 Dec. 1892.

[2] *MPR*, no. 38 (1907), pp. 1037–41, 1069–72.

[3] See above, p. 43.

sufficiently to support the scale of operations throughout the industry which his programme envisaged: the State had already provided against a foreign monopoly, and it could not command success in export markets, such as Germany, which were freely open to other sources of supply.[1] His reference to "unequal arms" curiously underestimated the already proven effect of state power through legislation and existing state institutions such as the C.F.R.

Four years later Vintila Brătianu (by this time Director of the National Bank) rehearsed these arguments in greater detail in a book written jointly with Constantin Hălăceanu, *Politica de stat în industria petrolului*.[2] The conclusions were similarly developed into a worked-out programme of state action, designed to determine the conditions for the immediate future of the industry and to promote national initiative. Measures under the first heading were those which the State as legislator could take at any time to protect the general interest and comprised (i) centralized control of all transport and storage; (ii) the completion of Constanța, and (iii) the construction of the main pipelines; (iv) and (v) limitation when necessary of all refining capacity to the level of the internal consumption of kerosine (to protect the small and medium refineries) with an eventual state monopoly of kerosine sales; (vi) a programme of prospecting organized by the State to establish the existence of oil deposits even on private properties; (vii) a cadastral survey of private properties in the oil area and the determination of minimum joint perimeters for exploitation; (viii) legislation to encourage industry and detailed regulations on administration making the widest possible use of the national element; (ix) the creation of an industrial credit bank, with Romanian capital, to finance installations, production, and the exploitation of lands found to be oil-bearing; (x) the development of professional and technical training; (xi) safeguarding the independence of Romanian exploitation from the influence of foreign control—though leaving open a possibility of co-operation with foreign groups for export sales.

Ideas for the promotion of national initiative aimed at the more distant future and envisaged the entry of Romanian capital into

[1] The Romanian Ambassador in Berlin argued Brătianu's case with greater relevance in 1934 at the time of exchange restrictions and import quotas: see below, p. 196 n. 1.

[2] Bucharest, 1911.

oil, above all in exploration, through the action of the State and through replacing the existing organizations. This was a matter of uniting the diverse interests of Romanians with those of the State, and of some of the foreigners who had already invested capital, as follows: state lands were to be reserved exclusively for Romanian capital, as a basis for a National Oil Company for exploration and production, which would collaborate in refining and marketing with some of the existing organizations. The State would contribute to the National Oil Company its own oil-lands and those bought back from companies wishing to participate in the venture. Romanian capital would be responsible for all installations and for future investment. This solution implied exclusively Romanian partners in the national concern: in fact, association of the State with foreign capital would be forbidden "for superior reasons" and because such a collaboration could be disguised as a concession, "which would exacerbate the abnormal situation in which we find ourselves today". This objection did not hold for refining, where collaboration between Romanian and the existing foreign capital—possibly involving the National Oil Company—would be permitted. As regards marketing, the only problem remaining in the internal market was one of organization: exports could continue as before, with participation in foreign sales organizations, or by direct selling, where eventually the interests of Romanian and foreign capital together would be "in a privileged situation". Collaboration between the three branches should be assured to attain the common end. The successive participation of the National Oil Company (for exploitation) in the two other spheres (manufacturing and selling) would form the bond of interest and the "happiest" means whereby the three could be transformed into a single unit. As the final result was due in the first instance to the selling price, a solution could certainly be found which, based on that price, allowed a profit margin to be allocated to each enterprise.

Brătianu and Hălăceanu concluded their exposition by referring to foreign capital already in Romanian oil, whose readiness to co-operate in the envisaged organization was held to be as necessary to the realization of the programme as action by the Romanian State, but their peroration restated the nationalist theme: the State would have to show more perseverance and concern itself in greater detail in all economic matters; "we should strive to convince Romanian capital and labour that their national and economic role is overwhelming and that, at the same time as [they earn] a brilliant return for their risks and their efforts, they will assist in the best possible way the work of economic emancipation for which we all strive".

Politica de stat în industria petrolului was a manifesto of economic nationalism in relation to oil resources. Some of its proposals, e.g. the pipeline and Constanța construction schemes (neither completed at the time), merely fitted particular technical necessities into a nationalist context and showed that they assisted towards "economic emancipation": others referred to developments such as the seaboard refineries, which never got beyond discussion. The novel feature was the emphasis on national ownership of raw materials, i.e. that they belonged to Romania whether they were exploited or not. Hence the concern in the first instance, for a National Oil Company for *exploration*. In this field, state power was to be exercised against foreigners not because, for example, they had infringed some law or other, but because they were foreign. Refinery and marketing could admit foreign capital under strictly controlled conditions: even here the Romanian voice, either state or private, would be decisive. The continued references to "part of" the existing foreign capital, and the unworked-out hints about the rationalization of various interests quoted above, seem to indicate the authors' willingness to come to terms with Germany (where in 1911 the agitation for a state kerosine monopoly was at its height) over joint working in parts of the Romanian industry in return for a guaranteed market—the "privileged situation" referred to at the end of the proposals for the future. Otherwise, Romanian oil was to be reserved for Romanians.

The arguments to this conclusion derived from an inaccurate interpretation of what Standard Oil had or had not tried to do in Romania, but Vintila Brătianu had already shown himself beyond reasoned discussion on this point. A more serious flaw was that they envisaged state action to realize a particular source of national wealth, the precise value of which at any given time depended on market factors which Romania could not control. Liberal opinion overestimated the importance of Romanian oil to the international oil trade. The failure of the attack by the Europäische Petroleum Union on the position of Standard Oil in Germany demonstrated that Romania was not a vital but merely a useful source of supply.[1] The agitation for a state monopoly also failed, and the guaranteed market did not materialize. By 1912 the Deutsche Bank was already seeking to sell its assets in Romania. Further, the increase in demand for gasoline and kerosine in Central and Western Europe was being met by production from the Netherlands East Indies and the United States rather than from Romania.

The programme implied political as well as economic results. State control of the industry would strengthen the Liberal interest: the manœuvres over the National Bank and other "patriotic undertakings" suggested that the industrial credit bank would be similarly manipulated for party ends. Moreover, the apparatus of state control would not only create jobs but, by the same token, multiply the number of Liberal Party dependants. This in fact happened. The next year, 1912, the Brătianus founded the "industrial credit bank", the Bancă Românească, whose operations became the mainspring in carrying through after the First World War the rest of the nationalization programme which, among other things, provided jobs for Liberal engineers and bureaucrats.

There is no evidence that these implications of *Politica de stat în industria petrolului* were widely discussed by the foreign capitalists[2] whose interests were threatened, though they reacted *ad hoc* on issues raised in the book such as the obligation to employ Romanian labour.[3] Foreign opinion noted that Romania's need of foreign capital was greater than the latter's need of Romania. This was, in 1911, correct, but did not remain so. Further, this reaction discounted the nationalist overtones in the programme. Seen from Bucharest, the oil industry not only offered a source of wealth and employment but symbolized Romania's emergence from tutelage and her aspirations to be a European power. It was noteworthy that the advocates of a rigorous interventionist state policy for oil were also those who promoted most vehemently the idea of *România Mare* ('Greater Romania') in which all Romanians should be united.[4]

[1] As it was, similarly, in the period prior to the Second World War: see below, p. 218.

[2] The book appears not to have been translated into French or German (the usual languages for works of this kind): summaries and excerpts appear in *MPR*, no. 32 (1911), pp. 1210–12; no. 34, pp. 1290–2; no. 35, pp. 1338–43.

[3] Henri Deterding of Royal Dutch-Shell and Vintila Brătianu quarrelled on this issue in 1909: see 5, also below, p. 78.

[4] This applied especially to Hungary, since the recovery of Bessarabia

Before the First World War these ideas[1] in Romania were confined to a small number of politicians, grouped around the two Brătianus; this in itself did not ensure that they would be carried out. The necessary opportunity was provided by the war itself, which enabled the nationalist programme in all its aspects to be realized.[2] For this reason the outbreak of the war provides a convenient juncture for a résumé of the industry as it had developed under the regime of foreign capital.

Total investment excluding the state contribution in the form of, for example, tank cars owned by the C.F.R. had reached Fr. fr. 440 million:[3] of this sum the individual contributions by nationality were as follows:

Germany/Austria-Hungary	31%
Great Britain	21%
Netherlands	20%
France/Belgium	14%
U.S.A.	7%
Romania	5%
Italy/Switzerland	2%

In terms of enterprises, German and Austrian capital financed 27 per cent of production[4] and 36 per cent of refining, Dutch

could only result from a dispute, possibly war, with Russia—as distinct from encouraging Vienna to exert pressure on Budapest. Further, Russian rule in the province (unlike Hungarian rule in Transylvania) involved no religious conflict, and did not permit the existence of national schools to be the focus of nationalist agitation and to train emergent leaders. There were signs as early as 1898 that Romania was prepared to "write off" Bessarabia: see Trotter's (chargé d'affaires, Bucharest) report, no. 47, 9 Aug. 1898, on the significance of King Carol's visit to St. Petersburg: and Kennedy's (minister's) dispatch on conversation with Sturdza, no. 53, 14 Sept. 1898: *FO Series 104*, vol. no. 136. On the subsequent developments in irredentist agitation, see R. W. Seton-Watson, *History*, Chap. XIII.

[1] Brătianu and Hălăceanu's study is additionally important outside the Romanian context, as being the first discussion of nationalist policies for the development of oil resources, which have since found a response in, e.g., Mexico, Iran, Iraq, and Indonesia, and have come to be argued in oil-consuming as well as oil-producing countries, in consequence of the greater involvement of government in all oil matters since 1911. The terms of the argument have become much more intricate but the basic attitudes and themes can be identified. See, e.g., J. E. Hartshorn, *Oil Companies and Governments*.

[2] See below, pp. 105–11.

[3] End-1913 figure: *MPR*, no. 3 (1914), pp. 116 and 118.

[4] 1913 figures: "refining" = % of 1913 total crude-oil runs, not theoretically available capacity.

capital 31 per cent and 29 per cent, French and Belgian capital 16 per cent and 12 per cent, British capital 5 per cent and 2 per cent, and U.S. 18 per cent and 17 per cent. In peacetime these participations merely indicated the predominant voice in management, the country of origin of some of its personnel, and the likely source of new capital. During the war, however, they took on a more direct political significance.[1]

Production in the first half of 1914 was running at a rate of 146,148 tons a month and was recovering from the drilling setbacks experienced during the two Balkan Wars of 1912 and 1913. New fields were developed at Bana-Moreni (Astra Română and Romanian Consolidated Oilfields) and Ţuicani (Steaua Română), both in Prahova; Fierbători (Romano Americana and Romanian Consolidated Oilfields) in Băicoi, and Ochiuri in Dămboviţa (Astra Română); exploration drilling at Ceptura, Valea Telegi, and Cricov showed positive results. Exports, likewise, averaged 80,604 tons a month and internal consumption 68,662: i.e. the stocks accumulated during 1913 were being liquidated. The course of the export trade had established the following pattern by destinations (1913):

Country	Quantity
U.K.	232,800
France	151,402
Germany	126,295
Egypt	121,642
Italy	118,643
Austria-Hungary	77,184
Turkey	64,682
Others	143,798
	1,036,446

The value of these exports to Romania was 100,773,850 lei, or 15 per cent of the total receipts from exports.

[1] See below, p. 72.

CHAPTER IV

The First World War

The industry during the period of neutrality, August 1914–August 1916

AT the outbreak of the First World War, Romania was still formally an adherent of the Triple Alliance, its connection having been reaffirmed as late as February 1913. The Romanian Government maintained successfully against its partners that the treaty was defensive in character and that, in the prevailing circumstances, it could not be invoked to compel Romania's participation. A Crown Council at Sinaia, on 27 August 1914, declared the kingdom's neutrality.[1] In consequence, for the next two years, Romania was subjected to various kinds of persuasion by both the Central Powers and the Allies to join in the conflict on their respective sides. The Romanian Government had no particular commitment to peace as such, if judicious participation would further Romania's nationalist aims. Ionel Brătianu's personal predilection was for France, but officially the choice between the two sides rested of necessity on what they could offer and on the military situation. The Allies could easily offer areas with Romanian population in Austria-Hungary but not Bessarabia, which was opportunely resurrected as a grievance and which Romania sought to regain not only on nationalist but also on strategic grounds. The Central Powers were quite open-handed about Bessarabia but were inhibited, by reason of Hungarian objections, from offering frontier rectifications in Transylvania that Romanian policy regarded as adequate. From either side Romania required war material.[2]

[1] See Seton-Watson, Chap. XV; and the report of Fasciotti, Italian minister in Bucharest, to the Minister of Foreign Affairs, Sanguitiano, dated 4 Aug. 1914, in *I Documenti Diplomatici Italiani*, Quinto Serie, 1914–18, vol. i, no. 63.

[2] See G. E. Torrey, *Roumania and the Belligerents, 1914–1916*; S. D. Spector, *Rumania at the Paris Peace Conference*, pp. 21–39; A. J. Rieber "Russian Diplomacy and Roumania", in H. L. Roberts (ed.), *Russian Diplomacy and Eastern Europe 1914–1917*.

In this situation Romanian assessments of the military position became the crucial factor. In 1915 the Allied failure at Gallipoli, the Central Powers' successes on the Eastern Front, and the final invasion of Serbia induced an orientation towards Berlin—as evidenced by the Romanian-German Trade Treaty of February 1916. Later in that year Brusilov's offensive in Galicia and the successful defence of Verdun seemed to tilt the balance of advantage in favour of the Allies, with whom Romania entered into a definite alliance on 17 August 1916,[1] joining in the war on 27 August, but only against Austria-Hungary and expressly for "the realization of her national unity". Germany, by reason of her alliance with Austria-Hungary, declared war on Romania the next day. Romania thus became a reluctant and suspicious ally of Russia.

The effects on the Romanian economy of this political and military situation as it developed over two years were far-reaching. The war created severe economic difficulties for Romania in that the course of hostilities deprived her of access to markets for her main exports and disrupted the main sources of external credit in Germany and Austria-Hungary for both state and private undertakings. The immediate effect was the accumulation of stocks and the restriction of credit within Romania. At the same time the State's policy of neutrality was called into question as a result of the commercial decisions of individuals; it was not a question of principle but as to where the balance of advantage lay in practice. Measures taken on cereals[2] and oil[3] demonstrate that irrespective of the formal legal position the character of neutrality was determined by the need to prevent the stifling of the economy, and entailed control by the State as *ad hoc* interpreter of the policy, initially in the form of prohibition but subsequently in positive direction.

[1] For text see Temperley, *History of the Peace Conference*, vol. iv, p. 516.

[2] The agricultural producers agitated for the removal of a ban on cereal exports, to overcome a storage crisis and prevent a complete financial collapse in agriculture. This accorded with the State's needs of increased revenue to support wartime expenditure, which implied earning foreign exchange and increasing taxation. Consequently, not only was the ban raised but export taxes were imposed in July 1915. However, the State was now involved, and sales were not left to individual exporters but were, from October 1915, made through a central bureau established in the Ministry of Agriculture. The proceeds of sales to Germany provided, *inter alia*, for the purchase of locomotives and other industrial goods not manufactured in Romania.

[3] See below, p. 73.

At the same time as "neutrality" as practised in Romania between August 1914 and August 1916 involved an extension of state power, the Allies imposed controls on their own nationals "trading with the enemy", as an instrument of economic warfare. These conditions posed new problems for the industry in that, through the nationality of shareholdings, pressure could be exercised on the decisions of managements of companies which were Romanian legal persons and therefore subject to the Romanian Government. Companies could therefore be classified in terms of their susceptibility to this sort of external pressure. According to an analysis made by the Romanian Government in 1915[1] thirteen companies representing 25·3 per cent of the total Romanian production (1914) were owned or controlled by nationals of the Central Powers, and sixty companies with 13·4 per cent of that production by nationals of the Allied powers. Among the neutrals were Netherlands, American, and Italian enterprises (with 31·4 per cent). The position of Astra Română, which produced 26·1 per cent, was recognized as Anglo-Dutch. In fact, the company functioned as a neutral, untouched by British legislation.[2]

In this way, therefore, economic warfare introduced the possibility of further tensions between the industry, or a particular section of it, and the State authority, on what were basically foreign-policy grounds—that is, in the sphere in which the primacy of the State was already acknowledged. Further, in the bidding for Romanian support, governments themselves went into the oil business, and arranged contracts through diplomatic representatives or accredited purchasing missions. After the closing of the Dardanelles in November 1914, which denied Russian output to the Allies, Romania became the sole European producer of gasoline, and as such, in the condition of contemporary warfare, of crucial importance to both sides. The Allies, however, had sources of production elsewhere, and their policy in Romania, through the

[1] Quoted in the *Report of the Anglo-French Commission to Roumania, 1920*, pp. 4a and 5a.

[2] In consequence of a decision taken in Dec. 1914 by the controlling Dutch shareholders, Deterding, whose sympathies were with the Allies, resigned from the board of Astra, whose General Manager, Jacobson, was instructed to inform the Romanian Government that the position to be adopted by Astra would be determined by Romanian Government measures. The British Government acquiesced, "trusting that Astra's policy would not be contrary to the interest of the Allies", but recognizing that "the company might have to dispose of products to enemy destinations". 6.

impossibility of shipment, could only be pre-emptive.[1] By contrast, the Central Powers, whose supplies were drastically curtailed by the Allied blockade, Russian enmity, and the fighting over the oil areas of Galicia, were the more dependent on Romanian production. As a result, the diplomatic contest between the two sides for Romania's participation issued in competition for petroleum products, principally gasoline, purchases being made from the companies owned by the respective nationals and from the open market. For such business, pressure was brought to bear on the Romanian Government to permit (or deny) the actual export. In these operations the companies were no longer principals but merely agents, and their autonomy was prescribed accordingly. Their strong point—the detailed knowledge of markets—became irrelevant to the purchase and sale of oil products.

The problems and new relationships created by law and diplomacy took shape and were worked out over a period of two years. In comparison the effect of the outbreak of war in Europe on the actual operations of the Romanian oil industry was instantaneous. In August and September 1914 the Romanian Government, in consequence of its policy of neutrality, forbade the export of crude oil, gasoline,[2] and fuel oil except with official permission. In November the closing of the Dardanelles denied access to about 70 per cent of the peacetime export market, and the Austrian operations against Serbia closed the Danube to traffic between Orșova and Belgrade. In 1914 and 1915 the loss of markets reduced exports from 55 per cent of refined production (1913) to 36·6 per cent and 25·6 per cent respectively, and created major refining and storage problems. Since it was necessary to keep refineries working to produce for domestic needs and for the permitted exports, principally kerosine and gas oil, an undisposable surplus of gasoline

[1] Large-scale buying of gasoline—84,000 tons in Dec. 1914 and 12,000 tons in Mar. 1915, at a cost of approximately £900,000—not only diminished quantities otherwise available for Germany, but assisted in solving the financial problems created by surplus stocks. Of the total purchases, 69,000 tons were stored at Constanța.

[2] The wisdom of this policy, as regards gasoline, was questioned by exporters on the grounds that it presented the Romanian market in Germany for this product, viz. 28 per cent of total requirements in 1913, to the benzol manufacturers, and this merely lost revenue to Romania without impairing German war potential: see *MPR*, no. 22 (1915).

and fuel oil choked the industry. This situation was only exacerbated by a successful outcome to exploration drilling during 1914.[1]

In default of an immediate increase in exports, the solution to the problem of excess stocks lay in discontinuing drilling programmes and cutting back producing-wells, but an equitable application of such a policy was ruled out by the way in which the producing sector of the industry was constituted: the disparities between producers were too wide. On 1 January 1915 of 988 wells in production 729 belonged to the nine major companies, while, at the other end of the scale, twenty-five companies owned one well each.[2] Additional complications ensued from the varying rates of flow.[3] Shutting back production was not technically difficult but entailed serious risks, such as the loss of a well through choking with sand. The major companies stopped the drilling of individual wells at levels short of the known oil-bearing strata and ceased exploration drilling. These straitened conditions, though serious for large companies, were more severe on the small ones, as they were less able to withstand the fall in crude prices[4] and the loss of their peacetime outlets, either brokers or the large refiners who had hitherto absorbed their output but were now preoccupied with minimizing runs. In 1915 the small producers petitioned the Government for a preferential allocation of the tankage constructed for the pipeline, and for permission to export crude oil to Hungary.[5]

The switching of the distillation process to maximizing the output of kerosine and gas oil (with demand for the latter stimulated by official price reductions) offered only marginal relief. In the end the problem of surplus output was met by accumulating the unusable products in reservoirs. Constanța became a storage depot and the tankage belonging to the Pipeline Service was similarly used. Individual companies built new tanks and in some areas

[1] See above, p. 69.

[2] See *Report of the Anglo-French Commission to Roumania*, p. 32 d.

[3] e.g. 1914

Company	Production (tons)	No. of wells
Astra Română	466,605	49
Romano Americana	420,531	48
Steaua Română	368,395	314
Concordia	81,965	146

[4] See *MPR*, no. 9 (1915), pp. 413–16; no. 14, pp. 631–2; no. 2 (1916), pp. 55–8.

[5] See *MPR*, no. 13 (1915), pp. 614–15.

dug pits; new construction, however, was restricted by the end of 1915 through the shortage of sheet iron: the substitute use of timber permitted heavy losses through leakage and increased the fire hazard. The conquest of Serbia (October 1915–January 1916), which reopened the Danube and allowed a greater rate of export, eased but did not solve the problem. Stocks, which in July 1914 totalled 300,000 tons, by the end of 1915 amounted to 800,000 tons—equivalent to 30 per cent of crude production. By the time Romania entered the war in August 1916, accumulated stocks in Wallachia, i.e. the main producing region and the area eventually occupied by the Central Powers, totalled 1,017,000 tons, nearly three times the level registered for the same area in the period 1912–14.

A transitory factor in the building up of excess stocks was the interruption in traffic to those markets still open to Romanian exporters—Germany, Austria-Hungary, Bulgaria, Turkey, and Switzerland. With the exception of Turkey, these countries were served by rail, but the system was inadequate for the tonnages involved, and the main export centre, the junction of Predeal, became jammed with tankcars. Conditions on the C.F.R. reflected also the chaos on the Hungarian railways, overburdened with war traffic, and the closure of the route via Czernowitz as a result of military operations in Bukovina and East Galicia. Trade with Turkey shrank because Constanța shippers declined the state war-risk insurance rates, and the Bulgarian authorities pleaded rail congestion against through traffic to Istanbul overland. For these reasons, by January 1915, the actual export traffic came to a standstill, with the exception of shipments to Bulgaria, where Baku kerosine was replaced by Romanian.

The export problem provoked a conflict between the industry and the Government; in April 1915 the latter took powers to seize, if necessary, all installations and products, on the grounds that only by so doing could the stagnant gasoline and fuel-oil export trade be revived.

This was the first time the State threatened control of the oil industry through seizure of physical assets as distinct from the control of phases of the industry's activity. In respect of the latter, control of exports exercised by the Romanian Government after the outbreak of war in Europe complemented that of the domestic market and virtually completed the structure of an oil monopoly, the achievement of which was part of the official Liberal programme. The lever was given to the Liberal Government as the interpreter of "neutrality". At the end of 1915 Vintila Brătianu in a speech to party members adduced a further reason to those given in *Politica de stat în industria petrolului* for giving this *de facto* state of affairs formal legal status, viz. that foreign capital had shown itself in some instances too readily identified with the interests of its own country of origin to the detriment of Romania's needs and interests.[1] He elaborated this view in the Chamber of Deputies, as *rapporteur* on the budget proposals for 1916,[2] arguing that the war had demonstrated how vital petroleum was to Romanian economic life, and that it was beyond question better that it should not be concentrated in a few foreign companies but in "the hands of the protector of all the general interests of the country—the State". In this way, wartime conditions in Europe and Romania furnished a telling argument of general rather than "technical" appeal for the nationalist programme.

The sequel was the introduction in May 1916 of a Bill to establish a state oil monopoly. The Bill virtually proposed the enactment of the recommendations of Brătianu and Hălăceanu's book, with the difference that the organization of the monopoly was to be left not to the State directly but to a mixed company in which the State and private Romanian capital should be represented. The Bill was not passed before the end of the session, and, by the time the next opened, Romania had joined in the war.

The industry during the period of belligerence and occupation, September 1916–November 1918

With the entry of Romania into the conflict, Vintila Brătianu became Minister of War. In this capacity he initiated measures to take over the management of "enemy" oil companies and to put the oil industry at the service of the State. The first course of action was secured by the decree of 28 August,[3] which defined as "enemy" those with any German or Austro-Hungarian element in their shareholding, no matter how small. Among the firms affected were Astra Română, in which the German shareholding

[1] *MPR*, no. 22 (1915), p. 1008.

[2] *MPR*, no. 7 (1916), pp. 312–14 (for criticisms see *MPR*, No. 6 (1916), pp. 245–8; and no. 8 (1916), pp. 325–8).

[3] See G. Pohl, *Die rumänische Petroleumindustrie während des Krieges, 1914–1918*, pp. 154–5.

was 4·11 per cent, and Orion, which had no German capital at the time, although at the founding of the company in 1910 one 2,000-lei share had been bought by a German national. On 7 October 1916 the Ministry of Justice ordered an exact evaluation of the property of the companies sequestered under the terms of this order. In this process their books and records were transported to Moldavia (and—as the course of the war turned against Romania—to Russia) for safe keeping. Employees of the companies were also removed to Moldavia, on security grounds.

The organization of the industry for war purposes derived from a decree of 10 September 1916, announcing the immediate control of all phases of the industry's activity (Article 1) and vesting detailed supervision in a supervisory commission set up in the Ministry of Industry and Commerce (Article 2).[1] The commission was appointed on 12 September under the presidency of Alexandre Radovici, Minister of Commerce. His colleagues included Alexandre Cottescu, the Managing Director of the State Railways, Professor Mrazec,[2] and, later, Dr. Edeleanu.

The commission's first task was to organize the industry: this it did by subdividing the oil regions into three groups, designed respectively to ensure essential internal consumption, to secure a three-months' reserve of stocks, and to cover the local needs of outlying districts. The first group—comprising the most important oilfields in Wallachia—was the most important since it assured government control of 90 per cent of the crude production at that time. Accordingly, this group was placed under a special inspectorate, subdivided further for the purposes of administration. Drilling was to be permitted here only under the express orders of the commission. The second group, which included the Arbanasi fields in Buzău and the fields of Băicoi, i.e. those the furthest away from the main refining centres, was held in reserve. New drilling could be undertaken in case of need by the operating companies. The third group, made up of several small scattered fields in Wallachia, was left unsupervised.

Control of crude-oil transport was also directly vested in the commission, who allocated crude supplies to the different refineries.

[1] *MPR*, no. 17 (1916), p. 706.

[2] Ibid., p. 707. Mrazec was a distinguished geologist who became adviser to Vintila Brătianu. The story that Mrazec promised to drink any oil found in Moreni testifies only to the hazards of finding oil.

Inland marketing was similarly centralized, being conducted by the Distribuția company, which had already been taken over by the State from its former owners[1] for that purpose. Foreign sales were to be conducted directly through the commission, which undertook to allocate orders among the refineries.[2] Finally, the commission instituted a complete survey of stocks to determine the quantities of crude oil and products available for the defence of the country. Such was the importance attributed to the commission that it was given special rights to requisition any geological or any other equipment necessary, and, at the same time, to supervise the requisitioning policy of the Ministry of War, particularly as regards transport, to ensure that production and distribution of oil were not adversely affected.

These decrees issued in consequence of Romania's entry into the war completed the structure of a state oil monopoly which Vintila Brătianu had introduced only four months previously: the treatment of Astra Română suggested that the Government hoped to make this emergency legislation permanent. On 5 September Jacobson, Astra's General Manager, presented evidence to the Ministry of Justice proving the Netherlands character of the company, and received assurances that it would be unmolested.[3] The next day, with the Orion, it was placed under sequestration. The Netherlands minister in Bucharest, van Vredenburgh, protested to Vintila Brătianu, to be regaled by a list of grievances against Astra, beginning with Deterding's opposition to the 'Romanization' proposals of 1909. He was further informed that, while Orion could be desequestered, Astra had to remain under official control on account of the German element in the company: gasoline and lubricants were vital to defence, and therefore the Government had felt itself obliged to take the companies under strict control.

Vintila Brătianu had added that, although the precise nature of the regime could not be foreseen, after the war the Government

[1] Astra Română, Steaua Română, and Româno-Americană.

[2] Since, in the circumstances, Russia was the only possible outlet, and transport facilities in that direction were wholly inadequate, foreign sales were negligible.

[3] Jacobson's efforts to hold Astra Română for the Netherlands were curiously paralleled by those of Rost van Tonningen during the Second World War, which involved him in trying to maintain a fine distinction between what was Dutch and what was German: see below, p. 237 n. 1.

would make a distinction between the companies: those that were tractable would be allowed to go on working, but those that were not could rely on the full opposition of the State. The Netherlands minister reported that the real reason for the maintenance of sequestration was the treatment to be meted out to enemy companies at the end of the war, when the price of the raising of the sequestration would be participation in some form of state-controlled enterprise. Sir George Barclay, the British minister, also protested to the Romanian Government and was informed that the sequestration was imposed because the Government suspected that the Astra was guilty of espionage in favour of Germany. Diplomatic interventions continued, and in December 1916 the sequestration on Orion was lifted: the Government refused to raise that on Astra, which was repealed only through the German occupation authorities annulling all such acts in the area they came to control.[1]

The role of the industry in the Treaty of Bucharest

This new relationship between the industry and the State was upset by the course of military events. The work of the supervisory commission began when Romanian troops were occupying Transylvania: it was cut short by a counter-offensive from the south and the west. By the end of October 1916 the southern and central Dobrogea had been overrun, and Constanța with its vast stocks of lubricants and fuel, including the 69,000 tons of gasoline belonging to the Allied Governments, captured intact. Furthermore, the German successes in forcing the passes of the Carpathians at the end of September opened Wallachia, the main oil-producing area, to imminent invasion.[2]

In view of the imperative need of the Central Powers for oil and other raw materials, the Allied Commands determined that some effective "scorched-earth" operation was necessary.

Allied notions on denying the oilfields to the Germans were first put to the Romanians in November 1916 by the French Military Mission in Bucharest, whose proposals envisaged the plugging of borings, the removal of essential plant, the transport,

[1] 7.

[2] The German novelist, Carossa, took part in the campaign as a medical officer and describes his experiences in his *Rumänisches Tagebuch*.

where possible, of existing stocks of products to Brăila, and the mixing together of what could not be so transported. In practice, however, the latter procedure was held up by the negotiations for the sale of products to Russia, and these stocks were run to waste.

Although these proposals had the backing of the Prime Minister, Ionel Brătianu, they did not, in the opinion of the British General Staff, go far enough, particularly in view of the fall of Constanța. In consequence, the possibility of effecting the thoroughgoing destruction of the oilfields was discussed by the British Cabinet War Committee on 31 October 1916, it was agreed:[1]

> that no efforts should be spared to ensure, in case of necessity, the destruction of the supplies of grain and oil, as well as of the oil wells. The latter are of particular importance owing to the German shortage of lubricating oils. . . . Lord Grey proposed . . . to press the Roumanian Government to take proper dispositions for the destruction of the supplies of oil and wheat in case of necessity, and for putting the oil wells out of action for some time to come. The War Committee approved of this proposal. . . . It was further agreed that the question of the best means of destroying the supplies should be referred for examination and action by the Chief of the Imperial General Staff. . . .

As a result, the Director of Military Intelligence, Sir George Macdonagh, was assigned the task of preventing the Germans from gaining possession of the corn and oil of Romania. The fieldwork was undertaken by the small party under the command of Colonel Norton Griffiths, who arrived in Bucharest on 18 November. By that time strong German and Austro-Hungarian forces had pushed through the Jiu and Olt passes and were turning south-eastwards across Wallachia, but the Romanian Government's opinion was overwhelmingly opposed to Colonel Norton Griffiths's mission, on the grounds that the German forces either would not enter the oilfields or would be quickly driven out if they did, and that anything approaching a wholesale destruction of such a valuable national asset would have a bad effect on Romanian morale. Colonel Norton Griffiths discussed the military situation with various Romanian general officers and concluded that, for the time being, the invaders' superior fire-power and tactics would enable them to overrun the oilfields. Thus he decided on drastic and immediate action.

Meanwhile the Romanian Government had appointed on 23

[1] 8.

November 1916—two days before the evacuation of Bucharest—a mixed Franco-Romanian commission to supervise the immobilizing of the petroleum industry on the lines originally urged by the French Military Mission. The rapid advance of the invading armies and diplomatic intervention from the Allies led to the official adoption of a policy of complete destruction. In the confused conditions prevailing, however, inadequate liaison between the Government and its local representatives led to opposition in the fields to acts of destruction or to participation in them only under threat of martial law. As a result, the measures ordered were not in all cases thoroughly executed. The destruction itself was carried out by the party already mentioned, together with Romanian engineers of the Franco-Romanian commission. In view of their imminent capture by the invading forces, Colonel Norton Griffiths himself concentrated his attention on the fields west of the Prahova river.

In spite of the areas to be covered, the state of the roads in winter, and the disadvantages accruing from military retreat, the destruction of the main centres proceeded rapidly. Târgoviște went up in flames on 26 and 27 November: installations at Moreni were destroyed on the following day, Ploești was gutted on 5–6 December, just before the arrival of the Austro-Hungarian and German forces. At Câmpina destruction was only partly effected. The refinery itself was situated in the middle of the town, which both restricted the use of explosives and entailed that important ancillary facilities, being located outside the town, could not be dealt with in time. Further, at the time of the operations, heavy rainfall soaked the lines of easily combustible material designed to carry fire into various parts of the refinery and made them useless. The effective burning area was limited, and such fires were the more easily dealt with since, among the German troops, whose entry into the town coincided with the withdrawal of the destruction party, there was an interpreter, Engineer Tappe, who had formerly been an employee of Steaua Română in charge of fire-prevention in Câmpina. Prompt action by the German troops under his direction ensured that fires were quickly brought under control and that various parts of the installation which had been soaked with gas oil that was inflammable only at a very high temperature never, in fact, caught fire. Thus it was that, of the major refineries and fields, Câmpina was brought back into production more rapidly than any other—the first output being of 332 tons for the month of February 1917.

Although fire was the principal and obvious method used to render the fields and installations inoperative, great damage was caused by the smashing of control apparatus and the dynamiting of derricks. Further, wells were plugged and pipelines squeezed by means of a mechanical press and a steam-hammer. Where circumstances permitted, the official destruction was supplemented by the quite unofficial but in their way no less thorough efforts of the local population. Altogether, Colonel Norton Griffiths estimated[1] that, as a result of his operations, about 800,000 tons of crude oil and products were destroyed. In other terms, the scorched-earth policy effectively denied the oil industry to the Germans, at a crucial stage in the war, for about five months.[2] From the German point of view the results of the destruction were mitigated by the capture of Constanța and more by the fact that, as German equipment and personnel had played an outstanding part in developing the industry before the war, the occupying forces had at their command both the technicians and the equipment most fitted for the reconstruction of the industry.

By 19 December 1916 the forces of the Central Powers had captured the whole of the Dobrogea and all Wallachia from east of a line from Focșani to Galați. The Court and Government removed from Bucharest to Iași. Military operations closed on 8 January 1917, with German troops on the Seret and Putna rivers. Included in the area of occupation were the districts of Prahova, Dâmbovița, and Buzău with their oilfields and the oil port of Constanța. Although responsibility for the occupation as a whole was vested in the Oberkommando Mackensen (OKM), the total occupied area was divided into four main zones—the fourteen districts of Wallachia being ruled by the *Militaerverwaltung*[3] (Military Administration).

[1] See *Report of the Anglo-French Commission to Roumania, 1920*; also *Report on the Anglo-French Destruction of the Oil Industry* by German Ministry of Foreign Affairs to Reichstag and Bundesrat, printed in *Petroleum* (1 and 15 Sept. 1918); and Karl Strauss (of Ploești), "Erinnerungen an die Zerstoerung der rumaenischen Erdoelanlagen", *Petroleum*, no. 36 (1922).

[2] The ultimate responsibility for the destruction, and therefore the source of compensation, became, in the general complex of post-war debts and claims, a much-disputed question between the Romanian Government and the French and British Governments.

[3] On the Occupation see Antipa, *L'Occupation ennemie de la Roumanie et ses*

The German High Command had given considerable thought to organizing the Romanian oil industry to serve the immediate strategic purposes of the war. Accordingly, an economic staff, the *Wirtschaftsstab*, including petroleum experts, had been attached to the OKM. The organization established itself in the offices of the Romanian Ministry of Agriculture on the evening of 6 December and immediately took charge of the whole of the economic life of the occupied areas. The petroleum section was subdivided into departments dealing with plant and refineries, distribution, consumption, finance and exploitation, and technical matters. The last two sections were stationed in Câmpina and eventually became part of the *Kommando der Erdoelfelder*,[1] while the other section remained in Bucharest under the style of the *Mineraloel Abteilung*.

The initial tasks of both organizations, in co-operation with other parts of the *Wirtschaftsstab*, were to survey and restore the oilfields, to seize any surviving stocks of crude oil or products, and to take a census of the materials available and necessary for repair work. The oil regions were isolated from the rest of the country and passports issued to the inhabitants. Reconstruction work, however, was hampered by the bad state of communications affected both by war and by the weather, the number of fugitives, and an outbreak of disease in the main oil areas of Câmpina and Ploeşti. Notwithstanding these difficulties, 65,000 tons of oil (apart from that captured at Constanţa) had been collected into German-controlled storage by the end of January 1917; production in Câmpina and the export of fuel to Germany began during the next month.

These measures were designed to meet immediate military requirements, but in addition the *Wirtschaftsstab* was charged with reorganizing the whole of the Romanian economy, in Germany's favour. Both the immediate and the long-term policy entailed a drastic reorganization of the oil industry. A beginning was made by

conséquences sociales; F. Sobotka, "Der Anteil Osterreich—Ungarns an der Militärverwaltung in Rumänien 1917–1918", in *Die Militärverwaltung in den von den österreichisch-ungarischen Truppen besetzten Gebieten*; K. Hesse, *Kriegführung und Kriegswirtschaft im Feinland: Lehren des Feldzuges in Rumänien.*

[1] "Oilfield Control", set up in February 1917 as a body with engineering and police sections to ensure the effective reconstruction of the industry. The Austro-Hungarian interest in this body was represented by the "K.u.K. Rohoel Kommando".

transferring plant and machine tools from companies owned by subjects of Allied or neutral countries to other firms who had made out a case to the *Mineraloel Abteilung* for their more effective use. In practice, this policy meant that Allied and neutral companies were deprived of much suitable equipment—gas engines, pumps, dynamos, etc.—in favour of German-owned companies. Shortly afterwards, in February 1917, the *Militaerverwaltung* ordered the various neutral companies to resume work in their fields and submit a programme of work to the *Kommando der Erdoelfelder* in Câmpina, under whose directions future operations were to be conducted.[1] This was followed by the publication of general regulations for the resumption of work throughout the oilfields on 3 April.[2] Companies with a predominant Allied shareholding interest were placed under the control of an administrator (*Zwangsverwalter*) who let out the property to contractors at rates designed to bring about the maximum increase of production in the minimum time. Among these contractors was the "K.u.K. Syndikat", which controlled various sequestered properties at Moreni and Băicoi and was formed to guarantee the Austro-Hungarian Monarchy's share in the exploitation of Romanian oil.

Considerable progress was made during 1917 in all aspects of the industry. The clearing of wells and repair of the fields in the occupied areas allowed a total production of 461,491[3] tons during that year. The concentration of effort in Prahova was demanded as much by its superior transport and refining facilities as by its already proven resources. The refining capacity of the Steaua Română and Vega companies was quickly put into action—the first products being made available at the beginning of February 1918.

Well-clearing went on during the whole period of the occupation: work was concentrated on those wells producing most efficiently before the invasion. On 1 July 1916, in the districts later to be occupied, there were 962 wells producing crude oil: in November 1918, when the German troops evacuated Wallachia, only 437 of these wells were in production, but from these wells about

[1] 9.

[2] See *Verordnungen fuer die Bevoelkerung Rumæniens im Gebiet der Militaerverwaltung*, no. 4 (3 Apr. 1917), item 29.

[3] Prahova, 419,476; Buzău, 14,639; Dâmbovița, 27,376. The same area yielded 1,219,093 tons during Jan.–Oct., inclusive, 1916, and 1,644,214 tons for the year 1915.

75 per cent of the total output of the area for July 1916 was obtained.

Overland transport difficulties persisted throughout 1917—principally through an acute shortage of rail tankcars, but the transport situation was somewhat improved by the reconstruction of Giurgiu and the linking of that port with the Prahova fields by means of a pipeline from Ploeşti. There the engineering branch of the *Mineraloel Abteilung* constructed a central pumping-station from which the oil flowed through the two five-inch lines previously laid for kerosine over the all-but-completed export trunk-route from Băcoi to Constanţa. The laying of these two lines from Ploeşti to Giurgiu took three months, and the first products—illuminating kerosine and diesel fuel—flowed through them in January 1918. Giurgiu was intended by the Germans to be the main Danube port for the river traffic of all the Central Powers, and, accordingly, not only were these pipelines constructed, but wharves, railways, and tankage facilities were also undertaken.

During the Occupation the distribution of products remained in the hands of the Distribuţia, but the undertaking was now directly controlled by the Germans. Crude oil prices were established by the *Wirtschaftsstab*, designed to benefit German, Austro-Hungarian, and neutral companies against those placed under forced administration—the former being paid until December 1917 150 lei a ton *loco chantier* and the latter only 80. After January 1918 these prices were raised respectively to 190 and 140 lei, at which level they remained until after the Treaty of Bucharest,[1] when one uniform price of 190 lei obtained for crude oil irrespective of the status of the company which produced it. The selling prices of products were similarly organized: sales to the Army of Occupation took place at a consistently lower level than those to the civilian population, for whom the price of kerosine was established at 400 lei a ton ex-refinery, against the 78 lei prevailing before 1916.

Control of the oil industry in all its aspects remained with the German authorities, who, given initially the task of reconstituting the disrupted industry for immediate strategic purposes, subsequently prepared the way for long-term German direction, with wider political and economic ends in view. This was initiated in advance of any settlement of the military and political situation and was marked by the liquidation of "Allied" companies and the

[1] See below, pp. 87 et seq.

establishment of the necessary German organs of control. The liquidation, carried out by the *Wirtschaftsstab*, was first announced on 1 June 1917 and extended to all companies with enemy-owned capital. The policy, however, was not put into effect until March 1918 (by which time formal peace negotiations were in train),[1] when, in two stages, a total of sixty-two enterprises, British, French, and American, were formally liquidated. This property devolved on the Erdoel Industrie Anlagen A.G., a company set up in Berlin for this purpose, for a total consideration of about 40 million lei. In return the Erdoel Industrie Anlagen A.G. took over all the rights and obligations previously vested in the individual *Zwangsverwalter*, including specifically all arrangements made by them with various independent contractors. The other aspect of this policy is represented by the setting up on 18 February 1918, by the *Wirtschaftsstab*, of the Oel-Laendereien Pachtgesellschaft (Oelpacht), to take over the exploitation of all Romanian state lands. This move issued specifically from the determination of the German military element (as distinct from their civilian colleagues, and their Austro-Hungarian allies) to secure its aims for fuel supplies and thus present the negotiators with a *fait accompli*.

Austria-Hungary's interests in Romania's oil were secured through the agency of the Petrol Block. This was constituted on 11 February 1918 from a number of small refining companies owned by Romanian, Austro-Hungarian, and Bavarian capitalists. In 1916 the different companies had jointly disposed of a refining capacity of 296,000 tons, though the actual throughput was less than half this figure (107,711). In addition, they had a small number of wells, but most of their revenue was derived from working-up crude on commission. After some quibbling by the *Militaerverwaltung* as to the precise ownership of the share capital of one of its member-companies, the Block was accepted throughout the occupation as Austro-Hungarian.

In this way the occupying powers clearly indicated their ideas about the future role of the oil industry and had organized the various bodies through which their policies were to be carried out.

During 1917 the front remained stabilized—the Romanian units reformed by the French Military Mission were effective in

[1] See below, p. 88.

preventing further advances, but could not themselves counter-attack by reason of the developing revolution in Russia. In the absence of effective Allied military operations, the Bolshevik *coup d'état* and the opening of Russo-German armistice negotiations at Brest-Litovsk prompted a similar Romanian-German armistice signed at Focșani on 9 December. The armistice paved the way for a peace settlement, negotiated between March and May 1918 and signed at Bucharest on 7 May. In the meantime, taking advantage of the situation in Russia, Romania formally annexed Bessarabia.

The foregoing summarizes a series of confused political situations—in unoccupied Romania itself, between the Romanian Government in Iași, Russia (both Tsarist and Bolshevik), and the Ukraine, and between the Iași Government and the Western Allies.[1] From the limited standpoint of this study, there emerge two factors of future relevance: (i) that the Treaty of Bucharest created feeling which complicated post-war relationships with Allied governments involving, *inter alia*, the oil industry, and (ii) that the attempts to keep Romania in the war entailed the setting up of the first Romanian diplomatic representation in Washington and the first active engagement of U.S. official opinion in the economic affairs and possibilities of the country.[2]

The political repercussions of this peace policy involved the resignation of Brătianu, who was prepared to make a tactical peace, and his ultimate replacement by Marghiloman, whose pre-war record as a Germanophil, it was hoped, would enable Romania to get lenient conditions.[3] These hopes were disappointed. The German-Austrian campaign had been undertaken not only to punish a faithless ally but—in a rather different order of motives—to exact specific economic benefits in the form of food and oil, for both civil and military requirements. There was some disagreement between the German military authorities on the one hand, and their civil counterparts and their Austro-Hungarian allies on the other as to what the pursuit of these benefits entailed, but none on the general issue. Further, it was held necessary to terminate operations as quickly as possible to release manpower

[1] See Spector, pp. 41–58, and references; Seton-Watson, pp. 498–520.

[2] See documents in *FRUS 1917*, Supplement 2, "The World War", vol. i, Washington, 1932; and discussion in B. S. Mamatey, *The United States and East Central Europe*.

[3] See Spector, pp. 50–3.

for the Western Front.[1] For economic and military reasons, therefore, the Central Powers were in no mood to wait. The terms of the treaty[2] reflected this mood and were severe, although Bessarabia was confirmed to Romania.[3]

The Romanian petroleum industry was prominently in the minds of all the major negotiating parties—Austria-Hungary, Germany, and Romania. Austria-Hungary was concerned mainly with effecting a political settlement to put relations with Romania back on their pre-war footing and with enrolling Romania as a third partner in an economic union with Germany. The petroleum industry entered into Austro-Hungarian calculations only in so far as deposits and existing workings were involved in frontier alterations.[4] The Germans, for whom immediate economic considerations were more important, had a more decided and comprehensive scheme for the Romanian industry, involving not only its production and refining aspects but marketing as well. As a direct con-

[1] See Ludendorff, "Notes for the Conference at Homburg of February 13th 1918", in *The General Staff and its Problems*, vol. ii, pp. 548–51.

[2] On the treaty, see G. Grätz and R. Schuller, *The Economic Policy of Austria-Hungary during the War in its External Relations*, Chaps. II–V (authors were economic advisers to the Austro-Hungarian delegation); Count Czernin (Austro-Hungarian Foreign Minister), *In the World War*, Chap. XI; F. Fischer, *Griff nach der Weltmacht*.

[3] During the dissolution of Tsarist Russia, local nationalists in Bessarabia proclaimed first its autonomy (May 1917) and then independence, as the "Moldavian Republic" (Jan. 1918), whose authorities, at the time of the Romanian–German armistice, called in Romanian troops to restore order. This was undertaken with French support and opposed by the Red Army. The resultant fighting was terminated in Mar. 1918 by the Agreement on Military and Political Matters, stipulating (Article 1) withdrawal of all Romanian troops within two months, except 10,000 specifically to protect Romanian warehouses and railways. This stipulation was not observed: the troops remained and on 8 Apr. 1918 the Bessarabian National Council voted for union with Romania. The Governments of Soviet Russia and of the Ukraine (Skoropadski) refused to recognize the validity of this vote: this created an issue which bedevilled post-war Romanian foreign policy (see below, pp. 100 and 102). See Seton-Watson, pp. 509–12, 519–20; Leonard Shapiro (ed.), *Soviet Treaty Series*, vol. i, no. 10.

[4] At the outset, in consequence of a projected frontier realignment in favour of the Dual Monarchy, the petroleum deposits at Okna and the frontier port of Turnu Severin would have fallen to Hungary. In the negotiations, however, these matters were not pressed and both places stayed in Romanian hands, though the shipyard at Turnu Severin was leased to Austria-Hungary for thirty years in return for a payment of 1,000 lei per annum and a guarantee that the yard and port workings would not be sublet to any other power. This stipulation was subsequently modified, with the consent of the Romanians, so that Austria-Hungary could grant to Germany a wharf-side and use of ancillary facilities in return for a 10 per cent share in the German shipyard at Giurgiu.

sequence of the latter consideration, German interest extended to the necessary ancillary features of the oil economy—railways, docks (particularly Constanța), river traffic, and the provision of materials. These divergent aims created dissensions between Germany and Austria-Hungary,[1] which were not fully overcome by the time negotiations opened. They also gave Romania an opportunity to defend her own assets, among which the petroleum industry was regarded as foremost. It was constantly urged from the Romanian side that Constanța was their only effective seaport, the loss of which would make Romania useless as a future economic partner. The Bessarabian coast provided no alternative outlet, since existing ports there were already silting up and, in any case, had poor communications with the hinterland. Furthermore, the Romanians argued that it would be dangerous to accede to Bulgaria's demand for the whole of the Dobrogea, which would place all the main communications between Central Europe and the Black Sea under the control of one power: any reserved rights in a Constanța–Cernavodă corridor would be only illusory.

In the event, these Romanian objections were partly sustained in the Treaty of Bucharest and its annexed economic agreements. The Dobrogea, as far as the St. George's, i.e. southern, arm of the Danube, was ceded by Romania to the Central Powers jointly, against an "assured trade route to the Black Sea via Cernavodă and Constanța". In the main, however, the treaty formalized the subjection of Romania politically and economically to the Central Powers, under the aegis of Germany. The occupied areas were to remain so for the time being, and the activity of the OKM was to be maintained until the ratification of the Treaty, although after it had been signed the Romanian Government had limited rights of interference. Certain communications, including railways, remained under military control, together with harbour works undertaken by the occupiers. Similarly, Romania's former status in the regulation of traffic on the Danube was abolished in favour of the hegemony of the Central Powers.[2]

The political aims of the Central Powers as set out in the terms of the full Treaty were supplemented by stringent regulation

[1] See Fischer, p. 449.

[2] The German plan provided for the replacement of international control of the Danube Delta, as established after the Crimean War, by a *Kommission der Donaumündung*, consisting of riparian states only. This plan was resurrected in 1939, and, *mutatis mutandis*, executed by the Russians in 1946.

of Romania's economic life in a series of agreements designed to secure to the victors Romania's "surplus" production, i.e. that which remained after the needs of the occupying forces and the rations of the civilian population had been accounted for. In the German view, German capital had played the predominant part in the creation of the Romanian oil industry and, in common with that of other foreign nations, had suffered severe restriction of action through the policies of successive Romanian Governments. A state monopoly such as introduced by Vintila Brătianu threatened the end of these assets, and the supervisory commission was regarded as a first step to the realization of such a policy. This diagnosis of the situation prevailing in the industry determined many of the issues included in the Petroleum Agreement, which, significantly, was negotiated directly with the Romanians by the Germans. The Agreement aimed not only at replacing Romanian by German authority as regards the limits within which all petroleum operations were to be carried on, but also at ensuring that the major share of the benefits resulting from such operations should accrue to Germany.

The key factor in the German position was the Oelpacht, which was to remain under the control of Berlin and to receive the sole right to exploit the oil resources of state lands in Romania, including those both in Wallachia and in the unoccupied area of Moldavia, for which in August 1914 no concessions had been granted. This right could be transferred to another company, if the Central Powers wished, and was to remain in force for thirty years—the period originally laid down for leases in the Romanian law of 1909.[1] Under the Agreement, however, the Romanian Government was bound to extend this by two further periods of thirty years, if required to do so, on five years' notice. Provisions of the 1909 legislation restricting the area of continuous land which could be leased were similarly done away with. In addition to its own rights on state lands, the Oelpacht was to receive a conditional right to territories already ceded to other companies when the concessions of the latter ran out, and the power to order the expropriation by the Romanian Government of privately owned land should the occasion arise. During the initial thirty-year term of its operations, the Oelpacht was to have complete tax-exemption and to enjoy all the favourable provisions of

[1] See above, p. 61.

existing Romanian legislation for the encouragement of industry: it was not, however, to be bound by the unfavourable conditions imposed by such laws. In order to ensure that the operations of the Oelpacht were not sabotaged or otherwise hampered, the company was given priority rights over all means of transport, private or state-owned, in the movement of crude oil or products. It was also to have extensive construction and maintenance rights over all ancillary facilities—roads, railways, telegraphs, telephones, and pipelines—and was empowered to demand for these purposes that the Romanian State furnish against payment the necessary timber from forests most suitably situated.

In return for its surrender of state lands, the Romanian Government was to receive a royalty of 8 per cent per annum on the market value of crude oil extracted by the Oelpacht, with deductions for refinery use and losses, 25 per cent of the ordinary shares in the Oelpacht, and participation in its profits.[1] The expected Romanian share was calculated on a basis of an 8–15 per cent dividend with an upward sliding scale should profits exceed this agreed rate. To secure the Romanian Government's expectations of income from the company, the Oelpacht was obliged to sink 100 wells every five years during the first half of its initial thirty-year term.

Under the terms of the Petroleum Agreement, the companies owned by "enemy", i.e. Allied, capital were dealt with as a separate entity. The Romanian Government undertook to recognize their compulsory liquidation by the *Mineraloel Abteilung* and the transfer of all their rights and obligations, which since the early part of the occupation had been held by the *Zwangsverwalter*, to the Erdoelindustrie-Anlagen G.m.b.H. (Petroleum Industry Investment Company). To this company was to be given the right to extend all concessions accruing to it as part of its assets for a period equivalent to the interval between the 27 August 1916 (i.e. the date of the outbreak of hostilities) and one year after the signing of a general peace.

The Oelpacht and Erdoelindustrie-Anlagen G.m.b.H. between them accounted for ultimate German dominance of the entire Romanian oil industry, except distribution. Control of the latter

[1] The capital structure of the company was such that the preference shares (representing not more than 10 per cent of the total share capital, but carrying all the voting rights, together with a fixed dividend of 8 per cent) were at the entire disposal of the Austro-Hungarians and the Germans.

was to be secured by the creation by the Romanian Government of a state petroleum monopoly, the rights of which were to be transferred to a commercial monopoly company, the Monopoly Trading Co. Ltd., to be founded by Austro-Hungarian and German financiers, under the direction of their respective governments. The capital structure was to be exactly the same as for the other two companies; the Romanian Government was to receive one-quarter of the ordinary shares. The commercial monopoly company was to control all sales of crude oil, for which purpose the Romanian Government was obliged to put at its disposal, on demand, all necessary refining and other facilities, including the free use of public highways and state wharves. Further, the new company could demand that the State expropriate assets belonging to third parties and transfer them to its own use. All new refinery construction was to be subject to its consent. The Monopoly was to control internal sales through Distribuția, in quantities and at prices to be established annually in consultation with the Romanian Government. However, all exports without exception were to be undertaken by the Monopoly alone, on payment to the Romanian Government of 3·60 lei a ton on crude and 4·0 lei a ton on products. Through these three organizations—the Oelpacht, Erdoelindustrie-Anlagen G.m.b.H., and the Commercial Monopoly—the Romanian oil industry was to pass under effective German control.[1]

The system of economic government envisaged by the Peace Treaty and its annexed conventions has been described as one of "military state socialism".[2] The retention of private property rights was made conditional on compliance with the instructions of the authorities, as regards both the management of the property concerned and the price at which its output was sold; failure to comply entailed either forced working by the military or outright expropriation. Irrespective of ownership, however, *all* industrial property in the occupied areas was regarded as being at the dis-

[1] The terms of the Treaty of Bucharest and its annexed agreements, taken in conjunction with the territorial and economic provisions of the Treaty of Brest-Litovsk with the Soviet Union of 3 March 1918, brought all the main European sources of oil under German control. The economic clauses relating to the oil industry were opposed in Germany by both the Deutsche Bank and the Disconto-Gesellschaft, who disliked the prospect of compulsory co-operation with the Oelpacht and rejected the idea of a trade monopoly. Neither the Deutsche Bank nor the Disconto-Gesellschaft were consulted by the German negotiators.

[2] By David Mitrany: cf. his *Effect of the War on South-Eastern Europe*.

posal of the occupying authorities or their agents. One of the results of this was the wholesale transfer of equipment belonging to any company (not merely to those with a predominantly Allied shareholding) to other parts of the country, as the authorities required. Only essential equipment for new plant was imported from Germany. In practice, however, the rigours of strict military government were relaxed when it was seen that a system largely based on requisitions failed to provide the required level of output. This in part provided a motive for the creation by the occupying authorities of the *Rumaenischer Wirtschaftsverband* (Romanian Economic Union), composed of occupation officials and Romanian representatives which determined the size of the "surplus" of any particular commodity and also had a share in determining prices. Furthermore, the *Wirtschaftsstab*, instead of relying wholly on its own military officials and experts, began to contract out its functions to middlemen, principally by granting monopolies for the sale of a single product in a given district. It was in this way that a "black" market in kerosine grew up in the towns. The production and bulk distribution of crude oil and products, however, remained under military control. Output steadily increased, the total for 1918 (1 January–31 October) for the occupied areas being 1,172,014 tons. Of this 827,295 tons was exported—Austria-Hungary received 194,442 tons, nearly all of crude oil, while Germany imported quantities of gas oil, fuel oil, and gasoline to a total of 608,543 tons; the balance of exports went to Bulgaria and Turkey.

The value of these oil products to the German war economy was attested by the Supreme Command which, meeting in October 1918 after the collapse of the Bulgarian front to consider future possibilities of carrying on the war at existing rates of consumption, should Romania also be denied to Germany, decided that both air and mechanized land warfare would cease absolutely in two months. However, limited land operations could be carried out on a strictly rationed consumption. Sea warfare could be maintained somewhat longer, for about six to eight months, but the outlook for home industry was very poor. Stocks of lubricants would be exhausted in six months and machinery running on oil would be without supplies after one to two months.[1]

[1] See the evidence of General von Kuhl given to the Reichstag Committee of inquiry into Germany's defeat, published in *Die Ursachen des deutschen*

The sudden deterioration in the German military position during the summer and autumn of 1918 presaged the end of German control of Romania and its oil industry. The collapse in the East began in Bulgaria. On 9 November Allied troops occupied Giurgiu. On the 10th Romania redeclared war on the Central Powers on the grounds that the Germans had increased the number of troops beyond those permitted by the Treaty of Bucharest, and Romanian forces advanced unopposed into Wallachia and Transylvania. The general cease-fire took place the next day on all fronts. The Germans evacuated Bucharest on 14 November but continued to occupy Buzău and Prahova. In these areas the transfer of power, which occupied a further week, reduced opportunities for pillage to a minimum: elsewhere, however, the absence of suitable precautions, and the internment by the Romanians of personnel left to hand over stocks and installations to the new authority, occasioned considerable losses through theft and severe damage to pipelines and storage facilities. In this way, the main lines on which the immediate movement of oil depended—those laid down by the Germans from Băicoi to Giurgiu—were made temporarily useless.

The collapse of German power in Romania entailed the end of the *Wirtschaftsstab* regime after nearly two years' operation and of the Petroleum Agreement, which had barely begun to be implemented. But both what the Germans did in regard to Romanian oil and what they planned to do had important residual consequences. The intended structure of permanent domination, the Oelpacht, E.I.A.G., and the commercial monopoly company, was recognizably the same as that which was the basis of Brătianu's monopoly proposals: it was therefore a clear demonstration of how they could be put into effect. Further, the German actions were widely considered as proof of the fate which awaited Romanian oil if foreigners gained the upper hand—a view which gained support for Liberal policies and was used subsequently to justify post-war Liberal enactments. Of more immediate importance, the Romanian Government had a lien on the demonstrably German sector as enemy property, and would necessarily be concerned in the administrative and judicial problems of "unscrambling" the assets rearranged by the occupiers to solve pressing technical problems.

Zusammenbruches. On the general question of oil supplies, see F. Friedensburg, *Das Erdoel im Weltkrieg*.

The war left the Romanian oil industry in serious straits. At the end of the Occupation, production in the areas under German control was running at about 105,000 tons a month. Immediately after the cessation of hostilities, the disorganization of the fields resulting from the departure of the Germans and the non-arrival of Romanian engineers from Moldavia caused production to fall to a rate of 45,000 tons a month. This output was obtained from 437 producing-wells. Stocks of crude oil products amounted to about 50,000 tons. A census of available transport showed that less than half the rail tank-cars available in Romania before the war were now available, and only 300 were found in Wallachia: even these could not be adequately employed, since there were insufficient locomotives for traction, the pre-war total of 1,500 having been reduced to a bare 150, mostly in need of repair. Tracks, bridges, and communication systems had been extensively damaged. Factories in Galaţi had to close down through lack of fuel, while available tankage at Buzău was completely full and production suspended, though these supplies and their prospective market were only seventy miles apart. Owing to looting, pipelines and tankage were in need of repair, and refining capacity had been reduced by destruction and dismantling to 1·5 million tons (from 4·5 in 1916). The repair of physical damage and renewal of facilities imposed a considerable burden on the administrative and financial resources of both State and industry in the post-war decade.[1]

[1] See below, pp. 136–7, 148–50.

CHAPTER V

Romãnia Mare

Changed relations between governments and the industry

THE First World War profoundly affected the relationships between the State and economic life in all the main belligerent countries. The necessity to mobilize the entire resources of the nation to support the efforts of mass-conscripted armies, the duration and intensity of the conflict, and its wide geographical extent, all involved detailed government or military regulation of the economy, even in states hitherto devoted to *laisser-faire*. Furthermore, the physical and financial damage in the industrialized areas of Europe was so extensive that reconstruction necessarily involved government. In the emergent countries the state authority was indispensable to their economic as well as their political organization. The effects of war direction persisted throughout the post-war period.

These general considerations are particularly relevant to the oil industry. The Allies had "floated to victory on a wave of oil"[1]—used not only as a fuel but also as a source of toluene and xylene for explosives. The evidence presented to the Reichstag Commission of Inquiry into the German defeat emphasized the importance of oil in terms of its absence. The implication was that no state, in future, could afford to leave the sources of these materials entirely unsupervised.[2] Additionally, the question of strategic

[1] Lord Curzon at British Government dinner to the (wartime) Petroleum Board, Lancaster House, 24 Nov. 1918.

[2] This view received its most authoritative public formulation from Senator Béranger, Chairman of the *Comité Generale du Petrole*, in his two books *Le Pétrole et la France* and *La Politique du pétrole*, which developed the theme of his inaugural speech to the *Comité* on 22 Aug. 1917. "La question du pétrole et de l'essence, moins que jamais, ne saurait être restreint a une simple question d'êpicerie privée. Elle devient chaque jour d'avantage une question d'utilité nationale." At the Lancaster House dinner he suggested that these wartime developments with regard to the industry, and the wartime inter-allied co-operation, should become the basis of "la politique interalliée du petrole realisé par l'accord des Gouvernements et des industries", as part of an international policy on raw materials.

control of sources of supply was strengthened by the further argument that the operations of the war itself as regards both the consumption of oil products and the physical destruction of producing areas had depleted the world's petroleum resources to a degree approximating to famine. This view was never more than plausible but nevertheless prompted governmental action to secure sources of supply. This in turn raised fundamental questions concerning the scope and nature of foreign participation in oil resources—whether it should be permitted and, if so, on what terms. The breakdown of wartime co-operation through the withdrawal immediately after the Armistice of the Americans from the Inter-Allied Petroleum Conference meant that the treatment of the problems referred to above was sharpened by rivalry between Britain and France on the one side and the United States on the other.[1] The result for the oil industry of this access of government authority was to place certain restrictions on its line of action, especially in requiring prior governmental consent to the sale of concerns to foreign nationals[2] and its use by governments as an instrument for the acquisition of concessions—evidence of a decisive change in the relationship between the industry and the State.

The search for "oil" security inevitably involved Romania as a source of supply whose potentialities had been increased by

[1] Mainly over rights in Iraq. The controversy terminated with the participation in principle of U.S. interests in the Turkish Petroleum Co. in 1923 (although the actual shareholding was not taken up until Oct. 1927). See *FRUS* and Gibb and Knowlton, Chap. 11; also H. Feis, *Petroleum and American Policy,* and J. de Novo, "The Movement for an Aggressive American Oil Policy Abroad 1918–1920", *American Historical Review*, vol. xvi, no. 4 (July 1956). On the other hand, during the same period, the pursuit of individual interests by Standard, Anglo-Persian, and Royal Dutch Shell in Russia led eventually to co-operation between the companies, once the intense suspicion generated by the Genoa Conference had been overcome (see Gibb and Knowlton, Chap. 12).

[2] e.g. British policy was set out in a letter from the Foreign Office to Lt.-Col. J. Boyle (accredited negotiator for U.K. oil interests concerning Russia), 29 Sept. 1921, ref. N. 10794/10242/38, para. 3, "Each company desirous of support by His Majesty's Government must undertake (1) to make such arrangements as will bring it under effective British control, (2) to guarantee that so long as His Majesty's Government accords to it the same diplomatic support as it accords to other British companies it will take no steps, either by disposal of shares or in any other way, which would affect the retention of British control without the previous consent of His Majesty's Government." On such a basis the British Government was prepared to offer all the support it properly could—it being understood that the Foreign Office decided in any particular case what constituted a "proper" degree of support.

the post-war frontier rearrangements.[1] Anglo-French policy towards Romania, as co-ordinated in the agreement of 8 April 1919 between the Rt. Hon. Walter Long and Senator Béranger, was specifically concerned with joint discussions with the Romanian Government for a participation in the distribution of enemy property and concessions on the state lands.[2] This agreement was withdrawn, for reasons concerning Iraq, but the policy regarding Romania was reaffirmed in slightly different form by an agreement between Béranger and Long's successor, Sir Hamar Greenwood, on 21 December 1919. The Governments no longer envisaged joint negotiations directly with the Romanian Government but only intended to support their respective nationals in any such negotiations. This policy was given the status of an international instrument in the Agreement of San Remo of 24 April 1920.[3] The future relationships between the two governments in regard to Romanian oil revealed one fundamental divergence of view. British concern stopped short at promoting the most suitable economic and financial conditions for the development of the oil trade and at protecting British investments. For France, however, Romania was not just a source from which, through investments and contracts, regular supplies of oil could be assured, but was part of the arrangements for containing Germany from the east that were developed as a substitute for the transfer of the left bank of the Rhine, which the Peace Conference did not allow.[4] The U.S. attitude towards Romanian oil was more akin to the British: the U.S. Government was concerned primarily to protect the position of the Standard Oil subsidiary, Româno-Americană, particularly in the matter of restitution for war damage. In July 1920 Bedford of Standard Oil and Colijin of Royal Dutch-Shell exchanged letters confirming their "accord as regards the policy of our Roumanian interests".[5]

[1] See below, p. 99.

[2] See below, pp. 107 et seq.

[3] On the development of Anglo-French policy, see *Documents on British Foreign Policy 1919–1939*, First Series, vol. iv, Chap. IV.

[4] On British and French policy in Eastern Europe and its determinants, see A. Wolfers, *Britain and France Between Two Wars*, particularly Chaps. VIII and XVII.

[5] This was not always reflected in relations between Româno-Americană and Astra Română, even though circumstances in Bucharest tended to throw them together. Till the end of their existence, the companies' policies clashed, partly through external factors, such as U.S. anti-trust legislation, partly through

Changes in the organization of the Romania state

The Romania which was the object of these governmental policies was vastly different from the Old Kingdom which had participated in the war. The crusade against Hungary for the possession of Transylvania had been successful: to this had been added Bukovina, Marmures, Crișana, and parts of the Bánát and the Dobrogea, though not without quarrelling with the Allies.[1] Bessarabia represented a bonus accruing from opportunism at the time of the disintegration of Tsarist Russia, though its formal status was not settled by the Peace Conference. These acquisitions transformed the Old Kingdom into *România Mare*—Greater Romania,—increasing its territorial area from 137,903 square kilometres to 292,244, and its population from 7·8 million to 16·9 million, of which about 28 per cent represented national minorities. All these acquired territories had to be grafted on to the Old Kingdom—a problem which was solved, formally, by the promulgation of a new constitution.[2] The territorial acquisitions

differing local practices in, e.g., production and marketing. There were certain identities of interest in times of crisis (see, e.g., pp. 126–9, 160–74), which prompted similar or agreed action.

[1] Romania's claims at the Peace Conference were based on the treaty with Great Britain and France of 17 August 1916, but by 1919 the original signatories found they had undertaken more than they could perform (since the new Kingdom of Serbs, Croats, and Slovenes also had extensive claims to the Bánát), while American policy rejected the entire policy of secret agreements. At Paris, therefore, the Allies argued that Romania, by making the Peace of Bucharest, had broken Art. 5 of the treaty of 17 August 1916 which was therefore no longer in force. They, the Allies, were thus released from the obligations under Arts. 3 and 4 concerning Romania's territorial acquisitions. Romanian claims that the peace treaty had remained unratified, and that Romania had not morally been at peace with the Central Powers, were countered by the argument that the Romanian Government had recognized the Peace of Bucharest, in that the renewal of hostilities had been justified on grounds of their infringement on its terms. Hence Britain and France considered themselves free to adjudicate between the rival claims to the Bánát and to determine the location of the Romanian-Hungarian frontier. Brătianu rejected outright (on the grounds of the infringement of national sovereignty) the proposed clauses for the protection of minorities within the areas to be ceded to Romania, which widened the breach between Romania and the Allies to a point where Brătianu broke off discussions and returned to Bucharest. Tension between the two sides increased after the Romanian invasion of Hungary, but Allied insistence on the withdrawal of Romanian troops and Romania's acceptance of the ultimatum of 1919 resulted in the resumption of relations. Brătianu had resigned on 27 September 1919, and the treaty was signed by a coalition government headed by Dr. Vaida Voevod. Romanian opinion remained convinced, however, that Romania had been the victim of the chicanery of Great Britain and the United States, and deprived of her just rights. See Spector, *passim.*

[2] See below, p. 112.

strengthened Romania in south-east Europe by extending its control of the Danube, though the Russian refusal to recognize the seizure of Bessarabia presented a continuing unresolved problem. Romanian foreign policy after the war became based on one simple proposition, viz. to maintain the new frontiers intact:[1] this suggested alliance with the other beneficiaries of the defeat of the Central Powers,[2] and with France, which replaced Germany and Austria-Hungary as the main source of external support. Romania's aims were also subscribed to by Italy, which emerged from the war determined to succeed to the former Austrian position in the Balkans, as envisaged by the wartime arrangements with the Allies, and accepted, in modified form, by the Peace Conference. In consequence, the Italian Government took the initiative in establishing political and commercial contacts with Greater Romania:[3] these developed to the point when, in 1923, the formation of a joint Italo-Romanian oil company was officially discussed.[4] Formal treaty relationships were established only in 1926—being retarded through Italian reluctance to ratify the Romanian position in Bessarabia, and by debt settlements.[5] Romanian relations with Great Britain had no such political or military basis. The British Government had recognized the acquisition of Bessarabia, but did not regard the settlements in Eastern Europe as irreversible, if agreed adjustments could be used to

[1] The single-mindedness with which this policy was pursued had disastrous effects during and after 1939: see below, pp. 223–5.

[2] From the Romanian standpoint the treaty with Poland (3 Mar. 1921) was designed to secure the frontier with the Soviet Union; that with Czechoslovakia (23 Apr. 1921) the frontier with Hungary; and that with Yugoslavia (7 June 1921) the frontiers with both Bulgaria and Hungary. These arrangements were part of the "Little Entente", and were supported by French policy in Eastern Europe, through the Franco-Polish treaty of 19 Feb. 1921 and the Franco-Czech treaty of 25 Jan. 1924. Discussions for a direct Franco-Romanian connection were initiated after the war but issued in a treaty only in Jan. 1926; see P. S. Wandycz, *France and her Eastern Allies 1919–1925*; R. Machray, *The Little Entente*. For treaties, see (i) Poland: *League of Nations Treaty Series*, Geneva, 1923. Wandycz, Appendix III, prints (reconstructed) military convention. (French policy in Poland was also concerned with oil supplies: see "Franco-Polish Convention of 6 February 1922", in *Journal Officiel*, 6 July 1922; also discussion in W. Lander, *Le Petrole en Pologne*); (ii) Czechoslovakia: *League of Nations Treaty Series*, Geneva, 1924; (iii) Romania: *Survey of International Affairs 1926*, London, 1928.

[3] See A. Adams, *Report on Economic Conditions and Prospects of Roumania at the Close of the Year 1919*, Cmd. 828, 1920, pp. 28–9.

[4] See *I Documenti Diplomatici Italiani*, Series 7/1923–25/2/pp. 16–19.

[5] See *Survey of International Affairs 1926*, pp. 157–9.

promote the general peace of Europe. Accordingly, until 1939, no specific political commitments were undertaken towards Romania, and, as mentioned above, official relations were concerned with economic and financial issues.

The international position of Romania, as ally of Poland and member of the Little Entente with the support of France and Italy, was much less strong than the formal structure of treaties made it appear. Effective co-operation with the other Danube states (Allied as well as former enemy) was hindered by the uncompromising nationalism which informed Romanian policy, and equally that of its neighbours.[1] What little provision was made in the peace treaties for economic co-operation was ignored, and wholesale restrictions were imposed on trade and finance. Tariffs intensified long-standing economic rivalries and were intended partly as weapons of external trade and partly as a screen for unstable currency situations.[2] Economic revival was hampered by inflation, the dislocation of traditional markets, and a railway transport system originally designed to serve main centres outside the new boundaries of the individual states and now deficient in vehicles and rolling stock. In Greater Romania, existing tracks in Transylvania served Budapest rather than Bucharest, and those in Bessarabia Odessa rather than Brăila and Galați; there was the further problem that the gauge in Bessarabia was Russian (seven feet) and had to be converted to standard.

The Danube ceased to offer a means of easy transport and communication: it now traversed seven different customs areas, with consequent multiplication of formalities and delays to shipping, and the riparian states differed over such essentials as the improvement of navigation—all of which stultified the internationalist intentions of the Peace Conference. Romanian policy shared in this: it consistently opposed the international control of the Danube and sought to limit its effect, denying access to particular waterways and discriminating against foreign shipping in cabotage (internal traffic) and use of port facilities.[3]

[1] "The foreign policies of the Eastern European states were directed by historical claims and memories to an extent that would have been paralleled in the west only if the basic object of British policy had been the recovery of Guienne and Normandy": Martin Wight, "Eastern Europe", in *The World in March 1939*.

[2] See Leo Pasvolsky, *Economic Nationalism in the Danubian States*.

[3] See W. D. Hinds, *Report on Danube Navigation, passim*; Sir Osborne Mance,

These external conditions prevalent in the Danube area after the war aggravated the difficult task of building up Greater Romania, which, although in political terms an extension of the Old Kingdom, had a markedly different economic structure. Considerable loss and damage resulted from the war and the occupation of Wallachia; the major part of the national gold stock had been transferred to Russia for safe keeping and had been seized by the Soviet Government in the course of the dispute over Bessarabia.[1] The financial situation was further bedevilled by the multiplicity of currencies in circulation and a special problem created by the occupation note issue.[2] In addition, the state finances were saddled with debts in the form of compensation payments to the owners of land expropriated under the various measures enacted between December 1918 and July 1921.[3] The creation of a unified currency, the ending of inflation, and the establishment of budgetary equilibrium were all prerequisites to the functioning of the new realm.

The internal politics of post-war Romania ensured that the problems of unification and reconstruction would be dealt with according to the theories of the Liberal Party.[4] With the final turn of the war, Ionel Brătianu, who had instigated Marghiloman's

International River and Canal Transport, pp. 51–70. For a Romanian view, see "Le Danube depuis sa source jusqu'à la Mer Noire", in *Correspondance économique* (official bulletin of Ministry of Industry and Commerce) (Jan. 1922), p. 22: for a summary of the technical problems, see "Le Réseau navigable de la Roumanie", in *Correspondance économique roumaine* (the same bulletin, restyled) (May/June 1925): on Romanian opposition to the European Danube Commission, see Permanent Court of International Justice, *Documents relating to Advisory Opinion*, no. 14, Series C 13, The Hague 1928; and E.D.C. secretariat's *Dix ans de régime international sur le Danube fluvial (1920–1930)*. For a comment after the Second World War, see below, p. 301 n. 3.

[1] The gold reserves of the Bancă Naţionala as at 31 Dec. 1920 totalled 495,458,997 lei, of which 315,179,980 was in Moscow, 80,000,000 in the Reichsbank in Berlin, 98,000,000 in the Bank of England, and, 1,700,000 in Bucharest. This gold stock was subsequently augmented by 14 million gold francs representing the Romanian share of the assets of the Austro-Hungarian National Bank: see A. Adams, *Report on Economic Conditions in Roumania*, to Apr. 1921, London D.O.T., 1921, pp. 17–19.

[2] These were in circulation during 1919 and 1920: "Bancă Generală" currency, issued by the occupation authorities; Austrian kronen in Transylvania, the Bánát, and Bukovina; "Lwow", "Kerensky", and "Bolshevik" roubles in Bessarabia; and Romanian lei issued by the Bancă Naţionala in the Old Kingdom: on currency and finance, see G. Cioriceanu, *La Dette Publique de la Roumanie*.

[3] See D. Mitrany, *The Land and the Peasant in Roumania*.

[4] See Manoliou, *passim*.

appointment,[1] now procured his dismissal on 5 November 1918. Brătianu and the Liberals returned to office on 14 December, and with consistent support from King Ferdinand ruled the country, but for two interludes,[2] until 1928.

The dominance of the Liberals in this decade rested on their being the "King's" party. Ferdinand had inherited them from his uncle in October 1914, and Brătianu's balancing act had been ultimately justified by the results. Liberal influence at Court, whether the party was in or out of office, was secured by Prince Stirbey, Administrator of Royal Domains and Brătianu's brother-in-law. Further, there was no Conservative party to which the King—if he wished to operate his uncle's system—could turn. The Conservatives were identified with the Treaty of Bucharest, and the basis of their power disappeared in consequence of the land reform. The party disintegrated. The Liberals' strength also derived from their hold on the bureaucracy and the banks—both of which were beneficiaries of Liberal policies.[3] This institutional factor was reinforced by opportunity. The Liberals maintained the wartime state of siege, and secured use of the administrative and election machinery, warding off emergent opposition groups by an election law whereby a party gaining 40 per cent of the total votes in a national election automatically received 70 per cent of the seats in the Chamber. For these reasons Liberal policies dictated the terms of Romanian politics until after the deaths of the King and Ionel Brătianu, when the party split on the constitutional issue and a coherent opposition[4] emerged.

[1] See above, p. 87.

[2] Brătianu resigned on 27 Sept. 1919 rather than accede to the Allies' ultimatum for the withdrawal of Romanian troops from Hungary. The odium of complying and of signing the peace treaties was left to a coalition under Vaida Voevod, whose Government, its mandate affirmed by relatively unmanipulated elections, was terminated abruptly by the King in Mar. 1920. It was followed by the Government of General Averescu, with Liberal support: this was withdrawn in the autumn of 1921. Averescu was eventually succeeded in Jan. 1922 by the Liberals under Ionel Brătianu, who remained in office until Mar. 1926. Averescu returned from 31 Mar. to 4 June 1926, and was succeeded by a proto-Liberal Government under Prince Stirbey, who gave way after eighteen days to the Liberals: they remained in office until Nov. 1928, by which time the deaths of King Ferdinand (20 Aug. 1927) and Ionel Brătianu (24 Nov. 1927) deprived them of their commanding figure and their main constitutional support.

[3] On methods, as seen by a contemporary business man, see G. C. Logio, *Rumania*, pp. 124–36.

[4] See below, p. 130.

Post-war Liberal policy aimed at creating out of the heterogeneous elements resulting from the peace treaties a centralized state with an industrial economy based on the extensive natural resources of the country[1] and the expanded internal market. Greater Romania was to be an expanded version of the Old Kingdom. Such a programme required considerable investment and was intended to be carried out with the greatest possible speed. Of possible external sources of funds, private foreign investment on the terms then understood in the main lending countries was anathema to Liberal opinion as leading towards "colonization": loans from foreign governments or the League of Nations were equally suspect, since they necessarily involved the giving of guarantees regarded as inconsistent with Romania's honour and an infringement of her sovereignty. The supply of funds within Romania, although adequate to serve a policy of selective industrialization, was insufficient to realize every one of the Liberal objectives. These funds derived from two main sources—from landowners whose estates had been drastically diminished by the expropriation, and who now sought an alternative outlet for their capital, part of which derived from compensation payments—and from the banks. The latter played a crucial role in Romanian economic life, and twelve large houses—all but one situated in Bucharest—formed the available source of credit. With the exception of the Bank of Romania Ltd., they functioned more as industrial investment trusts, the funds for which were derived from a policy of consistently high profits on banking transactions.[2] Furthermore, since the directorates of the banks were largely composed of leading members of the dominant Liberal Party, the banks[3] themselves became the channels through which the Government's economic policies were carried out.

[1] Of which oil was only one: *România Mare* also had bauxite, coal, chrome, manganese, copper, pyrites, lead, zinc, and mercury, with great resources of hydroelectricity.

[2] See V. Madgearu, *Evoluţia economii româneşti după războiul mondial*, p. 354.

[3] The principal Liberal banks were the Bancă Generală Română, the Bank Berkovitz, and the Bancă Românească. The latter, the Brătianu bank, increased its capital during 1920 from 60 million to 160 million lei: it opened branches in Bessarabia and Transylvania, acquired holdings in provincial banks, the Petrosani coal-mines, timber, the electrical industry, and other "commanding heights" of the economy. See D. Rottman, "Les Grandes institutions de crédit en Roumanie", *CER* (Jan. 1922), pp. 5–16.

The creation of a national oil industry

Their advent to office in the administrative and economic confusion prevailing at the end of the war presented the Liberals with an opportunity to realize their scheme, frustrated by the course of the war, for a state monopoly and the nationalization of the country's petroleum resources,[1] now enlarged by the addition of the Transylvanian natural-gas industry.[2] This opportunity derived from the dependence of the industry on the Government in such essential matters as re-establishing property rights and importing materials necessary for reconstruction, and from the right accorded to Romania to take over enemy assets in part-payment of reparations.[3]

The "National Oil Company" for exploration was founded in January 1919 as Creditul Minier by a group of Romanian engineers and geologists backed by the Bancă Albina, Sibiu, and the Creditul Tehnic, Bucharest. Its operations (which included exploitation of copper mines in the Dobrogea) were heavily financed by the Bancă Naționala and the Bancă Românească, whose representative, Stefanescu, was appointed to the Board in October 1919. In conjunction with the Bancă Românească, Creditul Minier later participated in other oil companies of which the most important was the Industria Română de Petrol (I.R.D.P.) founded in June 1920, again in conjunction with a group of Romanian engineers.[4] Minority participations were conditionally offered to other companies.[5]

Hoisescu, Secretary-General of the Ministry of Industry and Commerce, became the Director-General of the new company, which was originally designed to operate state land concessions and a state export monopoly, to run the Giurgiu and Constanța pipelines, and have sole charge of inland distribution. This latter attribute was withdrawn after the controversy generated by the relevant decree law[6] and in consequence of Distribuția's refusal to

[1] See speech by Osiceanu at Câmpina, 18 Jan. 1920: *MPR*, no. 3 (1920), pp. 81–90.

[2] Which passed directly into the hands of the Romanian State as "enemy" property: see below, p. 135.

[3] See Treaty of Versailles (Article 279), Treaty of St. Germain (Article 249), Treaty of Trianon (Article 232).

[4] 10.

[5] 11.

[6] See below, pp. 139–41.

transfer the former Deutsche Bank share interest to the I.R.D.P.[1] The company was therefore given 100 hectares of state lands in Runcu, Moreni, and Ochiuri as compensation. From the outset these two companies, as "friends of the administration", received preferential treatment in the matter of concessions on state lands and became the organs *par excellence* of Liberal oil policy. Guest[2] commented subsequently on their position: "Their progress was not looked upon as a matter which concerns their management and shareholders but as something in which the pride and prestige of the whole nation were involved."

The prospect of the State's being able to take over the German and Austro-Hungarian sector, raised in Allied discussions of post-war policy, implied that the Romanian Government would dispose of shares in thirteen principal undertakings engaged in all phases of the industry and accounting for about 30 per cent of the total nominal share capital and 25 per cent of total crude oil output. The Romanian Government anticipated the probable outcome to the peace treaty discussions by appointing sequestrators to all companies with any "enemy" element in the shareholding.[3] This included not only such German companies as Steaua Română (to which an Inspector at the Ministry of Finance, Constantin Osiceanu, was appointed),[4] but also Astra Română, which as an officially neutral concern still had two German nationals on its board, and whose sequestrator was Constantin Hălăceanu,[5] co-author of *Politica de stat în industria petrolului*. This circumstance, and his exercise of the widest powers, encouraged the belief in the Astra shareholders that this was a Liberal party manœuvre designed to "nationalize" the concern. The Astra management appealed to the courts, who pronounced the sequestration illegal, and Deterding inaugurated official interventions with Ionel

[1] See below, pp. 139–41.

[2] Astra Română General Manager after Jacobson. 12.

[3] This implied not only the appointment of officials to the firms but also the prohibition of stock-exchange dealings in their shares.

[4] See *MPR*, no. 10 (1920), p. 329.

[5] Hălăceanu had been appointed to Astra as Military Director, i.e. representative of the Ministries of War and Industry and Commerce, under the Decree of 9 Nov. 1918 reimposing the type of control inaugurated by Vintila Brătianu in Sept. 1916 (see above, p. 77). On assuming office, Hălăceanu addressed a series of demands to the management, including disclosure of all the company's books, prohibition of foreign languages in correspondence, reports on oilfield operations and production and the imposing of his own censorship on all correspondence and payments.

Brătianu (then in Paris), who in January 1919 instructed that it should be lifted. Hălăceanu remained in the organization as 'military director', with virtually the same extensive powers as before and with salary charged to the company until February 1920.[1]

The state controllers, however their authority was derived, were the means of promoting the acquisition of enemy assets, and, as a first step, on 15 May 1919, the Ministry of Industry and Commerce ordered the depositing with these officials of all shares of companies controlled by them, whether held in Romania or abroad, for scrutiny and stamping.[2] The shares were to be accompanied by certificates proving ownership after 28 August 1916; shares found to be in the hands of enemy nationals after this date were to be confiscated. The decree set a time-limit for registration but said nothing about the return of shares to their owners—an omission which gave rise to the suspicion that the result would be "a temporary sequestration rendering the shareholders' capital immobile".[3] By this means German and Austro-Hungarian holdings in banks, brewing, milling, and other economic activities passed into Romanian hands.[4]

The most important oil enterprises, however, escaped confiscation. Post-war changes in the nationality of shareholders converted former citizens of Austria-Hungary into respectable members of the Succession States whose property could therefore not be touched: on these grounds the Petrol Block, which during the occupation had been treated by the authorities as an Austro-Hungarian interest, escaped liquidation after an appeal to the

[1] 13. The abolition of the military directorate offers an interesting comment on the Romanian political situation. Guest reported: "As the diplomatic intervention did not appear to have any result, I took the step of enlisting the services of Osiceanu, and obtaining the assurance from him that the Liberal Party had no objection to offer to the removal. I conveyed this intimation (together with further full explanation of the legal position) through Mr. Zanne to the Minister of Industry and Commerce. I think it highly probable that Mr. Osiceanu's help was useful, not only because of the tendency I have noticed in this Government to try to avoid giving offence to the Liberals but also because I attach some significance to the fact that Osiceanu was well informed of the decision some days ago, i.e. before we had heard of it."

[2] See *PT* (28 June 1919): also *MPR*, no. 5 (1919), p. 162; no. 8 (1919), p. 266.

[3] 14.

[4] The final law for the liquidation of enemy property as held in Aug. 1916 was passed on 13 June 1923: see *MPR*, no. 12 (1923), p. 887.

Romanian courts.[1] Similarly, it was found that the most important German interest—the shareholding of the Deutsche Bank in Steaua Română—had been deposited by the bank with a Swiss company, Eos S.A. of Glarus, which had further deposited them in Zurich and Stockholm as a pledge for debts. The enemy-assets provisions were therefore inoperative, and the Romanian Government lacked the foreign exchange to redeem the pledges. Standard Oil, however, had the necessary resources and attempted to buy the shares: this failed in face of a threat by the Romanian Government to refuse to ratify the transation by stamping the share certificates. In July 1920 an Anglo-French financial group acquired an option to buy the shares and negotiated with the Liberal banks to put up the necessary capital to acquire them.[2] The Romanian group had not sufficient foreign currency to purchase its proportion of shares, and the funds for so doing were subscribed by a loan from their Franco-British partners, with the shares themselves as security. This transaction was conditional on the approval of the three governments concerned, which was given during July and August. In consequence, the Romanian Government removed the sequestration from Steaua Română and her affiliated companies on 18 September 1920 and duly ratified the changes in ownership by stamping the share certificates on 7 October 1920.[3] The cost of the shares was Sw. fr. 77 million (about £3·7 million at the current rate of exchange).[4] The manage-

[1] *MPR*, no. 1 (1919), p. 20; no. 2 (1919), pp. 63–5.

[2] Agreement of 9 Aug. 1920, between

Anglo-French Group		*Romanian Group*
Banque de Paris et des Pays Bas	Stern Bros.	Bancă Românească
Miraband et Cie	James Dunning & Co.	Bancă de Credit Român
Societe d'Études et d'Entreprises		Bancă Marmarosch Blank
A group represented by M. Mercier		Bancă Generală Română

The Romanian group had 50·1 per cent of the interest on the option, the French group 24·95 per cent, and the British group 24·95 per cent.

[3] See *Monitor Oficial*, No. 132 (17 Sept. 1920).

[4] 15. This transaction was later held up by their opponents as an example of the Liberals' operation of government to suit party ends, by which they had privately settled for 51 per cent of Steaua Română instead of the 100 per cent to which the Romanian State was entitled. Many of the details remain obscure, but in essence it appears that, after the sale of their assets to Eos was concluded, the Deutsche Bank acquired an option on them for resale, which they eventually transferred to the Banque de Paris et des Pays Bas. The option was first offered

ment of the company became predominantly Romanian, and the Romanian banking group nominated the State Administrator, Osiceanu, as General Manager.

In addition to the Deutsche Bank's interest in Romania, that of the Disconto-Gesellschaft also reverted to Allied ownership, although the process did not require a series of complicated and interlocking negotiations between the governments, banking groups, and oil companies of the three countries concerned. The Disconto interest, held through Deutsche Erdoel A.G., in the assets of the Concordia, Vega, and Creditul Petrolifer, was acquired by a Belgian group, "Sirius", financed by Romanian capitalists and by the Cie Financière Belge de Pétrole. Sirius obtained the desequestration of these assets on the grounds that during the war they had become vested in a Swiss company, "Terra", from whom Sirius had bought them.[1]

Although control of the former German companies was shared between Romanian and other nationals, the Romanians filled the positions of authority on the boards of the various companies and had direct personal links with their country's politics.

Thus, by the end of 1920, as a result of the war, the pattern of ownership in Romanian oil had radically changed. All German and Austro-Hungarian capital, which had been largely instrumental in building up the industry, was now excluded, and Romanian nationals controlled six of the largest enterprises in the industry, representing 30 per cent of the crude oil output. Foreign-capital enterprises, of which the main representatives were Astra Română and Româno-Americană, were politically isolated and faced by competitors in the industry who at any juncture were assured of government support. These two concerns

through Lazards to Royal Dutch-Shell, but the price was considered excessive and Royal Dutch-Shell stipulated for control of Steaua Română, which was unacceptable. The foreign shareholders were later embodied in two holding companies, respectively Steaua Française (now "Omnium Français de Pétrole") and Steaua Română (British) Ltd., in which the shareholders are Stern Brothers, Whitehall Securities, BP, and Shell International Petroleum Company.

[1] Following the decision of the tribunal, Concordia and Creditul Petrolifer held on 1 July 1920 general meetings to alter the distribution of share capital so that a majority would be held by Romanian nationals: cf. *MPR* (1920), pp. 365–7. In the case of the Vega this process was held up by a dispute between the French and Belgian shareholders in the Compagnie Financière Belge de Pétrole (MPR (1920), p. 501); for the further consequences of this deal, see below, pp. 119 and 120.

came under incessant pressure to "nationalize". Guest reported on 12 November 1920:

> The Nationalisation of the Petroleum Industry is getting more and more intensively discussed. Our friends at Brașov [Photogen] report that they have been practically threatened with confiscation unless their capital comes under Roumanian control by the 31st December.... The latest gentleman to tackle me on the subject was Mr. L. Berkowitz, the Banker. As others have done before him he told me it was essential for the Astra so to rearrange the holding of its capital that it should be in appearance at any rate predominantly Roumanian. Like all others who speak in this way he appears to be an agent of the Liberals. His suggestion was more businesslike than that of Osiceanu, whose proposition is briefly 'give us your shares or otherwise you will not have any peace'. Berkowitz suggested that the Liberals should form a Syndicate which should obtain a concession on State lands which they would bring to the Astra in exchange for a block of shares. Further he suggested, though rather obscurely, that the shares could be held under such trusts and agreements as would prevent the Roumanians, even if in a majority, from exercising any real control.

This appears to be the first recorded approach of a number made to Astra in the course of the period 1920–5, with much the same ends and inducements:[1] they were all turned down. Româno-Americană was similarly approached. The ability of the two companies to withstand such pressure rested on their size and efficiency relative to the industry as a whole, and their international connections.

Astra's basic policy was set out by Kessler, reporting a joint view of Deterding and himself, in a letter to Guest of 17 August 1920:

> It is worse than useless and entirely wrong to fight anything made by the Roumanian Government which they consider in the interests of their country. If their decision is wise and sound we should support them instead of fighting them and if their decision is unsound it is no use fighting it because such a decision will never stand and anything which is economically wrong will disappear after a short while. We do not wish to mix in politics in any way and we want to look after our own business and do all we can to develop our company into a prosperous and sound one for the benefit of the country in which it is working, its shareholders, employees and workmen.[2]

This declaration of intent was written apropos the foundation of I.R.D.P. It remained the point of departure for Astra's policy

[1] See also below, p. 125. [2] 16.

but came to be very much modified. The setting up of a party oil concern was a strictly internal matter: the attack on "acquired rights"[1] belonged to a different order of activity. However much the management of Astra wanted to look after their own business, the fact remained that Vintila Brătianu wanted to look after it too—which inevitably forced the company to "mix in politics", not in the sense of subsidizing particular Romanian political groups, but in requesting external support from the governments of its shareholders on issues affecting its interests, where international law allowed. Deterding's view of business derived from his previous experience was essentially international, which, together with his extreme practicality, led him at this stage to underrate the extent to which nationalist political practice could maintain in being what he regarded as economically unsound.[2] (There were, however, no precedents in this respect.) He tended to think of the Liberals in terms of Horatio Bottomley[3] and misjudged the compelling emotional attraction of "Romania for the Romanians", particularly amid the euphoria of the peace treaties. The Liberal schemes for oil-exploitation failed—but not through economic factors alone.[4]

The main contribution of the Romanian State to the national-capital companies was twofold—administrative permissions and allocations of areas of state lands as working territories. Nevertheless the exploitation of the lands as a whole extended beyond the immediate competence of any one concern. The Romanian Government, therefore, sought to use them as a means of raising a loan, or as a source of crude oil to supply against raw materials which the country needed. At the same juncture, as explained above, the British and French Governments were particularly interested in Romanian oil and were very aware of Romania's foreign exchange stringency. The joint action agreed by Long and Béranger issued in discussions with Ionel Brătianu in Paris. Discussions on the same two issues—state lands and the enemy

[1] See below, p. 121.

[2] Deterding's post-war career, till his retirement in 1936, can be viewed as one large-scale attempt to recreate the situation which had been disrupted in 1914, in which the easy exchange of goods and services across frontiers was an essential component.

[3] 17: Guest commented "the real difficulty here is that . . . the Bottomleys are rather apt to rise to the top". His own view was that only the peasants could sweep away the Brătianist system: see below, p. 124.

[4] See below, p. 130.

oil companies—were continued with the Vaida Voevod administration by Anglo-Dutch companies supported by their governments.[1] Foreign governments were also necessarily involved in the Steaua Română deal—which, in one aspect, represents the success of the joint Anglo-French policy of assuring sources of supply. Romania's pressing need for materials also dictated negotiations in Paris, under the aegis of Titulescu, Minister of Finance in the Averescu Government,[2] between Romanian officials and members of a group of French metallurgical interests in January 1921. The proposals envisaged the use of Astra Română, in association with the State, to drill state lands for oil, a proportion of which would be supplied by the Romanian Government to the French metallurgical industry against a loan and deliveries of steel products.[3] The scheme was not put into operation. The financial problem remained,[4] and so did the possibility of using state lands as a solution. Ionel Brătianu and the Italian Minister, Aloisi, discussed a possible concession to an Italo-Romanian oil company in April 1923.[5]

The Constitution of 1923 and the Mining Law of 1924

In *Politica de stat în industria petrolului*, Brătianu and Hălăceanu had suggested that the scope of state lands should be widened and that all subsoil resources not already concessioned should become state property. The opportunity to realize this part of the programme arose in connection with the new constitution passed in a special session of Parliament on 29 March 1923.

The newly acquired provinces of Greater Romania had their respective legal systems rooted in Austrian, Hungarian, Bulgarian, and Russian law—which posed the problem of their relationship with that of the Old Kingdom. The Romanian Government's immediate response at the time of the *de facto* union of the new territories was to guarantee a certain legal and administrative autonomy. After their accession to office in January 1922, the Liberals elaborated a new constitution,[6] designed to provide for

[1] 18. [2] See above, p. 103 n. 2. [3] 19.

[4] See Vintila Brătianu, *La Situation financière de la Roumanie*.

[5] *I Documenti Diplomatici Italiani*, 7th Series, vol. ii, no. 10: for the outcome of this policy, see below, p. 130.

[6] See *BFSP*, vol. 121 (1925): also D. Mitrany, "The New Roumanian Constitution", in the *Journal of Comparative Legislation and International Law*, 3rd Series (Feb. 1924).

the assimilation of the post-war acquisitions to the Old Kingdom. The first Article declared Romania a unitary state and the national territory indivisible. The traditional provinces were suppressed, under Article 4 the country being divided into districts and subdivided into communes, under prefects. These articles[1] prompted the opposition parties, containing representatives of the territories, to boycott the proceedings. The Constitution therefore became a party document which reflected *inter alia* its authors' views about property ownership and the rights of foreigners in Romania.

Article 18 restricted the holding of rural property (under which rubric the mining areas came) in full right to Romanian and naturalized Romanian subjects; Article 19 appropriated mining rights to the State in the following terms:

> Without prejudicing the rights acquired by the State on the basis of previous legislation, mineral deposits, as well as subsoil resources of all kinds, are the property of the State. . . .
>
> A special Mining Law shall establish the rules and conditions for exploiting these assets, shall fix the royalty due to surface owners, indicating also the proportion of their participation in the exploitation of this wealth.
>
> Acquired rights shall be respected as far as they correspond to an exploitation of the subsoil in compliance with the provisions which shall be provided under the special law.
>
> Leases for mining exploitation established or granted in compliance with the law in force shall be respected for the period for which they are granted, while existing mining exploitations worked by owners shall be respected so long as they are exploited by them. No perpetual leases shall be granted.
>
> All leases and exploitations mentioned under the preceding paragraph must comply with the rules which shall be fixed by law, stipulating also the maximum duration of these leases and exploitations which shall not exceed, however, fifty years from the promulgation of this Constitution.

The Article thus envisaged a uniform system of subsoil ownership throughout Romania, including the territories newly acquired from Austria-Hungary and Russia.[2] All newly discovered subsoil

[1] This method of dealing with the problem put the more developed territories—the former Austro-Hungarian provinces—under the less developed—the Old Kingdom—with a consequent fillip to the national pride and administrative opportunities of "ethnic" Romanians.

[2] In Bessarabia: On lands belonging to the State, exploitation rights were attributed to the discoverer of mineral deposits. On private lands, following

resources became state property[1] and their exploitation would involve the payment of royalties to the State, which undertook to reimburse surface owners. The reference to the "new mining law" was only one of a number of important details left to be settled by subsequent legislation—a feature which excited generally much criticism because of the uncertainty entailed.

That the policy of nationalization of the subsoil had particular reference to the oil industry is evident from the speech of Constantin Dissescu, *rapporteur* on the draft constitution to the Chamber of Deputies: "The present and future economic life of Greater Romania depends on what the new constitutional dispositions will be. The petroleum regime depends on the regime for the subsoil." Dissescu quoted from Ludendorff's memoirs on the reasons of oil supply prompting the attack on Romania in 1916, and rehearsed the activities of the Erdoel Industrie Anlagengesellschaft, which were aimed at creating "a vast monopoly of exploitation and distribution, only suppressed in 1918 by the double victory of the Allies in east and west". He also argued from "the great struggle for petroleum supremacy", repeating the standard Liberal explanation of the current phase of the world oil industry, and concluded that "the national interest and the possibility of the exhaustion of petroleum demanded its nationalization".[2]

The appropriation of the subsoil by the State was not in itself prejudicial to the oil companies, whatever the complexion of their shareholding, so long as exploitation continued on a royalty basis. Guest reported to Deterding:

> In the new Constitution the subsoil is "nationalised" in the ordinary sense that is to say, it becomes the property of the State. This change did not directly concern us, and though we naturally study the attitude

Russian law, the surface owner had the right to the subsoil. The Bessarabian land reform of 13 Mar. 1920 declared (Article 11) all mining subsoil in expropriated areas to be state property. In Bukovina and Transylvania: The Austro-Hungarian law of 1854 separated subsoil from surface rights where mineral deposits were found to be of economic importance. The law of 7 Jan. 1911 (Article 31) reserved all petroleum exploration and exploitation rights to the State, though these could be conceded to third parties. It was by virtue of this law that the Transylvanian natural-gas industry was carried on by the Hungarian State: as such it passed direct to the Romanian State, under the terms of the Treaty of Trianon (see below, p. 135).

[1] At the time of the promulgation of the Constitution, 33 per cent (502,973 tons) of Romania's crude output was derived from lands already owned by the State.

[2] See *MPR*, no. 5 (1923), pp. 392–3; no. 6, pp. 407–10.

of the new subsoil proprietor with interest and even concern, questions can arise, in the first instance, only between the State and the original proprietors. We, however, as holders of concessions are not entitled to protest for their sakes.[1]

This was formally correct, but the declared aims and known actions of the Romanian Government and its subordinate authorities did not give grounds for expecting that the State, as owner of the subsoil rights, would be content merely to collect royalties.

The Law was originally drafted in October 1923 by a committee consisting of Vintila Brătianu, Professor Mrazec, Tancred Constantinescu (former Director-General of the C.F.R. and current Minister of Industry and Commerce), I. N. Anghelescu (Secretary General, Ministry of Industry and Commerce), and I. Tănăsescu (former Secretary General, Ministry of Industry and Commerce). The draft was then submitted to a Parliamentary Commission, comprising members of both the Chamber of Deputies and the Senate. This commission, which was a constitutional innovation, met in February 1924 at the Ministry of Industry and Commerce, and, after a long exposition from Tancred Constantinescu, the draft was slightly amended. In this form it was then discussed by a Mixed Commission, comprising the authors of the Bill, Members of Parliament, and representatives of the industry. The latter were Osiceanu, Ficșinescu (President of Colombia), and I. Ghica (Director-General I.R.D.P., and President of the Association of Mining Engineers). The industry representatives had technical criticism to offer but none affecting the clauses of ownership. At the end of April, Constantinescu announced that, owing to the congestion of the parliamentary time-table, the draft law would be held over for consideration in the autumn session. Guest gave his personal opinion that the motive was "to create a favourable atmosphere and generally remove the question from the area of urgent discussion during the royal visit abroad, during which it is probably desired to solicit favours abroad, e.g. support on the Bessarabian question". He added a warning that it was "highly probable that the real intention is to rush the Bill right through Parliament during the short residue of the Session which will remain after Parliament reassembles on the return of the Sovereign".[2] This in fact happened. On 14 May the Council of Ministers decided

[1] 20. [2] 21.

to submit the Bill to the Senate and to prolong the parliamentary session till 25 June in order that the Bill should be passed.[1] The Senate passed the Bill with one major, and several minor, amendments,[2] but none affecting the principle of the Bill, by 83 votes to 4 on 28 June and the next day the Chamber passed it, as thus amended, by 138 votes to nil, the Opposition abstaining. The King gave his signature on 3 July and the Law was promulgated in the Monitor Oficial on 4 July, when it became effective.

The Mining Law of 28 June 1924 defined the new relationship between the Romanian Government and the oil industry. The text[3] consisted of 271 articles, and its underlying concepts were indicated by an *exposé des motifs* published over the signature of Tancred Constantinescu. The exposé outlined the official view of the petroleum policies in force elsewhere and of the early history of foreign capital in Romania's oil industry: it concluded that the time had now come for a Romanian state policy "to protect the country's economic interests and to safeguard political rights", and hence "to put subsoil wealth into exploration and to stabilize permanently vital national interests".

These considerations made for great changes in the temper and complexity of the law. In 1895 the State exercised only minimal functions in order to encourage investment in petroleum and, as landowner, enjoyed no exceptionally privileged position. By contrast, thirty years later, the State's requirements had become more detailed and its domain extended by the constitution to all minerals (except those held as "acquired rights")[4] regardless of the legal identity of their surface owners. The new law set out comprehensively the terms on which these new rights were to be enjoyed—and by whom. The State, represented by the Ministry of Industry and Commerce and the Romanian Geological Institute, assumed a strong directing role, reserving explicitly to itself a right to undertake exploration work. Further, it supervised other explorers by requiring them to demonstrate, among other things,

[1] 22. [2] See below, p. 118 n. 3.

[3] This analysis of the Law extends only to those provisions relating to the rights of the State and of private property owners, both Romanian and foreign. The Law also contained technical provisions concerning exploration and exploitation. For Romanian and French texts of the Law, see *MPR*, no. 14 (1924), pp. 1107–65.

[4] i.e. concessions already held on privately owned or state-owned lands which were exempted from the operation of the law, subject to their being validated, by Articles 235–58.

their technical and financial capacity as a condition of being granted a licence to prospect.

The terms on which minerals could be exploited were also drastically revised in favour of Romanian nationals. Concession rights on all lands not validated as "acquired rights" were reserved exclusively to Romanian mining joint-stock companies and co-operatives. All such companies were to have their stock in nominative shares[1] which were not transferable except on the authorization of the board of directors. At least 60 per cent of the share capital in these companies, i.e. those to be organized in future for exploiting state lands, was to be held by Romanian citizens, although for existing companies which in a period of ten years undertook to "nationalize" themselves this percentage was reduced to 55. Further, the president and two-thirds of the members of the board of directors had to be Romanian citizens. Existing companies which did not fulfil this particular condition were to transform themselves in accordance with the Law or lose the right to concessions.[2]

The concessions to which this general scheme applied were to be of three kinds[3]—those on a straight royalty basis, those involving a working association between the State and the lessee, or those that were an amalgam of both systems. The participation of the State entitled it to share in the profits in proportion to the assets supplied. The surface owners, whether private persons or the State itself, had a right to share in the capital of a concessionaire up to 10 per cent of the capital devoted to the exploitation of the perimeter concerned. Such participation had to be in the form of nominative shares, which were transferable only among Romanian nationals. The rights of these citizens were upheld not only as investors but also as employees: under the schedule of the concessionaire's rights and obligations, it was stipulated that not less than 75 per cent of all the categories of employees, as determined by the Minister of Industry and Commerce, should be Romanian nationals.[4]

[1] See above, p. 61.

[2] Article 32: it was modified on 6 Dec. 1925 by the addition of a clause admitting "to the same right" Romanian limited companies employed in mining, which had Romanian capital and board members in a majority, and which fulfilled the conditions regarding nationality of employees. This clause was designed to benefit two Romanian mining companies with capital less than 55 per cent Romanian—the "Lupeni" and the "Petroşani", which mined coal in the Jiu valley and with which Tancred Constantinescu had been connected.

[3] Articles 35–40.

[4] Article 80.

These provisions applied to all mining deposits: Part II of the Law laid down special provisions concerning the exploitation of particular minerals, including "liquid bitumens" (petroleum),[1] and introduced that distinction between various categories of perimeter which became fundamental to Romanian oil law and which secured more favourable prospecting terms for "national" than for "foreign" companies. If a foreign company struck oil it was allowed the right to only 50 hectares in the area but if a "national" company was successful it was granted from 80 to 100 hectares.[2] The rest of the area declared as oil-bearing was to be divided into three groups; the first, to be chosen at the State's discretion for reserves, and the other two to be leased only to "national" companies constituted in accordance with the conditions of Article 33. Article 181 limited the number of perimeters to any one undertaking and permitted "Romanian mining companies of great importance" one extra perimeter.[3]

Special provisions[4] put the construction of pipelines and tankage under the control of the State, reserving explicitly in the case of existing pipelines the State's right of expropriation against payment of compensation. Finally, under Article 198, the State declared that priority was to be given in the operations of the oil industry to the satisfaction of internal consumption, at prices determined by the State and through a form of distribution organized by the State.[5]

On the side of the industry, the reactions to the promulgation of the bill and its passage into law were expressed through the "Asociație Industriașilor de Petrol din România" (A.I.P.R.),[6] whose advice was sought by the Government.[7] The draft law was considered at a meeting on 24 March, under Osiceanu's chairmanship, at which it appeared that all parties, including the representative of Astra Română, were in favour of "nationalization".[8] The "foreign-capital" companies were invited to submit

[1] Articles 175–80. [2] Article 179.

[3] This provision was absent from the original draft law and was inserted during the Senate debate to cover the position of Steaua Română.

[4] Articles 195–8. [5] Article 198. [6] Founded 1922.

[7] The A.I.P.R. was asked to nominate eight representatives, of whom the Government would choose four: in fact it chose two, Banciu (Creditul Minier) and Luca (Româno-Americană), who, however, had no power of decision, which was reserved to the Association. Osiceanu and Ficşinescu (Colombia) represented the Association of Mining Engineers in the discussions. 22.

[8] See *MPR*, no. 7 (1924), p. 567: de Bruyn's remarks appear to have been

their complaints for consideration so that the Association could arrive at a formula to suit all interests. Further discussion was prevented by reason of the announcement of the postponement of the Bill to the autumn. While the Bill was as yet in the Senate, the Association sent its proposals to Constantinescu, making technical criticisms about the conditions of exploration, size of perimeters, permit procedures, etc., and, on the basic "political" issue, urging that the necessity rapidly to develop Romania's oil resources made nationality irrelevant as a criterion, and that acquired rights, i.e. those guaranteed by earlier legislation, should be observed.[1] The memorandum was originally to be subscribed by all the constituents except Creditul Minier and I.R.D.P., but at the last minute the representatives of Steaua Română, Concordia, and Vega withdrew their signatures—according to Guest "on pressure from Osiceanu, who came in from a long conference with Tancred Constantinescu and Stefanescu, the Manager of the Banca Românească".[2]

The "national" companies[3] countered with their own joint memorandum to the Government, offering technical objections but stressing that the nationality qualifications concerning capital-holding did not go far enough to even out the inequalities between "national" companies and "foreign-capital" companies. The memorandum attempted to justify the preferential treatment for "national" companies, and to rebut the arguments against "nationalization". In particular, the alleged lack of Romanian funds to buy participation in "foreign-capital" companies within a time-limit, which the signatories noted with surprise had been extended from

misinterpreted, either by Osiceanu or by the Minister, Constantinescu, who was managing the party press. 23.

[1] See *MPR*, no. 12 (1924), pp. 931–50. Guest reported: "the Association is putting in a rather good protest—which I have had a bit of trouble to get into shape. At first it was very difficult to get any Roumanian to make a stand against 'nationalisation' but by degrees we have got quite a creditable protest on that point drafted by a prominent Roumanian Engineer who has now been dropped from the Council of the Engineer's Association on account of his attitude (this is Ficşinescu, manager of Colombia)." 24.

[2] 25. On 8 June Guest had reported, "as a sidelight on Vintilist methods", that Marinescu (Romanian), Manager of Concordia, was being victimized for his "vigorous attitude" over the law for Commercialization of State Enterprises (see below, p. 120 n. 4) and the Mining Law, in that, "among other things, steps are being taken to deprive him of a Bank directorship from which he derives a not unimportant part of his income".

[3] Including Creditul Minier, I.R.D.P., and Redevenţa, but not Steaua Română. 26.

five years to ten, was a mere pretext: an exchange of shares would have made it groundless.[1] Concordia submitted a separate memorandum, admitting "nationalization" but limiting the obligatory Romanian element in management to a simple majority (which already qualified the company as Romanian and therefore entitled to the consequent privileges): acquired rights should be recognized: only non-consolidated rights should need validation.[2] The abstentions from the A.I.P.R. memorandum significantly robbed it of much of its force and demonstrate the paralysis of the official representative body of the industry *vis-à-vis* the State on the most important issue to both—its legal basis.

Although formulated at different stages with varying degrees of emphasis, the case of the foreign-capital companies—principally Astra Română and Romăno-Americană—who took the initiative in organizing opposition to the Mining Law was as follows: the "nationalization" provisions, combined with the appropriation of the subsoil by the State under the Constitution, deprived them of their right to expand their operations without admitting Romanian capital and thus deprived them of the fundamental prerequisite of their future existence—viz. the right to explore new areas. The ultimate alternative to "nationalization" was forced liquidation. In addition, although imposing a rearrangement of share capital, the Law made no mention of compensation for the loss of the rights of foreign shareholders, and the situation on the existing capital market was too stringent to allow of public large-scale buying of foreign-owned shares.[3] Accordingly, the "nationalization" clauses were tantamount to unilateral alienation of private property by the Romanian Government.[4] Similar fundamental

[1] See below, pp. 128–9. [2] 27.

[3] Estimates of the current market value of foreign-owned shares varied from £37 million to £50 million.

[4] This view of nationalization derived additional weight from the "Law for the Commercialization of State Enterprises", announced by Tancred Constantinescu in Jan. 1924, which, under the provision of running some state enterprises on commercial lines, implied a state-run inland and export trade in oil, ignoring acquired rights. The draft law was attacked by the A.I.P.R.—mainly on the grounds that the State's claim to distribute oil products was unconstitutional. There were no grounds for external protest, since the Romanian State was entitled to create a monopoly in principle—particularly one which affected all companies equally. The Law was passed in June; the question thereafter being whether it would be applied to oil. It had not (except for the foundation of a methane-gas company) by the time the Liberals left office. The National Peasants elaborated new rules for the conduct of state enterprises. See 28 and *MPR* references.

criticism attached to the clauses relating to the validation of acquired rights: as enacted,[1] they implied that property rights acquired in good faith and due form were again to be questioned and the owners reduced to the position of litigants. It was always possible that rights definitely and legally acquired might not be validated and therefore, under Article 253, annulled, for purely technical reasons in the administration of the law. The wording of the Law was sufficiently imprecise, and the vagaries of Romanian administrative tribunals sufficiently well known, to make it certain that validation would be no simple process.[2]

Apart from the necessity to prove again rights already legitimately gained, with consequent delay and expense, the Law was held to be inconsistent in that it abrogated certain rights already acquired by operating companies, notably those relating to the expropriation of private pipelines (Articles 71 and 195), the obligatory

[1] *Article 235*. Rights of any kind that come under the provisions of this law and which were acquired in the various provinces of Romania before the promulgation of the Constitution on 28 Mar. 1923, by any manner or by any act of definite date, will be recognized and legalized according to the ruling of the present law and under the following conditions:

(*a*) The rights must be acquired in conformity with the laws, decree laws, and regulations, and any other dispositions of this kind which were in force at the date of their acquisition.

(*b*) The rights claimed must not affect the rights which the State has under the law of 13 June 1923 for the liquidation of assets of foreign subjects who were former enemies, as well as those which the State has through other laws, decree laws, regulations, ordinances, or by international deeds.

(*c*) The rights must correspond to the value of the subsoil in accordance with the distinction shown below.

All rights which are not legalized become void.

Rights which are consolidated, or are in the course of consolidation in accordance with the law of consolidation up to the promulgation of this law, are exempted from the obligation of recognition and legalization on lands that fulfil the condition of sub-paragraph (*b*).

Article 236. Any person who, before the promulgation of the Constitution, acquired a right of any nature which comes under the provisions of the present law, on the basis of any deed, must, within one year's time from the promulgation of the present law, request the recognition and legalization of his rights.

[2] Validation was carried out by District Commissioners nominated *ad hoc*. In his *exposé des motifs* at the presentation of the Mining Law to Parliament, Tancred Constantinescu stated that the verification of acquired rights was necessary only to ascertain that such rights and properties had not been acquired during the war and immediately afterwards without respecting the rights and claims of Romania, and that for established companies validation would be a matter of form. In fact, the Romanian Government contested *all* validations, which both in principle and in practice became a major issue between the companies and the State.

pooling of plots (Articles 113–15, and 190), the annulment of reserve lands in areas not hitherto exploited, if certain conditions imposed by the law were not complied with (253, 254), the exploitation of gas (Articles 193 and 199–202), and the rights claimed by the State to installations on expired concessions (Articles 118 and 150).

The provisions regarding staff also represented an excessive degree of state interference in the rights of management. By Articles 75 to 81, the Government retained extensive powers regarding the appointment, employment, and qualifications of personnel. The Minister of Industry and Commerce had the right to fix the categories of work on which the 75 per cent quota operated and also had the right to dismiss a General Manager (Articles 141 and 142).[1]

These legal objections were supplemented by a general political criticism that the main beneficiaries of the Law would be the Liberal-financed companies, which had escaped the main burden of exploration in Romania and the need to renovate war-damaged fields and installations, and at the same time had already received favourable concessions on state lands. On this view the Law had really the character of an expropriative measure in favour of private interests. Further, it was feared that interference by the Romanian State in the accepted rights of management would increase the scope for further arbitrary government action, taking advantage of the imprecision of the Law in both general and technical provisions.[2]

"Nationalization" that was construed as "uncompensatable" expropriation of private property and the infringement of acquired rights gave, under international law, the governments of the foreign shareholders affected grounds for intervention on their behalf. This right was exercised by the British, U.S., French, Belgian, and Netherlands representatives between the appearance of the draft law in late March 1924 and its enactment at the be-

[1] State policy on employment also had a personal appeal: "Officials in the Ministry of Industry and Commerce seem to entertain high hopes of Clause 80. One of these informed me [Guest], through a friend of his, that it would be insisted that three-quarters of the managers were Roumanians at the same pay as existing managers and was kind enough to offer his services." 29.

[2] The U.S. minister in Romania summed up this view of the situation in his report to the Secretary of State of 17 July 1924: see *FRUS* 1924, ii.

ginning of July.[1] In general, the grounds of protest were enumerated by the local companies concerned both to their parent concerns and to local legations who submitted them for consideration to their respective Foreign Ministries for authority to intervene. The diplomatic wrangle is illustrated in the reports of the U.S. Minister, Peter Jay[2] (which also show how the Mining Law issue became caught up with others between Romania and the United States, such as war debts and an outstanding extradition treaty). His representations advanced the arguments mentioned above, together with one specific to Româno-Americană—viz. that its re-establishment after the war had been based on certain written assurances of the Government as to equitable treatment which the Law now controverted.[3] The Romanian Government's response—as set out in its note to Jay of 24 June and more fully developed in the memorandum to the U.S. State Department of 21 July—asserted that the modifications made to the original draft of the Law had removed any "scope for justifiable criticism" and that the Romanian Government alone had "the right to decide the future disposal of its petroleum lands", as "an essential and legitimate attribute of its sovereignty". The memorandum denied that the Law established any difference between the petroleum enterprises of the various nationalities: on the contrary, it allowed "all companies without distinction to benefit by all its advantages under certain most acceptable conditions, as is shown by their acceptance by the Franco-British, British, French, and Belgian companies".[4] Mr. Jay's colleagues met with the same rejection. The British interests affected took advantage of King Ferdinand's visit to London—during which he spoke at the Guildhall about co-operation with British capital—to intervene with him for the abandonment of the Mining Law and the Law for the Commercialization of State Enterprises.[5] Subsequently, on his return, the King was given a memorandum from Guest summing up the main case against "nationalization".

[1] The first diplomatic inquiries had been made in Nov. 1923, registering as a precaution the possibilities which the foreign companies would find objectionable. Tancred Constantinescu was reassuring: see Jay to State Department: telegram no. 60, 13 Nov. 1923; and letter no. 495, 10 Dec. 1923, ibid., pp. 597–8.

[2] *FRUS* 1924, ii. It is also discussed by J. C. McKenna, S.J., in *Diplomatic Protest in Foreign Policy*, Chap. V.

[3] Jay to Duca, 26 June 1924: *FRUS* 1924, ii.

[4] See below, p. 124.

[5] 30.

All these interventions secured only minor amendments, though of importance in principle, such as the provision of compensation for the expropriation of pipelines,[1] and, after the passage of the Bill, the deletion of the clause prohibiting drilling within thirty metres of a boundary of a concession[2] on the grounds that, by rendering existing concessions or parts of concessions useless, the clause constituted a violation of acquired rights. It would have taken a complete political reversal in Bucharest for these representations to have achieved more: "nationalization" was the essence of Liberal industrial policy, and validation was one kind of inducement to "nationalize". Furthermore there was, as Guest recognized, in the given domestic political situation no possibility of a government really independent of the Liberals ("unless perhaps the 'peasant' and 'national' parties unite") and therefore one which could do anything other than "slow down the realization of Vintila's programme".[3]

Furthermore, foreign capital was not united in opposition. Judged by its results, the strategy of Vintila Brătianu was to persuade those foreign-capital companies which were already vulnerable either through an existing large Romanian shareholding, or through being too dependent on short-term leases, to accept "nationalization" as involving little more than a minor readjustment in order to qualify for new concessions. At the same time, diplomatic representations were to be met with assertions of Romania's legal competence to insist on "nationalization" as a pre-condition of exploiting the country's oil resources, which now belonged to the State, with assurances that acquired rights would be respected. In this way, the opposition in France and Belgium dissolved. The representatives of Concordia and Vega withdrew their signatures from the A.I.P.R. memorandum on the Mining Law.[4] Steaua Française, representing the French interest in Steaua Română, accepted Brătianu's policy; and, after enactment, Wenger, of Petrofina, put the view that intervention was a mistake[5]

[1] Article 195.

[2] Article 190. In justification, the Romanian Government advanced the desirability of preventing the extraction of the resources of neighbouring plots. In practice, the distance at which wells affect each other, even in the same fields, varies according to the geological features of the subsoil and the levels at which oil deposits occur.

[3] 31. In November Guest found himself approached by Averescu, whose return to office was (erroneously) thought imminent, to discuss revision: 32.

[4] See above, p. 123.

[5] 33.

and published a favourable article on the Law in *La Roumanie Économique et Financière*, which was republished in full in the Liberal paper *Viitorul* on 3 October 1924. The objections of the British Phoenix Oil and Transport Company were similarly overcome.[1] These manœuvres enabled the Romanian Government to use the "acceptance by the Franco-British, British, French, and Belgian companies" as a means of rebutting U.S. official criticism[2] and spreading the notion abroad that foreign opinion was satisfied with the Romanian measures,[3] while at the same time isolating Astra Română and Romăno-Americană within Romania.

The modification of nationalist policies—the Mining Law of 1929

Nevertheless, the Romanian Government made last-minute efforts to persuade Astra Română to "nationalize". Before the passage of the Law, Guest was approached by Gheorghe Dumitrescu, through Zentler, General Manager of Distribuţia, with a proposal that Astra Română should form a company in which it held 45 per cent of the shares—the rest being held by Romanian nationals. Behind this dummy company, Astra would be able to continue its operations undisturbed and would be assured of state lands. The reply was that the proposal could only be taken seriously if made by Vintila Brătianu in person.[4] After the Bill was passed, a similar proposal was put to Guest by Dumitrescu directly.[5] At the same time, Vintila Brătianu, at his request, discussed possibilities with Sir Henri Deterding and his associates in London, from which subsequently emerged a suggestion that, if capital for a large company were forthcoming, the Law would be managed so that control remained with the foreigner.[6] The foreign-capital companies decided that they would be exposed to the same political risks as before: the proposals came oddly from one who publicly

[1] Apparently Sir Edward Manville, Chairman of Phoenix, who had initially taken a hostile attitude, came to a private arrangement with the Romanian Government: see *Annual Report Phoenix Oil and Transport Co. for 1924*; also 34 and 35; Manville's circular to shareholders, *Financial News*, 4 July 1924.

[2] See Romanian memorandum of 21 July 1924 to State Department: *FRUS*, 1924, ii, p. 621.

[3] Tancred Constantinescu in *Argus*, 19 Feb. 1925.

[4] 36. [5] 37. [6] 38.

denounced them as preventing Romania from occupying the place in the sun to which she was entitled.[1]

The first concessioning of state-owned lands under the new regime was announced in November 1924 but was completed only in March 1926. Oil-bearing territory of 495 hectares were involved. One-half was to be allotted to foreign companies "considered as complying with the regulations regarding Romanian ownership and representation" or those "who agreed to make the necessary modifications within ten years". The other half of the land under concession went to Romanian concerns.[2] It was to take advantage of this opportunity that Steaua Română, 51 per cent of the shares of which were held by Romanian nationals, increased that percentage to 55 per cent. In return Steaua gained the right under Article 181 of the Mining Law to participate in the concessioning and also the renewal of existing leases on state-owned lands due to expire within a short period. In theory, Steaua thus became a "national" company like Creditul Minier: in practice, it could not become so until the shares which had been mortgaged by the Romanian banking group to its British and French partners in 1920 as part of the purchase arrangements of the Steaua shares had been redeemed.

The Concordia group also became "nationalized" in order to preserve existing concessions on state lands due to expire in 1925 and 1927 and to qualify for further plots. Since the concessions were held in the name of Internationale, to which Concordia was successor in title, the Franco-Belgian shareholders attempted to ensure that only this company was "nationalized". The Romanian Government insisted that all the companies were involved and by the agreement of 9 July 1925 it acquired share interests to a value of 200 million lei (approx. £200,000) in Internationale, Vega, and Concordia. Further, Romanian citizens were given the right to buy shares in Vega and Concordia at below the ruling market price.[3]

As a result of these arrangements, Steaua and Concordia were admitted to operating on state lands, on the same footing as I.R.D.P. and Creditul Minier. Româno-Americană and Astra Română refrained from putting in a bid, since they decided not to

[1] See also below, p. 129.

[2] *MPR*, no. 1 (1925), p. 49; no. 5, p. 396; no. 8, p. 603; no. 21, p. 1769; no. 23, pp. 1954–5; no. 6 (1926), p. 563.

[3] For these arrangements, see *MPR*, no. 14 (1925), pp. 1210–12; and no. 21, pp. 1758–9.

accept the conditions for "nationalization". Astra Română had the greater area of private concessions and had already some concessions on state lands taken over with Regatul Român in 1910. All these came under the acquired-rights clauses of the Mining Law and were subject to validation. Romăno-Americană sought to increase its output by deeper drilling. This technique, which also began to be used by Astra, involved the use of rotary drills and presented special legal problems of its own.[1]

In so far as the Mining Law generated new relationships between the State and the oil industry, and reserved the initiative in the future development of the country's oil resources to the State, Vintila Brătianu's policy succeeded. The instruments of this policy, I.R.D.P. and Creditul Minier, had been created prior to the Law: the Law now ensured their scope and restricted that of their foreign-capital competitors. The "national" sector was enlarged: 26 per cent of the oil industry was controlled by Romanian capital, as against the 4–5 per cent of 1914. Vintila Brătianu, in his public speeches, liked to portray this result as the reward for valiant defiance by a small state of the great world trusts that had wished to keep Romania in subjection. This echoed the "unequal-arms" theme of fourteen years before[2] but was at variance with the facts: neither the State nor the industry was in the position implied in Brătianu's rhetoric. In the given situation the State, as a recognized sovereign, which could determine the law and commanded the means of enforcement, was by far the stronger party. Romăno-Americană and Astra Română were Romanian nationals and therefore bound by the Romanian municipal law. However much they considered that Brătianu's policy was prejudicial to Romania's interests, as they saw them, or however much they resented the local chicaneries to which they were subjected, both were part of the political context which they could not themselves alter. The Mining Law when passed was a factor to which they had to adjust themselves: they did so by taking one of the options

[1] The rotary system involved the use of drilling-fluids under pressure which acted as coolants and permitted faster drilling. Moreover, it required only half the well-casing used in percussion systems, enabled the flow from the boring to be more easily controlled, and reduced the fire hazard. Legal obstacles, deriving from pre-war legislation against the flooding of drillings, prevented until 1928 the use of the rotary system to "bring in" a well, even though it had been used to reach the oil sands: at this stage some dry system was compulsory.

[2] See above, p. 63.

available to them, i.e. by declining to "nationalize" and accepting the consequent restriction on their own future. Their foreign shareholders could not contest the Romanian State's right to act according to its own interests: they could invoke the support of their own governments only when these actions gave rise to a definable dispute in international law. Even then oil problems were only one of a number of different questions at issue, and, moreover, Romanian supplies were not essential to British oil policies, still less to American. Accordingly, both the companies in Romania and their parent concerns were at the time in a weak and essentially defensive position with regard to the Romanian State. Their hope was for a revision of the legislation, either enforced by the demonstration of its disadvantages in practice or through the breakdown of the Brătianist system: time was a prerequisite of either alternative.

In the circumstances, the chances of a *rapprochement* between Vintila Brătianu and the two companies were negligible. The points fundamentally at issue were the conduct of their business and the scope given to Romanian elements in capital and management. Duca summed this up to Guest when complaining that Astra did not drill, held back land from exploitation, and employed only foreigners, allowing Romanians to come in as domestics. These charges indicated a total difference of outlook in the oil business. Both companies pursued more conservative dividend policies than their competitors, and, while they had always made a point of employing Romanians, refused to accept Brătianist nominees—which "nationalization" implied. Personal distrust of Vintila Brătianu and his associates entered into their reaction to events.

In the event, the signs of breakdown were evident even before the concessioning of state lands was completed: its beneficiaries, the main "national" companies, were unable to sustain the role designed for them. As early as 1925 I.R.D.P. and Creditul Minier encountered operational difficulties[1] which eventually forced them to make supply and processing arrangements with foreign-owned companies and then with foreign lenders. In 1925 Române-Americană was running in its refinery at Teleajen a large proportion

[1] These also forced the abandonment, in 1926, of the favourite Liberal scheme, resurrected by Tancred Constantinescu, for a single Romanian company, with a capital of 200 million lei, of which 10 per cent would be contributed by the State and the rest by the national companies, and to which the State would also contribute oil-bearing lands.

of Creditul Minier's crude-oil output and marketing it abroad. At the same time, negotiations, ultimately abortive, were in hand for Români-Americană's parent company, Standard Oil of New Jersey, to take an interest in Creditul Minier. In 1927 Creditul Minier made arrangements involving French capital for the formation of a subsidiary company, as a market outlet in France.[1] Similarly in August 1925 it was rumoured that I.R.D.P. were negotiating with Petrofina (Cie Financière Belge des Pétroles) for the sale of I.R.D.P. shares and the working of I.R.D.P. crude oil in the Vega refinery. In November 1925 proposals were made to Astra Română that either that company or the Royal Dutch–Shell group of companies should take a substantial interest in I.R.D.P.—leaving the Banca Românească as the only other large shareholder—and that in consequence of this arrangement I.R.D.P. crude should be run in Astra's refinery at Ploeşti and the resultant products exported through Royal Dutch–Shell outlets.[2] This scheme was not adopted, and in 1926 I.R.D.P. solved its financial difficulties by selling to a specially created French company, the Societe Industrielle des Pétroles Roumains (S.I.P.E.R.), 300,000 shares, which were subsequently traded on the French market.[3] In this way the quintessentially national companies became responsive to a degree of foreign direction.

Finance was necessary not only to overcome the operational shortcomings of the national companies but to strengthen the Romanian economy as a whole, particularly to enable the currency to be stabilized, and the railways renovated.[4] In this situation the shareholders in Astra Română and Români-Americană were strongly placed. In August 1924 Titulescu (Romanian minister in London), negotiating for Vintila Brătianu, was informed that Romanian legislators were not to be trusted[5] and no shareholders money would be risked in Romania until the legislation was altered. The companies saw to it that their attitude became known in banking circles in London, Paris, and New York. From May to July 1925 Standard Oil tried to use Romania's indigence to sponsor

[1] *MPR*, no. 3 (1928), p. 195.
[2] See 39.
[3] *MPR*, no. 17 (1926), p. 1737.
[4] See below, p. 148.
[5] This may be illustrated by a comment of Guest's: "I am told that had I given Tancred and the Secretary General [Anghelescu] 20 or 30 million apiece during the winter, things would have gone much more smoothly. I am of course certain that these gentlemen would have accepted this or any other bribe with alacrity, but I doubt whether it would have produced any lasting result." 40.

a U.S. loan to Romania, in return for a repeal of the objectionable features of the laws. This attempt was unavailing, since the State Department refused to allow any loans to Romania until outstanding war-debts questions were settled,[1] an episode which emphasizes the fact that the attitude of Royal Dutch–Shell and Standard Oil was not the only, or even necessarily a major, deterrent to foreign investment in Romania. The instability of the leu and the refusal of the Brătianus to concede adequate physical guarantees were also observable factors—and the British and French Governments had their own debt problems with Romania.[2] Nevertheless, the Mining Law was held up as an example of Romanian malpractice. Except for a short period, the Romanians found the international money market closed to them till 1927, when Vintila Brătianu at last undertook to give the requisite guarantees.[3] The exception occurred during the short-lived Averescu ministry of April–June 1926, which unexpectedly brought to a conclusion negotiations with the Italian Government for loans and the recognition of the Romanian occupation of Bessarabia. In consequence Italian capital was admitted to a controlling interest in the companies Prahova, Petrolul București, and Atlas Petrol, which were given concessions of 350 hectares on state lands in Colebași and Rasvad.[4]

The prospects for foreign capital, and therefore ultimately of the foreign-capital oil companies, were improved by political and constitutional changes. In October 1926 the opposition which Guest had postulated as necessary to any meaningful departure

[1] *FRUS*, 1928, iii, p. 814.

[2] Particularly in regard to the settlement of obligations arising from the 1916 destruction.

[3] The incentive was perhaps the disastrous harvest of 1927 as much as the arguments of foreign bankers. A French mission under Professor Charles Rist arrived in Bucharest in July 1928 to study the conditions for a foreign loan. By the time its report was published, Vintila Brătianu was out of office.

[4] For details of these concessions see *MPR*, no. 4 (1927), p. 277. The formation of an Italo-Romanian company for oil was originally projected in conversations between Ionel Brătianu and Signor Aloisi, Italian minister in Bucharest, in 1923 (see above, p. 112). During negotiations in the spring of 1926 for the loan of 200 million lire it was arranged that annuities on the loan could be paid in oil products or crude oil (this provision was in itself contrary to Article 196 of the Mining Law). The loan was negotiated through the "Azienda Generale Italiana Petrolii" (AGIP) set up by decree of 3 Apr. 1926, with 60 per cent of the capital directly subscribed by the Italian Government. For details of these negotiations see *Survey of International Affairs 1926*, R.I.I.A., London, 1928, pp. 157–9; and "Italy and Roumania", *Petroleum Times*, 20 July 1926.

from the Brătianist system came into existence, when the Romanian National Party of Transylvania, led by Vaida Voevod and Maniu, and the Peasant Party, whose leaders included Mihalache and Virgil Madgearu, fused into the National Peasant Party (Ţărănesc). Its economic programme at that time stressed the necessity to attract foreign capital, both for renovations and for development.[1] In practice, this implied equality of treatment between national and foreign capital, the suppression of special taxes on foreign money, and the abolition of nominative shares and the fixed ratios of foreign to national capital in undertakings. More generally, a healthier economic order should be introduced, doing away with administrative chicanery, exercised under the pretext of state control. Reform of the Mining Law necessarily followed from such a policy.[2] The opportunity to carry out this programme was created by the death of King Ferdinand on 20 July 1927. His successor was his grandson Michael, since the immediate heir to the throne, Prince Carol, Michael's father, had, as a solution to filial and matrimonial problems, renounced his rights to the throne and retired to Paris. Michael being 5 years old, the royal authority was exercised through a Regency.[3] It was no longer committed to the Liberals, who also began to divide on the succession question, i.e. whether the parliamentary decision of 4 January 1926 confirming Carol's renunciation should be overturned, which became the central issue of Romanian domestic politics, and which the conduct of the Regency did much to promote.[4] The death of Ionel Brătianu, in November 1927, removed the authority in the party capable of keeping these divisive tendencies in check. The leadership devolved on Vintila Brătianu, who had the same *idées fixes* as his brother, but was politically less adroit.

During 1928 the harvest failed for the second successive year—at a time when the emergence of the National Peasant Party as an alternative government encouraged foreign lenders that their hopes for a more reasonable economic order were likely to be fulfilled. Vintila Brătianu turned to the Liberals' former financial

[1] See H. L. Roberts, *Rumania: Political Problems of an Agrarian State*, pp. 139–56.

[2] See Hoisescu, *The Revision of the Economic Legislation of the Liberal Party*: Report to National Peasant Party Congress, 26–7 July 1928.

[3] See Henri Prost, *Destin de la Roumanie*, Chap. V.

[4] For the outcome, see below, p. 181.

haven—Berlin—but, in the course of attempting to strengthen his domestic political position through tying the outcome of the negotiations to a guaranteed renewal of office, submitted his resignation, which the Regency accepted. The National Peasants, who had meanwhile demonstrated their public strength in a series of rallies and conferences, formed the new Government under Maniu on 10 November 1928. One of its first administrative decisions was to raise the state of siege and end the censorship which, since the war, had been the basis of Liberal election-manipulation. The elections, unregulated by the police and the prefecture, on 15 December gave the National Peasants 78 per cent of the popular vote. The new Government secured foreign assistance, enabling the currency to be stabilized and various reconstruction projects begun.[1] The loans benefited the oil industry in that they enabled government departments—and particularly the C.F.R.—to pay their outstanding debts for oil supplies.

The Government also opened the question of the Mining Law, asking both individual companies and the A.I.P.R. for proposals, in January 1929.[2] The draft law was submitted to Parliament on 12 March and given to the A.I.P.R. for discussion on the next day. The title of the draft was significant—"Draft for the modification of the Mining Law of 4 July 1924",[3] and the only substantial alterations, apart from those of a technical nature, concerned nationality qualifications and the rights flowing from them—which were abolished. The main feature of the old law, validation of acquired rights, was not only retained but became more stringent. The financial obligations, particularly with regard to royalties, which were of general concern to the industry as a whole, were unmodified. The draft also retained the basis for the erection of a state monopoly, as Madgearu explained to Otto Stern (Guest's successor in Astra Română), "as a kind of Damocles' sword over the head of the petroleum industry, so that the Government could make use of these rights in case a private monopoly should be exercised in the country to the disadvantage of the owners of the oil—i.e. the Government—or to the disadvantage of the consumers

[1] See Madgearu, *Roumania's New Economic Policy: Economy Survey of Roumania* (articles reprinted from *The Economist*): particularly S. Radulescu (Romanian Minister of Foreign Affairs) on "The National Debt" and Popovici (Minister of Finance) on "Public Finance".

[2] 41.

[3] *MPR*, no. 6 (1929), p. 461.

of oil products".[1] The Bill became law on 27 March 1929. Article 30 permitted concessions to be granted "to any company either Romanian or foreign which complies with the requirements of Romanian legislation, as well as to juridical persons of any kind"—which allowed foreign-owned companies to acquire concessions and prospect for oil on the same basis as "national" ones, and thus guaranteed their continued existence. Additionally, nationality qualifications for employees exempted higher management, and in any case were to be applied to 75 per cent of the staff as a whole, not by categories.

The grounds for dissatisfaction with the modifications to the Mining Law were enumerated by Otto Stern for the British minister, at the latter's request.[2] They were: the incidence of taxation and royalties (which even extended to accidental losses in production); the State's reservation of crude-oil export rights, the right to expropriate field pipelines, and the reassertion of the State's right to construct trunk lines—which, combined with the necessity for state authorization to build refineries, left open an easy way for a state oil monopoly; the vagueness of clauses governing the transfer of equipment and installations to the State (e.g. tank-cars); the application of compulsory employment of Romanians to the oil industry only; the absence of provisions on the formation of land reserves by companies; and the new conditions for validation, in particular the stipulation that constant exploration was indispensable to it. Stern added one final caveat:

"Only from the way and the spirit in which [the law] will be applied will it be possible to judge if this law is really an improvement on the law of 1924."

The reason for this situation appears to lie in the Government's wish to achieve something quickly, as a demonstration of effectiveness and good faith to foreign lenders, and to establish sound regulations for leasing new perimeters.[3] At the same time there were factors in the internal political situation limiting the scope of the Government. Radical reform of the 1924 Mining Law implied reform of the Constitution, and, although they had, as separate parties, abstained from the one and boycotted the other, the National Peasants'[4] immediate concern was electoral and

[1] Ibid. [2] 42. [3] 43.

[4] The Law eventually became an issue in the trial of the National Peasants' leadership in 1947: see below, p. 302.

administrative reform to dismantle the apparatus set up by their opponents. Moreover, at the beginning of its period of office, the Government was faced with Brătianist supporters entrenched in the banks and the Civil Service. The Liberals abandoned resistance to the 1929 Law but threatened to reintroduce the 1924 Law in full rigour on their return to office.[1] This, however, took place only after the Depression and at a time of constitutional crisis.

[1] 44.

CHAPTER VI

Romania and her Markets after 1918

The resources of post-war Romania

IN 1918 the prospects for the Romanian oil industry seemed highly favourable once the residual war damage had been repaired. During the Occupation the recognized oil-bearing zone had been extended eastward by discoveries at Pacureți and Ochiuri. These new sources of oil were augmented by the acquisition under the Treaty of Trianon of the natural-gas industry of Transylvania.[1] The marketing outlook was also regarded as promising: the potential internal market had been enlarged by the addition of population and by new outlets such as the iron and steel industry of Reșița, in the Banat of Timișoara. Export potentialities were improved by reason of the removal of two pre-war competitors from the Central European oil market.[2] The realization of the new potentialities depended on an improvement in European trading conditions, particularly in the area served by the Danube (conditions in which are described above),[3] on the overhauling and

[1] Natural gas (99 per cent methane) had been discovered in the area around Sármás (Sărmăşel), east of Kolozsvár (Klausenburg, Cluj), in 1908, as a result of prospecting by the Hungarian Geological Institute. The gas-bearing lands were immediately taken over by the Hungarian State, which, with the aid of Austrian and German drilling firms, exploited the area for eight years. In 1916, however, part of the lands had been leased for fifty years to a private concern, the Ungarische Erdgasgesellschaft, set up by an agreement between a banking group, headed by the Deutsche Bank, and the Hungarian Ministry of Finance. In return for leasing these exploitation rights, the Hungarian Government retained a 15 per cent interest in the company. The gas was used for lighting and by industry, and by the outbreak of war had begun to be piped to various places within the Principality. Production in 1913 totalled 12,063 cubic metres, but by 1918 had increased to 55,588,570 cubic metres. As a result of the war, the Romanian State took over directly the Hungarian State property, which it managed until 1925, and sequestrated the assets of the Ungarische Erdgasgesellschaft as enemy property.

[2] As a result of the Revolution and disorders in the producing areas, Russia ceased to be an exporter until 1928, and the Galician oilfields, which had been severely damaged both during the First World War and the Russo-Polish War of 1919–20, were located within the frontiers of the resurrected Poland, where their declining output was consumed.

[3] See above, p. 101.

reconstruction of Romania's systems of transport, and on an increase in storage capacity. These prerequisites depended on state initiative. Railway transport had suffered generally in the war-affected areas: track and installations had been extensively damaged, and the number of locomotives and rail tank-cars had been cut respectively to 10 per cent and 35 per cent of 1916 totals. Furthermore, for all but the sections between Bucharest and Ploești, Ploești and Câmpina, and Ploești and Buzău (203 kilometres), single tracks prevailed. New construction was required to link Transylvania with the Bánát and to connect Bessarabia with Cetatea Albă rather than Odessa. For these tasks, both finance and raw material were lacking. The immediate deficiencies were partly remedied by a credit of £100,000 granted by the United Kingdom, and by arrangements such as that between Romania and Austria of December 1919 whereby oil products were exported in return for Austrian railway material and locomotives. Notwithstanding these measures, the repair of locomotives was protracted through the multiplicity of types in service and official indecision as between policies of repair or replacement. By 1922 the number of locomotives in service had been increased to 1,650, with about 3,000 in need of repair. Recovery was hampered by lack of funds: C.F.R. revenue derived from passenger and freight charges and subventions *ad hoc* by the State, but, nevertheless, would only cover normal running expenses; any fundamental improvements could only be financed by loan.[1]

The pipeline situation was equally unsatisfactory. During the Occupation, part of the Constanța line had been taken up and relaid from Băicoi to Giurgiu, where tankage and loading facilities had been constructed so as to make Giurgiu the main Danube port of oil traffic. Thus, there were at the end of the war three lines, one of 9–10 in. from Băicoi to Constanța and two of 5 in. from Băicoi to Giurgiu, one of which was used for unrefined kerosine and the other for gas oil and crude. One Giurgiu line was in working order but the other two were in need of repair. The shortage of materials entailed the reconstruction of the Constanța line with sections only 5 in. in diameter, and these joints therefore governed

[1] On post-war railway facilities see "Roumania" in the *General Transport Situation in 1921* (statements submitted to the Barcelona Conference on Communications, Mar.–Apr. 1921), League of Nations, vol. ii, pp. 351–9.

the performance of the whole line. The operation of all lines was unsatisfactory, pumping losses were high and delays frequent.[1] It was proposed that the two existing lines to Giurgiu should be supplemented by a third to run gasoline: work was begun in 1920 but suspended within a few months for lack of funds. The law governing the Pipeline Service made over pumping charges to the C.F.R. as part of its revenues: the Pipeline Service was therefore debarred from this means of raising funds, which by the same law had to be voted in Parliament by special credits.[2] Constanţa itself, which was the only harbour open all the year round and which had been neglected during the Occupation in favour of Giurgiu, was in need of renovation and extension. In 1919 its unloading machinery comprised one hand-operated crane. In the oil dock, tankage comprised 120,000 tons, of which about half was ready for immediate use.

Inadequate and unreliable transport caused more than half the 1919 production (920,437 tons) to be consumed as fuel, either directly as crude or as residual fuel oil. Exports were confined to the Danube and—by reason of wartime losses and post-war confiscations of oil vessels by the Succession States—had to be shipped in converted grain vessels. The situation improved somewhat in 1920. Higher production in Moreni and restoration of damaged wells permitted an increase in output to 1,034,123 tons, but transport again restricted refining[3] and export possibilities. A prolonged freeze, low water in the summer, and lack of tugs and fuel restricted the employment of the limited shipping available. Constanţa was used only for shipments under the Boyle and Standard Oil contracts.[4]

The restoration of the industry was also hampered by fluctuations in the rates of exchange. The course of the leu in post-war years is shown in Figure 3. The rapid movements in the rate adversely affected the purchase of drilling and other materials from abroad and realizations from the crude-oil output.

[1] For a detailed description of the technical problems of the Constanţa line, see Cornelius Toroceanu, Director of the Pipeline Administration, "L'Hiver et les pipe-lines de l'Etat", *MPR*, no. 3 (1925), pp. 197–203.

[2] See report of conference of 8 Apr. 1921 between Pipeline Service, Ministry of Industry and Commerce, and users in *MPR*, no. 8 (1921), p. 376; and 45. Until 1 Apr. 1921 the Pipeline Service was part of the C.F.R. but after that date came directly under the Minister of Communications.

[3] A sulphuric-acid shortage resulted from the lack of locomotives to transport pyrites from Transylvania to the Old Kingdom.

[4] See below, p. 143.

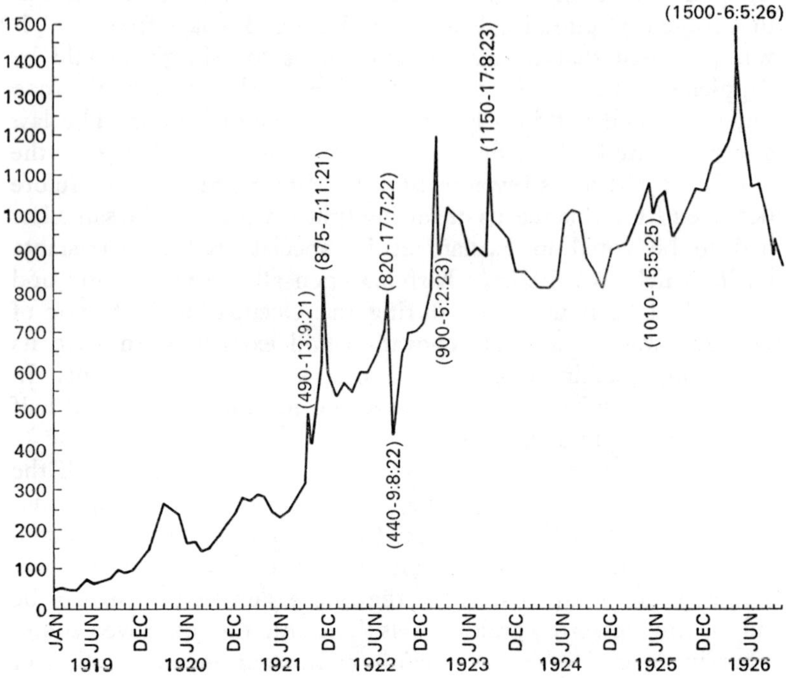

Fig. 3. Leu exchange rates.

Relations between the State and the industry

Post-war policy in both State and industry referred to this situation. The question of its improvement, however, was bound up in the attempts to create a "national" industry and to extend the area of direct state control. Even if the dominant political party had not evolved an *étatiste* programme for the industry, the assistance of the State would have been necessary to its reconstruction, to offset the advantages gained by the "enemy" concerns whose plants had been restored and operated by the occupying authorities. In 1919, for example, the Government enforced an agreement on crude-oil producers and refinery companies whereby crude oil could be processed in return for a fee—the resultant products remaining the property of the producing company. Liberal policy, however, went beyond such necessary *ad hoc* arrangements and attempted to use the opportunity to create a complete monopoly of marketing, both import and export. This isolated Astra Română and Romano-Americană who, by reason of their international connections with U.S., British, and Dutch finance, were best placed to further the re-entry of private foreign capital into the industry and to market its exports.

* * *

The end of the Occupation and the adherence of the new territories suddenly enlarged the internal market but its legal basis remained vague: it was far from certain what local administrative autonomy was to mean in practice and therefore whether the Distribuţia organization would be permitted to operate, or, if so, on what terms.[1] It appears that the various *Consiliu Dirigenti*, the responsible councils in each territory, made their requests to the Ministry of Industry and Commerce in Bucharest for the necessary permissions[2] but treated directly with suppliers on contract and delivery terms.[3] Conversely, the Distribuţia partners supplied outside their arrangement.[4] The administrative centralization of the country did away with this situation, since it established that authority ran only from the Ministries in Bucharest. In the domestic market the inland sales monopoly necessarily followed: The Steaua Română share in Distribuţia appeared to offer a means whereby it could be exercised. The Romanian Government attempted to sequestrate the Deutsche Bank's indirect interest in Distribuţia through Steaua Română and make it over to I.R.D.P.[5]

[1] 46. [2] 47. [3] 48.

[4] "So far we have not declared the shipments made to Orşova to the Distribuţia as we consider this as business with the Bánát and it is not yet settled whether the Bánát will form a part of Roumanian territory. . . . We may expect considerable discussion in the Distribuţia regarding the Transylvanian business, especially as the Orion refuses to declare its sales to Transylvania and Bessarabia": 49.

[5] Decree Law No. 2491 of 10 June 1920. In his report to the Council of Ministers (no. 13540 of 7 June 1920) Octavian Tăslăuanu, Minister of Industry and Commerce, set out the reasons for the Decree Law as:

(*a*) the extraordinary speculation in petroleum products deriving from the absence of organized internal consumption and the irresponsibility of innumerable intermediaries,

(*b*) the existence of different regimes of distribution in the Regat and the new provinces,

(*c*) the necessity, demanded by the higher interests of the country, for the

on the grounds that Distribuția had been compromised by its activity during the Occupation and that the transfer to I.R.D.P. was necessary to put distribution on a national basis. The Government's intentions regarding the inland oil trade were set out as follows:

Article I: All petroleum products necessary for inland consumption . . . which have been sold hitherto by individual companies will be sold in future by the State with the aid of organs controlled by the Ministry of Industry and Commerce.

Article II: All means of transport and storage such as pipelines, rail tank-cars, lorries, reservoirs, and all installations used for the sale of petroleum products within the country are declared to be at the absolute disposal of the State. To this end, they will pass into the domain or the usufruct of the State, according to which it will order their purchase or their hire.

The vagueness of the wording of the Decree aroused much speculation: it was not clear what was to happen to installations brought into Distribuția (even if the Romanian Government succeeded with its plans for liquidation), whether the Constanța line and the state tank-farms would be included—in which case they would surely provide the basis for an export monopoly—

"urgent, uniform, and systematic organization of distribution of all products throughout the country—an organization which could at any moment come under the direct control of the State, governing the quality and manner of distribution, ensuring that consumers were not exploited by intermediaries, and creating for the State an important source of revenue",

(*d*) the fact that the present company, "which is in course of liquidation", could not in consequence partake in a development appropriate to the situation in Romania,

(*e*) that through the Proration Law of 1908, and during the war years, by exclusive right the State had reserved to itself the right to dispose of all oil products: post-war, the different measures taken by the State for sharing and fixing the prices of petroleum products demonstrated the imperative necessity of state intervention for regulating the trade in those products to which economic life was so closely bound—therefore, for all times, the right to distribute products for internal consumption should be reserved exclusively to the State: because, however, a state monopoly was difficult to effect through "the rigid organization of a state administration" it was necessary to concede the working of the monopoly for fifty years to a Romanian company in which the State would have the facility of reserving for itself part of the initial capital, against either cash or storage and transport facilities which it possesses or would possess after the liquidation of companies with German capital. There should be a special convention between the Ministry of Industry and Commerce and the concessionaire. See *MPR*, no. 12 (1920), pp. 377–80.

or whether producers would be compelled to supply products to the monopoly. Further, Article II was widely held to be confiscatory in its implications, since, in practice, it would not be feasible to determine which equipment was used for the manufacture, transportation, etc., of domestic, as distinct from export, products.

The agitation[1] throughout the industry over Article II drew a disclaimer from the Ministry of Industry and Commerce at the beginning of July by which it was to be construed to relate only to "installations necessary to furnishing petroleum products to the interior of the country and over the major portion of which the rights of redemption by the State is recognized in conventions concluded with individual companies". It was added that the measure did not extend to the means of transport or storage that served oilfield exploitation or refineries, which continued to remain fully under the control of their respective companies.

Furthermore, the concession of the former German interest in Distribuţia to I.R.D.P. (Article V) was disputed by Steaua Română on the plea that as a recognized Romanian company it could not be deprived of any of its holdings on the grounds that they were enemy property, that it was contrary to Romanian law to nationalize *part* of a shareholding in a company, and that the Decree law had never been ratified. On these grounds Steaua Română refused to transfer the shares and to permit representatives of the I.R.D.P. to attend annual general meetings. The I.R.D.P. therefore brought a lawsuit against Distribuţia to oblige the latter to transfer 58 per cent of its capital, standing in the name of Steaua Română.

Although the chosen method of effecting the Government's policy was diverted into a long-drawn-out legal dispute, the policy itself was reaffirmed in the Law for Commercialization of State Enterprises and the Mining Act of 1924 (Article 198). In the following year the lawsuit was abandoned in consequence of the "nationalization" of Steaua Română, which formally admitted a "national" company into the business of inland distribution: this was the extent of the Liberals' achievement as far as the question of a state monopoly was concerned—the attempt to vest it in a specifically Liberal company was abandoned. In the 1929 Mining Act

[1] On 16 June 1920 the diplomatic representatives of the countries whose nationals owned interests likely to be affected protested against the decree on the grounds of its confiscatory character: 49.

(Article 200) inland trade was declared free, i.e. not state-run, and left in the hands of Distribuţia and its suppliers.

The Romanian Government's attempts to carry through schemes for a monopoly of oil exports were similarly unsuccessful. As devised immediately after the war, state policy on exports[1] comprised three elements: (i) goods could be exported only under some form of state control, (ii) they were a primary means of earning foreign currency for the State, (iii) and they were a source of tax revenue. The implications for the industry were that exports would be directed, through a licensing procedure, to officially approved destinations, and that the resultant foreign currency would accrue, in whole or in part, to the State, which would give the exporter his return in lei. Whether any export duty handicapped the Romanian exporter or not would depend on the rate and the situation in the markets.

At the end of the Occupation, the Government imposed a ban on exports: this was lifted on 7 June 1919—exports were permitted under strict government control, and the payment of a 20 per cent tax *ad valorem*. In order to maximize the yield in terms of both taxes and foreign exchange the Government set up a *Bureau d'exportation* (known as "Birex"), which divided among those refiners who agreed to operate under its terms orders in proportion to their stocks. All receipts of foreign currency were taken over by the State, which paid the individual exporter one-half of his proceeds in foreign currency and one-half in lei at a rate fixed by the authorities.[2] The State export tax was established at a flat 2 per cent, as distinct from the 20 per cent levied on non-Birex exports. The Bureau was not a corporate entity but merely a common sales office established at the premises of Steaua Română, at the time

[1] This paragraph abstracts the main elements from detailed correspondence. There was no explicit statement of policy by the Government as such: 50. In Sept. 1925 the Romanian correspondent of *The Petroleum Times* put the breaking-point between \$4 and \$3 a barrel for Pennsylvania crude, as against 3,000 and 2,200 lei a ton for Moreni crude: see "Recent Developments in Roumania", 19 Sept. 1925, p. 469.

[2] The rates differed from the free market rate. On 10 June 1919 they were fixed as follows (free market rates in parentheses):

£ sterling	44·25 (cheques)	48·65 (drafts)	(255)
Fr. francs	150·0	165·0	(465)
U.S. \$	9·35	10·0	(65)
Lire	116·15	127·87	(330)
Sw. francs	186·5	204·0	(1175)

51.

still under Constantin Osiceanu as government sequestrator. Apart from Steaua Română, the Petrol Block, Concordia, Aquila, Vega, and Internationale participated in the Bureau. In general, however, the work of the Bureau was conducted by Steaua Română —all correspondence being conducted under its letter-head and signed by Steaua officials. An important factor in the strength of the "Birex" was its control of state transport and storage facilities, use of which was restricted to its members:[1] one consequence was that the Constanța pipeline functioned much below capacity through lack of regular intake.[2]

The "Birex" arrangements remained in force until 30 April 1920. During the year of the Bureau's operations only two large contracts were signed, under pressure from the Government. Of the two contracts, one was between the "Birex" companies and Standard Oil, concluded in September 1919, for 70,000 tons of products, and the other between the Ministry of Industry and Commerce and Lt.-Col. J. W. Boyle, representing the Royal Dutch–Shell Group, also for 70,000 tons of products.[3] Owing to transport difficulties the deliveries could not be effected until the year following, by which time an improvement in oil prices had made the terms of the contracts unfavourable to the sellers. The "Birex" regime failed to maximize exports since the Government could not induce either Astra Română or Romàno-Americană to participate, or to do business with it, although both courses were discussed.[4] They in turn objected to the 20 per cent *ad valorem* tax, payable in foreign currency, during the period up to August 1919, when the tax was suspended, and were deprived of access to pipeline transport, which was used as an inducement for them to enter the "Birex". Accordingly they made no exports until after the "Birex" arrangements had been formally terminated.[5]

The regime which followed was more complicated but designed to serve the same ends. An arrangement on exports was promulgated on 10 April 1920[6] which distinguished two categories—those products exported by the State directly or through consortiums under the direction of the Ministry of Industry and Commerce, and those exported by individual companies: the latter

[1] 52.
[2] 53.
[3] For details see *MPR* no. 4 (1919), p. 121; no. 6 (1919), pp. 188 and 202.
[4] 54.
[5] *MPR*, no. 21 (1920), p. 734.
[6] Published in *Monitor Oficial*, no. 11 (18 Apr. 1920).

could be carried out only under the supervision of a commission for exports composed of representatives of agriculture, industry, and the Ministry. Under the first category, the State had the right to take over its export requirements amounting to 20 per cent of the total export quota at the maximum inland prices, and to use the difference between those and the export price to finance compensation business or purchases for state undertakings in order to do away with extra-budgetary credits. The basis of officially approved private business—the second category—was to be established by a price commission which both fixed the relevant inland price and calculated the relevant export price. The 20 per cent *ad valorem* duty was abolished in favour of a sliding scale payable in lei, based on the difference between the inland and the export price.

Româno-Americană and Astra Română took the view that this scheme was designed to perpetuate the "Birex" type of control by other means, and intervened respectively with the U.S. and British ministers. The Româno-Americană's protest (23 April 1920) was the more comprehensive, stressing that state control or monopoly necessarily prohibited freedom of export for private companies, and that the Law deprived American capital from obtaining a commensurate return on investments made on the understanding that exports would be free. Astra Română's protest confined itself to expressing the fear that the law was specially directed against companies with foreign capital and that there was no guarantee that the producers would receive anything like the real value of their products. Both memoranda requested action as the recipient thought appropriate.[1] The difference in tone and scope reflected partly differences of approach, since Româno-Americană were opposed to all export taxes, while Astra admitted them in principle but questioned how they were to be computed and administered.[2] Diplomatic intervention on 29 April elicited reassurances[3] from the Romanian Minister of Foreign Affairs, Zamfirescu (particularly that export permits would always be granted and were devised only as a basis for levying the tax), and the protests were withdrawn. The Law duly came into force. On 21 May the ministers of the United States, France, Belgium,

[1] 55. The companies' decisions also took into account the likelihood of the creation of a state monopoly for inland trade (published 13 June 1920).

[2] 56.

[3] 57.

the Netherlands, and Great Britain entered a further *note verbale*[1] for consideration, setting out the fears of their nationals and suggesting that the legitimate aims of the Romanian State might be met by allowing freedom to export, with a flat export tax related to the market. This last suggestion was carried into effect in the law of 27 July 1920, which established the bases on which oil exports were to be made, as follows: the export of crude oil, residual fuel oil, heavy lubricating oils, and paraffin wax was prohibited (Article 1); the export of all other products was unrestricted, but only after inland requirements had been met (Article 2); exports were subject to taxes and a 2 per cent commission (Article 3); the right to export was accorded only to producers also supplying the inland trade (Article 4); and the State's right to purchase for export at inland prices was restricted to 10 per cent of the quota exported. This law,[2] together with a law of 27 August (published on 8 October 1920) confirming the State's right to levy taxes but limiting them to 50 per cent of the value of the goods exported, laid down the general lines of state export policy[3] followed by successive Romanian governments, and marked the end of attempts at control through a specific export office. State intervention was restricted to defining products to be exported and taxing them. Thereafter, the arguments of the industry were concerned with widening the definition and with decreasing the taxes, i.e. with commercial rather than political issues. The development of exports under the various post-war regimes may be seen in Figure 4.

This change in emphasis was more marked in regard to exports than to the inland trade. The latter by definition lay within the authority of the State and a state monopoly was both a cherished idea of the Liberals[4] and a reserve defensive measure on the part of their opponents.[5] This factor persisted in the argument between the industry and the State, and the companies' vulnerability influenced their response on domestic matters. Nevertheless, even in this field there was scope for disputes on economic or commercial questions rather than on the "superior interests of the State". It is noteworthy that in these disputes, the dichotomy between "national" and "foreign-capital" companies tended to disappear, since the problems were not of the kind on which

[1] 58. [2] See *MPR*, no. 15 (1920), p. 477.
[3] See policy statement by Argetoianu (Minister of Finance): *MPR*, no. 21 (1920), p. 730. [4] See above, p. 140. [5] See above, p. 132.

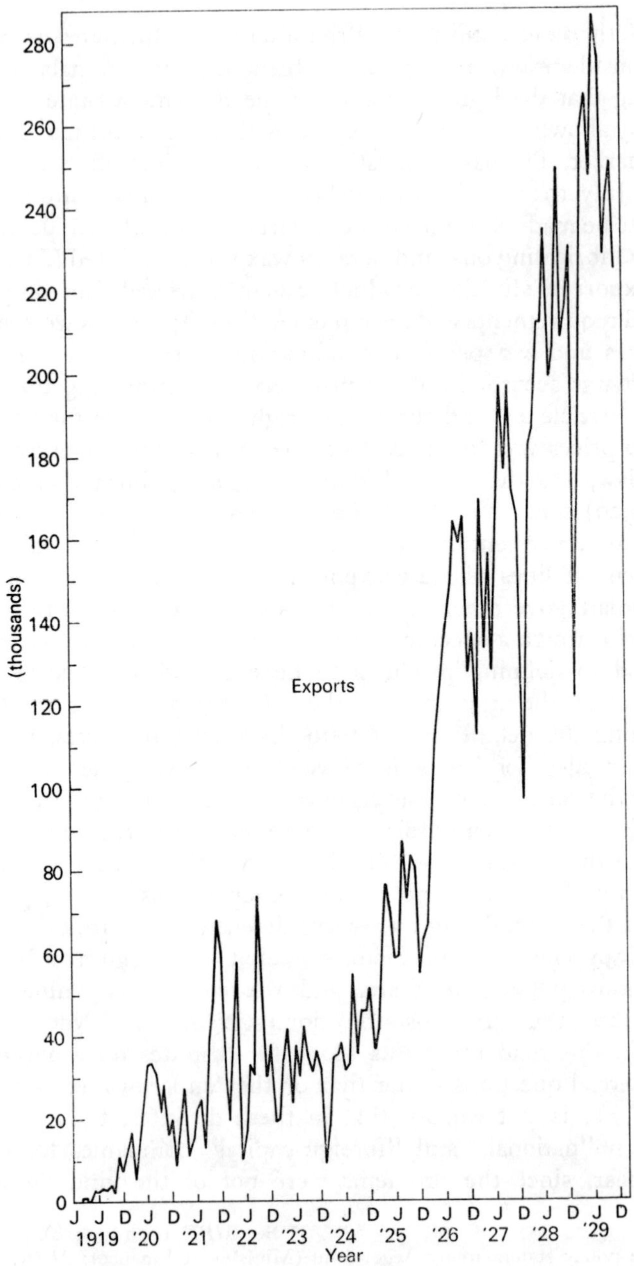

FIG. 4. Exports.

shareholding interest had any bearing: they arose specifically from the incidence of state price and tax policies, and from the operation of transport, which affected all companies alike. Complete price control was reimposed by the Government as part of its reassertion of a Romanian authority over the industry at the end of the Occupation. As exports were banned, there was no compulsion to take notice of external yardsticks in setting prices, which reflected strictly administrative considerations and were, it appears, below costs of production.[1] This was the ground of the first objection by the industry after the war, which opened a continuous dispute between the two sides. Throughout the State insisted that the inland market had priority, and that it had the right to set maximum prices for that market and to extend them to supplies not only to all state institutions, e.g. the armed forces and C.F.R., but also to all industries set up under the laws to encourage industry. In the Government's view the difference between these nominal prices inland and an economic return was to be made up by the receipts from export. This activity, when not forbidden to private companies, was also taxed for revenue-raising purposes. Conversely, the industry conceded the points about the State's right to set prices (although demanding that they should be high enough to cover costs and risks) and about supply to state institutions (but complained that they were dilatory in settling their accounts). The industry objected to subsidizing the fuel costs of private Romanian manufacturers and stressed that export taxes restricted Romania's competitiveness in the market, since the ruling prices there reflected the costs of supply from alternative sources. Additionally, the administration of these taxes was a separate factor, since the levying of the tax took no account of the situation in the market and adjustments tended to lag.[2] These general arguments were supplemented by disagreement on specific issues, such as the claim by the State to take over either all or a percentage of the refiners' export output at inland prices in order to provide for the State's own barter business.

[1] See notes in *MPR*, nos. 5–7 (1920), pp. 175–6; no. 11 (1920), p. 339.

[2] In 1922 the Government decided to use the revenue from export taxes to create a sinking fund for the consolidation of bonds to be issued in exchange for Treasury bills. For this reason export taxes on petroleum products were increased—including those on gasoline—at a time of steadily falling world market prices.

Successive representations by the A.I.P.R.[1] brought about modifications in the tax regime and even the abolition of the export tax in 1929. Nevertheless, the effectiveness of industry objections to fiscal policies was limited by the State's need[2] for revenue, enhanced by the disruption of agriculture.

Its pre-war history had demonstrated effectively that transport[3] was in the hands of the State. No attempt was made after the war to controvert this position. The restoration of railways and pipelines depended on state finances, but the very improvements which would increase the possibilities of raising revenue demanded money which the State lacked. A special foreign loan for the reconstruction of the railways therefore became one of the objects of Government policy,[4] but it remained suspended through the reluctance of the Brătianus to give guarantees. Meanwhile, for political reasons, such new construction as took place was devoted to binding the new territories to the Old Kingdom. Essential work, as, for example, the doubling of the track from the oilfields to Constanța, could not be undertaken. In 1925 the journey of an oil-train at harvest time from Ploești to Constanța—less than 200 miles—could still take up to a fortnight. Poor performance derived from defective physical assets[5] and indifferent

[1] See, e.g., *MPR*, no. 15 (1922), p. 889; no. 1 (1924), p. 75; no 19 (1924), pp. 1565 and 1567; no. 10 (1925), p. 794; no. 21 (1925), p. 1774; no. 5 (1926), pp. 453 and 575; no. 23 (1926), p. 2256; no. 13 (1928), p. 1269.

[2] Not only that of the central Government: after 1922 communes such as Ploești and Constanța taxed the movement of products through the areas under their control: see *MPR*, no. 9 (1922), p. 508.

[3] Here considered only as railways and pipelines: road transport also presented considerable problems, but these had no *particular* bearing on the oil industry. The root cause was that the adequate road system of the Old Kingdom did not suffice in Greater Romania, and that there was insufficient money for renovations and extensions. Astra Română's viewpoint on the matter was: "We experience already considerable difficulties in obtaining payment from the Government for our fuel supplies, and, as asphalt supplies for road construction would mean an increase of these difficulties, we are not in favour of interesting the Government in improving their roads" (59). It is reasonable to suppose that other suppliers to the Government shared this view—which explains, *inter alia*, the continued reliance on oxen for draught purposes: see "Transportation in Roumania", article by Romanian correspondent, *PT*, 20 Nov. 1926.

[4] Railway reconstruction was one of the objects of the projected American loan to Romania during the war (Mamatey, p. 167). Sir Albert Stern informed the writer that his participation in the purchase of the Steaua Română shares developed from the discussion of a possible railway loan.

[5] See references in various reports published by the Department of Overseas Trade on economic conditions in Romania, particularly Apr. 1921, pp. 58–60; Mar. 1923, pp. 36–40; 1926, pp. 16–17; Mar. 1928, pp. 27–8.

administration,[1] both directly related to lack of funds. From the oil exporters' standpoint the Pipeline Service was on the same footing as the C.F.R., since whether the Service was part of the C.F.R. or separate under the Ministry of Communications, charges were fixed so as to rule out competition with rail transport.

The financial stringency which held up not only necessary technical renovations but also the payment of salaries due to Pipeline Service officials inhibited the full use and further development of pipeline transport in Romania. Inadequate repairs and maintenance entailed pumping losses exceeding the permitted 2 per cent[2] and periods of reduced working—thus raising the cost per unit of product transported and consequently tending to increase the tariff.

For the oil operators, railways and pipelines were two aspects of the same problem—i.e. the movement of products throughout the country for the inland trade and to points of export in a way consistent with keeping refineries and wells operating at the maximum possible capacity and with fulfilling delivery schedules, particularly those to tankers at Constanța. Delays in either respect increased the unit cost of operations—in manufacturing, through accumulating stocks or diminishing the rates of refinery throughput, or, in export, through incurring demurrage charges. To a lesser degree there was also the problem of moving drilling and other equipment from Constanța and Galați inland to the oil fields.[3] These motives therefore prompted intervention by operators, through the A.I.P.R. and individually, with the authorities to improve the service of the C.F.R. and the Pipeline Administration. A series of formal discussions between representatives of the Government and industry took place during April 1923, when a number of outstanding questions, e.g. transit losses and tariff charges, were solved. The improvement of facilities[4] presented

[1] e.g. the balance sheets for 1919–28 were drawn up in 1929: the railway was overstaffed; officials (like other civil servants and government departments) had free use of facilities; at the same time they were poorly paid. David Mitrany —in *The Land and the Peasant in Roumania* (p. 465)—reports "The memorandum submitted by the Society of Engineers in March 1923 stated that the nominal wages of railwaymen were nineteen times above pre-war level but the salaries of railway inspectors only 3·15 higher, which meant that their real salaries were 89 per cent lower. The latter figure was probably applicable to all civil servants."

[2] 60.

[3] Transport from the ports was complicated by the precedence given to all government departments in the allocation of available trucks.

[4] 61.

greater difficulty. At the end of May a draft pipeline scheme was drawn up, reaffirming the State's exclusive rights and envisaging a 45 per cent tariff reduction to "national" companies. The Government representatives were trying to combine political aims with technical solutions and no progress resulted. The Giurgiu lines were patched up after the looting at the end of the Occupation and functioned adequately after mid 1922.[1] At this stage, improvements to the Constanța line, replacing the 5-in. joints and adding the pumping capacity necessary to increase its capacity from 600 to 1,000 tons a day were estimated to cost 84 million lei. The increase in output during 1924 and 1925 enforced further consideration of the transport problem as an alternative to shutting back production, with consequent loss to the industry and the State. In June 1926 the Averescu Government appointed a commission[2] to investigate the construction of an additional line to Constanța. The commission decided in favour of a scheme put forward by Mrazec and Osiceanu[3] whereby a line should be built at the cost of the A.I.P.R., which would obtain materials on credit from foreign manufacturers who would be paid from the resultant revenues. When the last credit was liquidated the line would revert to the State under the Commercialization Law. For this purpose the oil companies would form a company[4] with state participation to work all pipelines, including the new one, for thirty years. The companies, through the A.I.P.R. entered into discussions, which were held up through disputes on the valuation of existing installations.

The inadequacy of existing facilities was emphasized when the larger companies began shutting back production during the last quarter of 1926, while the urgency of dealing with the problem was demonstrated by the re-emergence of Russian oil in Germany in 1927 and the prospect of more serious competition from Iraq. The Averescu Government resigned in June 1927 and was succeeded by the Liberals. It appears that Vintila Brătianu had ideas

[1] 62.

[2] Consisting of Osiceanu, Pretorian (General Manager, CFR), Popescu (Ministry of Industry and Commerce), Mrazec (Geological Institute), Toroceanu (Pipeline Service).

[3] 63. Guest reported in April that Mrazec's scheme originated with Steaua Française.

[4] The draft Articles of Association were published on 1 July 1926: *MPR*, no. 13 (1926), p. 1278.

of using the pipeline scheme for the benefit of the "national" companies by converting it to the supply of crude oil to a joint "national" refinery at Constanța.[1] Whatever the reason, it was only in January 1928 that a definitive form was given to the projected bill, but no further action took place during the parliamentary session. Further progress had to wait on the formation of the National Peasant Government, which enacted a law establishing a regime for the running of state pipelines in July 1929.[2] But at the end of the decade the extra line to Constanța remained unbuilt, and the companies were still petitioning the Minister of Industry and Commerce over transport tariffs.[3] In default of an additional pipeline, Romania's gasoline exports depended on the performance of the railways. Representations to improve the C.F.R. were no more effective than those on the pipeline, since the sums involved for repairing and modernizing the system could be obtained only from a large loan, the basis of which could not be agreed between the Romanian State and possible lenders until 1927–8. Additionally the status of the C.F.R. within Romania itself was a political issue.[4] The oil industry had to restrict itself to suggestions for improving traffic, arguing about freights, and demanding payment for the supplies of fuels and lubricants it was obliged by law to supply to the C.F.R. at nominal prices.[5]

Changes in export markets and their effect on the industry

The changes and attempted changes affecting both inland and export markets so far discussed were those which resulted from the initiative of the State and the reaction of the industry within the new Romania. In addition there were during the post-war decade other changes, mostly outside the country, to which both state and industry had to adjust themselves.

[1] 64; see also report by a special correspondent, "A New Era in Roumania", *PT* (24 Nov. 1928), p. 930.

[2] See *MPR*, no. 13 (1929), pp. 1154–5; no. 14 (1929), pp. 1217–26 (text); no. 15 (1929), p. 1336; see also 65.

[3] 66.

[4] The Liberals introduced changes in June 1925 which were reversed by the Averescu Government in Feb. 1927 and re-enacted by the Liberals in November of the same year.

[5] See, e.g., A.I.P.R. discussions and representations in *MPR*, no. 20 (1923), p. 1405; no. 23 (1926), pp. 2256–7; no. 21 (1929), p. 1990.

The rapid development of the production and use of automotive vehicles confirmed the importance of the trade in gasoline, as distinct from that in kerosine, which declined as an illuminant and heating agent in favour of electricity and gas. The predominance of gasoline was reflected during the 1920s in the standardization and improvement of the product, in consequence of which a number of processes were devised to increase gasoline yields by cracking heavy distillates and residuals and by recovery from refinery gases. Blending was introduced to improve volatility and anti-knock qualities.

The emphasis on gasoline encouraged in Europe the commercial development of its production by the hydrogenation of coal,[1] but the cost of synthetic gasoline remained above that of the petroleum-derived product and all manufacture required direct government subvention or duties and excise taxes on gasoline from other sources to make it economic. Benzol became another coal-based competitor. In its pure state, although effective, it was too expensive for widespread commercial use, but blended with gasoline, began to be sold as a motor fuel. Other non-petroleum-based blending agents, such as alcohol and methanol, also appeared: these were less technically successful than benzol largely due to the susceptibility of alcohol to atmospheric humidity and the lower calorific value of the mixture. Their introduction was in each case enforced by governments to assist agriculture.[2]

Even in Germany, where conditions for the production of synthetic gasoline were the most favourable, its development took

[1] Research into and manufacture of synthetic fuels were carried furthest in Germany, where synthesis was the local version of that search for "oil" security which pre-occupied the U.S., French, and U.K. governments as described at the beginning of Chapter V. Germany was well supplied with raw materials, possessing large deposits not only of bituminous coal, but also of lignite, which was obtained by surface working and offered certain technical advantages that made a lignite-based process cheaper. The first manufacture took place in the Leuna Werke in 1927. On developments in synthetic fuels in Germany see *Report on the Petroleum and Synthetic Oil Industry of Germany*, Ministry of Fuel and Power, 1947; and W. Birkenfeld, *Der synthetische Treibstoff, 1933–1945*.

[2] In Feb. 1923 the French Government imposed compulsory blending of alcohol to aid the potato, sugar-beet, and vine growers. This example was followed in 1926 by the German Government to assist grain production, and successively by other states of Europe with an influential agricultural sector or where the production of alcohol was a state monopoly—including, among Romania's customers, Italy, Austria, Hungary, Czechoslovakia, and the Balkan states.

time, so that during the 1920s it remained a potential rather than an actual threat to the petroleum-based product in the market. Compulsory blending, however, diminished the gasoline market with immediate effect: blending ratios ranged from 10 to 15 per cent of gasoline and for short periods in certain states were even higher. Moreover, the fact that both developments were state-directed was a new element in the market situation.

Concurrently with the increase in demand for and quality of gasoline there took place a similar development in lubricants. After the mid 1920s improved engines demanded a product free from the tendency to form gum and carbon deposits, and the existing method of ordinary distillation followed by acid treatment no longer sufficed: neither did the output of paraffinic crude oil yielding the higher-grade lubricating oil. It therefore became necessary to experiment with other crudes hitherto considered unsuitable. Improved refining methods (Trumble plants) yielded a better distillate, and the application of de-waxing and solvent extraction techniques gave an extensive range of lubricants to more exact specifications.

These changes were necessarily reflected in the operations of the Romanian oil industry. Crude oils with a high gasoline content were in greater demand, but the increases in Romanian production during this period were predominantly in paraffinic crudes, leaving in ordinary distillation a high proportion (up to 50 per cent) of residues containing so much paraffin as to remain solid up to 35 °C. This was unsuitable for the fuel-oil market, but could be treated to provide gasoline. Romanian refining practice then changed so that on first distillation the paraffins were eliminated, leaving a residue suitable for cracking. Colombia initiated the change in 1927, and by 1930 Romanian industry included eleven cracking plants. These changes in refining evidenced a response to the demands in the market where profits were to be earned: against this, however, in any consideration of the future, were to be set the increasing gas–oil ratio[1] in the known areas from which the annual increases of output were derived, and the virtual absence of wildcatting,[2] so that vast tracts of possibly oil-bearing

[1] The quantity of gas per m³ from 1 ton of crude oil.

[2] i.e. of the drilling of exploration wells to "prove" a suspected oil-field: if successful, "wildcats" are followed by (i) appraisal wells, to establish the limits of the structure, and (ii) exploitation wells: if unsuccessful they remain as "dry holes".

territory remained unsurveyed. This failure to explore was the result partly of the high cost of drilling[1] and partly of the drive for "nationalization". Creditul Minier and I.R.D.P. were designed to provide a short-run return: they therefore operated on state lands in the traditional areas granted to them on favourable terms. The companies best equipped in terms of finance and technique, Româno-Americană and Astra Română, desisted from wildcatting, since, after the Mining Law of 1924, success implied the retention of only fifty out of a possible thousand hectares explored—with the rest being taken over by the State for itself and for "national" concerns. Both companies therefore concentrated on further exploitation of areas held as acquired rights. These factors, it must be stressed, were relevant to future operations and their significance depended on a revised distribution of the rewards to prospectors and the use of improved and cheaper drilling methods. Both eventually involved alterations in legislation—in which respect beginnings were made at the end of the decade. The elimination of nationality qualifications under the 1929 Mining Law removed one hindrance to wildcatting, and general permission to use the more efficient drilling rotary system to "bring in" wells (instead of having to discontinue its use at the oil layers) helped to inhibit the rise in extraction costs—as well as, for example, decreasing the fire risk and waste of natural gas.

The contemporary situation, seen within Romania, was more satisfactory. Successful exploitation at Gura Ocniței in 1924 and 1925 by Concordia and Phoenix, and penetration of Meotic strata in Ochiuri by I.R.D.P., Creditul Minier, Astra Română, Phoenix, and Steaua Română between 1925 and 1927 increased the importance of Dâmbovița as a producing area. In Prahova the exhaustion of the shallower deposits and inflow of water in the old-established Moreni field was compensated by discoveries at deeper levels at Pleasa (Româno-Americană), Țuicani, and Ceptura. These operations account for the increases and acceleration in production after 1924 shown in Figure 5. The drilling and production effort giving these results was shared between "national" and "foreign-capital" companies: the former were concerned to consolidate their legally privileged position, the latter to offset their disadvantages by maintaining and improving their share of Romanian output.

[1] See "Rising Drilling Costs in Roumania" *PT* (5 Feb. 1927).

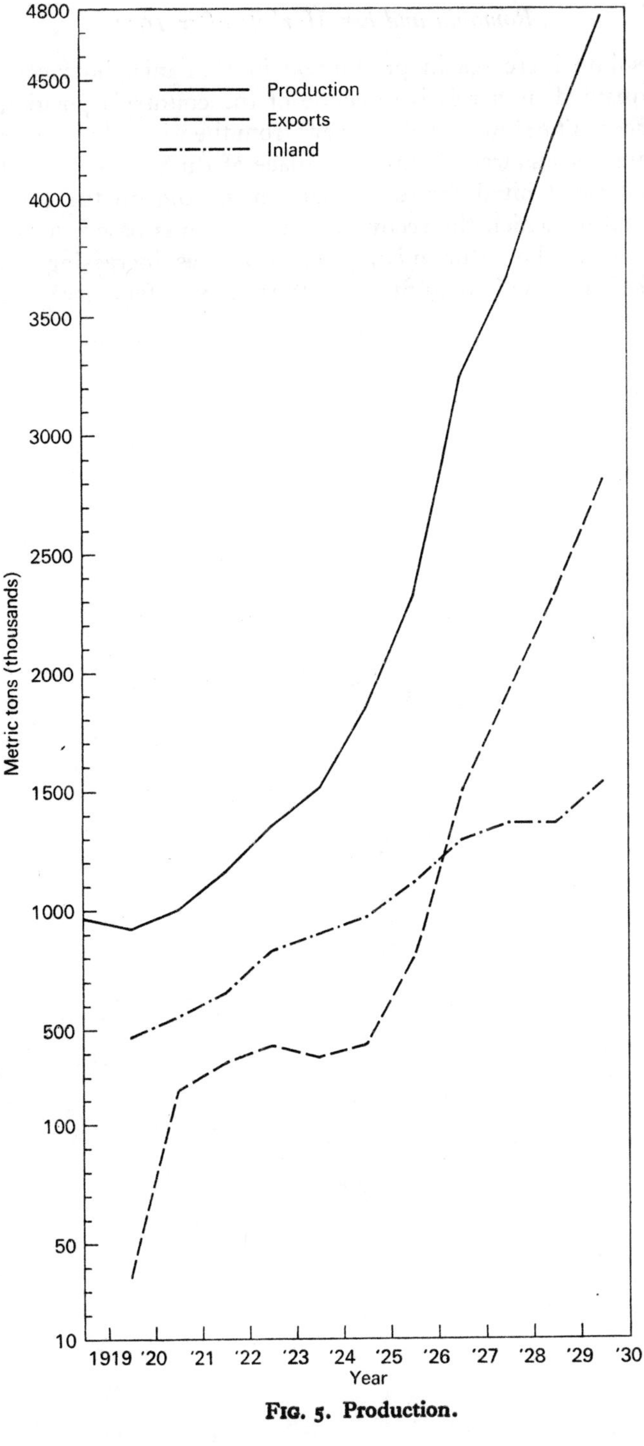

Fig. 5. Production.

Absolute increases in production in Romania, however, were accompanied by a relative decline of the country's position as a producer. The slow rate of recovery from the war failed to establish Romania as a source of supply in place of Russia and Poland, and allowed the United States to share in making up the supply of crude oil on which the recovery of Europe was based. Moreover, after 1925, when Romanian production was increasing rapidly, so was that of the United States, so the relative inferiority remained.

CHAPTER VII

Crisis and Depression

The effect of the world oil crisis of 1929–1933 on the Romanian industry

As was shown above in Chapter V, the main post-war preoccupation of oil-producing states and international oil companies was securing sources of supply against an expected decline in output in consequence of the exhaustion of reserves. From 1925 onwards, however, the discovery of large flush-pools in Oklahoma and Texas combined with the emergence of Venezuela, and, to a lesser extent, the Middle East, to create a crude-oil surplus, with a consequent depressing effect on crude-oil prices and a cumulative reaction throughout the industry.[1] In these circumstances the main problem became not one of supply but one of conservation, and various unsuccessful attempts were made in the United States to restrain the production from the prolific new fields.[2] Since the United States was still the main single source of supply for Central and Western Europe, the repercussions of American over-production on the Romanian oil trade were severe, and only areas not open to American seaborne trade remained safe markets

[1] An excess of crude oil necessarily created a surplus of fuel oil which both depressed the general fuel market to uneconomic levels and promoted the manufacture of gasoline through cracking. In consequence the gasoline surplus caused a fall in gasoline prices, cracking operations ceased, and the fuel-oil surplus continued to depress the price of fuels to even lower levels.

[2] Until May 1933 conservation was the responsibility of the governments of the individual producing states: there was, in consequence, no correlation between any state's permissible production and the total domestic and export demand of the United States. This correlation was ultimately enforced by Federal action under the Industrial Recovery Act. On the conservation problem in the United States, see the *Report of the Federal Oil Conservation Board* (from 1926 onwards), Washington D.C.; *The Legal History of Conservation of Oil and Gas*, published by the Section of Mineral Law of the American Bar Association, New York, 1938; and "Petroleum Development and Technology", in the *Transactions of the American Institute of Mining and Metallurgical Engineers*, vol. 103, published by the Institute, New York, 1933.

for Romanian products.[1] In general, however, the prices commanded by Romanian exports were drastically reduced, and so, in consequence, was the financial return to the oil industry and the State of Romania. Furthermore, during the period of U.S. over-production, drilling activity in Romania carried on by the "national" companies in their newly acquired areas, as well as the penetration of deeper layers by improved methods by the major "foreign-capital" companies, resulted in the discovery of prolific wells in the Meotic strata of the old Moreni field and increases in Ochiuri, Ceptura, and Gura Ocniței. As in the United States, the legal situation discouraged conservation: since, from the beginning, the emphasis in oilfield development had been on the fulfilment of drilling obligations under leases, there was no basis in Romanian law for their postponement.

In consequence of these developments, the Romanian oil industry, by the middle of 1930, was faced with an output of crude oil much in excess of its refining capacity[2] and extremely low prices in markets.[3] This bore more heavily on the smaller companies than on the larger, since the latter in the circumstances merely discontinued buying crude oil and refined entirely their own oil. This reaction displaced the crude oil normally supplied by the smaller companies, who were deprived by law of their obvious alternative, i.e. selling for export through brokers. The producers without refineries, therefore, began to press through their Association for the removal of this restriction. The A.I.P.R. appointed a committee[4] to examine the measures necessary to reduce crude-oil output, which reported in May 1930 in favour of restrictions

[1] In 1931, e.g., the U.S. supplied 34 per cent gasoline, 36 per cent kerosine, and 32 per cent lubricants to the U.K.; 23 per cent, 10 per cent, and 22 per cent to France; 16 per cent, 9 per cent, and 17 per cent to Germany. See R. Schwarz, *Petroleum Vademecum*, 9th edition. See also summary of lost outlets in the representation of the A.I.P.R. to the Minister of Industry and Commerce: *MPR*, no. 22 (1931), pp. 1075–6.

[2] This had not been experienced since the 1880s, except for the war years.

[3] The effect on Romanian export prices for gasoline is shown in Figure 6; for a discussion of prices during the Depression see editorial in *MPR*, no. 2 (1934). The market situation was additionally complicated in Germany by the reappearance of Russian supplies through a Soviet-owned concern, Die Deutsche Vertriebsgesellschaft fuer russische Oelproduckte ("Derop"), established in 1927, which sold gasoline through retailers and in bulk to benzol producers. In consequence, the Russian share of the market, which had since 1924 averaged 6·5 per cent, increased in 1928 to 14·4 per cent.

[4] Consisting of Matasaru (Steaua Română), Ficşinescu (Columbia), and Tacit (Creditul Minier): see *MPR*, no. 4 (1930), pp. 821–3.

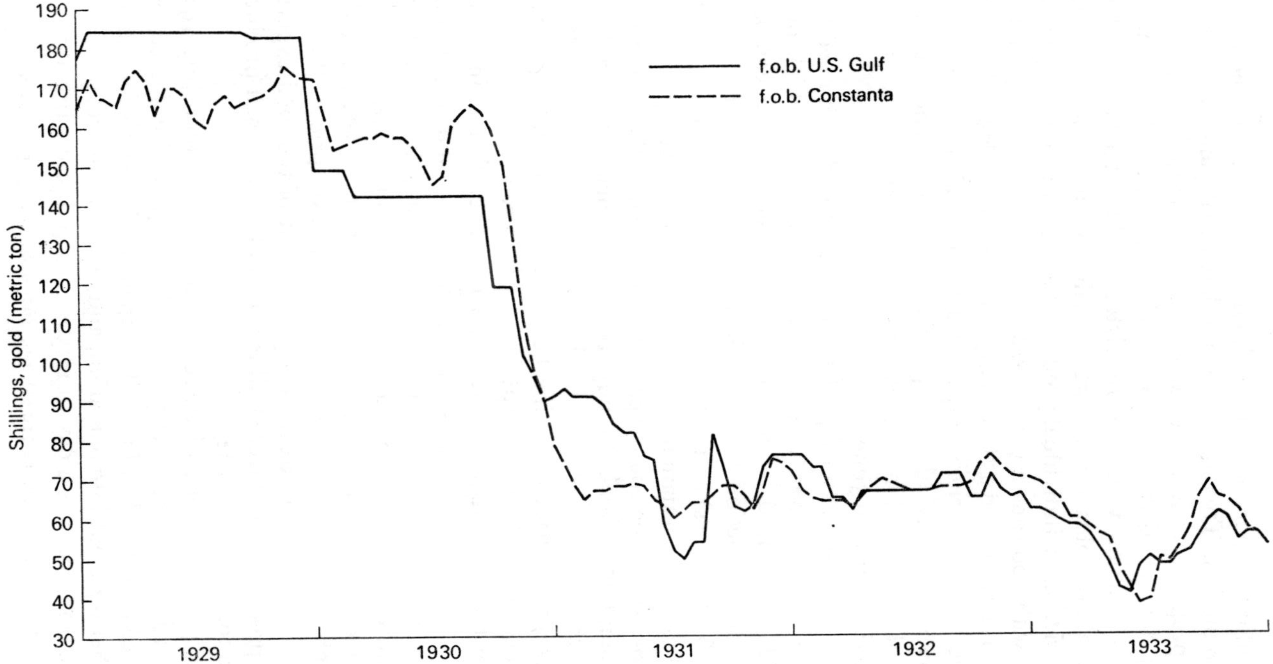

FIG. 6. Gasoline prices.

on drilling proportionate to each company's share in the increase in production. This approach was left in abeyance in view of a joint proposal by Romăno-Americană and Astra Română (4 July 1930) for a "Conservation Convention" whereby Romania's total production would be restricted to the total of the internal consumption plus the exports to adjacent countries. The level of restriction on individual producers was to be assessed by output; those with 100 tons a day or less would be unaffected; all others were to be permitted a "free" output of 100 tons a day and restricted on the balance of their output in a ratio equal to the proportion between the individual company's daily output and Romania's internal consumption and exports to adjacent countries during the period April to June 1930.

This scheme was designed to be put into effect immediately: for the future it was suggested that the scale by which output should be reduced would be the ratio between the company's potential production (to be estimated by letting all wells flow freely for two days) and potential consumption as already defined. Changes in the situation were to be met by repeating tests every six months. As the counterpart of both schemes, the two proposer companies undertook to buy from Romanian producers as a whole the same quantity of products as for 1929, viz. 67,000 tons gasoline, 235,000 tons kerosine, 68,000 tons gas oil, and 105,000 tons fuel oil, on condition that the proposals were accepted by 95 per cent of the oil industry by 19 July 1930. The scheme, in essence, offered to the participants an opportunity to find a guaranteed and profitable outlet for a proportion of their output and thereby an inducement to restrict their production to that level, while the proposers hoped to stabilize conditions in the European oil markets through a curtailment of supplies at "distress" prices: the stated totals comprised approximately 31 per cent of the total export in these products and covered all exporters but the small producers of 100 tons per day or less.

The proposals were discussed in the A.I.P.R. on 9 July 1930,[1] approved by the Ministry of Industry and Commerce, and accepted by virtually all the oil companies.[2] Measures for immediate relief were put into operation, and, in consequence, production fell from 17,000 tons a day to 13,500 tons and prices began to rise.

[1] See *MPR*, no. 14 (1930), pp. 867–970; no. 15, pp. 941–2.
[2] 67.

While these measures were being implemented and details of the long-term arrangements being worked out, there appeared a possible rival solution to the Romanian marketing problem, deriving not from the industry but from the Government. On 12 August, it was announced in *Le Temps* that following discussions between Tardieu and Mironescu an inter-governmental oil agreement was about to be concluded. The next day the French Government issued a *démenti*, and the agreement[1] was signed on the 27th. Each government undertook to set up organizations within the petroleum industry, respectively to supply and receive 400,000 tons of products and 400,000 tons of crude oil residue for cracking and an undetermined quantity of fuel oil. The Romanian group was to consist of all companies exporting directly or indirectly to France, and its supplies would be restricted to those French importers who were in the corresponding group. In return the Romanian Government was prepared to accord "Romanian national" treatment to "French-capital" companies in Romania. A state-guaranteed market for more than twice the quantity envisaged under the conservation scheme was correspondingly the more attractive and, at the proposed scale of exports, implied an increase in production rather than a restriction to 1929 levels: the financial returns to the industry and the State could be maximized through the sale of large tonnages at low prices.

A fortnight later, on 12 and 13 September, the flow test was carried out to establish the potential production of each of the participating companies. The authors of the scheme recognized that the test might not in all cases represent any one company's maximum production, but urged that in the time available it was the most satisfactory working basis. In the event, the test was not wholly successful: by the time it took place many small wells had already been shut in for some months and others working on restricted flow, so that in these instances, during the actual test, the operating gas pressures were above normal and the rate of flow correspondingly increased. Furthermore, some wells were allowed to run without a choke.[2] Nevertheless, the results showed the Romanian potential output was 31,000 tons a day: allowable October production was fixed at half this figure, and the maximum

[1] Text in *CER* (1931), pp. 130–56.
[2] 68.

percentage production of the major companies for the period 30 October 1930–31 March 1931 was established as follows:

Astra Română	15·3%
Concordia	13·3%
Phoenix	13·2%
Steaua Română	12·5%
Creditul Minier	10·5%
Romăno-Americană	7·7%

These results gave rise to discontent among those producers—notably Steaua Română—whose maximum allowable production was less than their expectations. Astra Română and Romăno-Americană sought to meet the acknowledged shortcomings of the tests by increasing the frequency of future tests from six months to three months and proposing a drilling-restriction scheme, details of which were published on 17 October 1930.[1]

Before their proposals could be adopted, the proration scheme broke down through the withdrawal of Steaua Română, notified on 24 October, on the grounds that the flow test had worked against its interests and that its production and refining output during 1928 and 1929 had been such as to satisfy a higher quota.[2] Steaua's action was discussed at a meeting of the A.I.P.R.,[3] during which Osiceanu refused to continue to operate the convention while modifications to meet his company's difficulties were being agreed, on the grounds that its technical problems could be overcome only by a higher production. He further attempted to use these arguments to justify Steaua's withdrawal from the scheme from the end of October, but the meeting insisted successfully on his observing the obligatory period of notice; the withdrawal therefore took effect from 30 November 1930. The action of Steaua, the second-largest producer, was decisive, and subsequent attempts to establish an agreed conservation policy failed.[4] On 1 December Astra Română and Romăno-Americană officially withdrew their joint offer.[5]

On the same day as Steaua Română gave official notice of withdrawal, it became known that Pallas, one of its associates,

[1] On the relationship between the potentiality test and the drilling proposals, see 69: for proposals see 70.

[2] For Steaua Română's letter of withdrawal see *MPR*, no. 21 (1930), p. 1259.

[3] For A.I.P.R. discussion, see *MPR*, no. 23 (1930), pp. 1249–51, 1353–4.

[4] See *MPR*, no. 24 (1930), pp. 1363–7.

[5] For the performance of the major companies under restriction, see Figure 7.

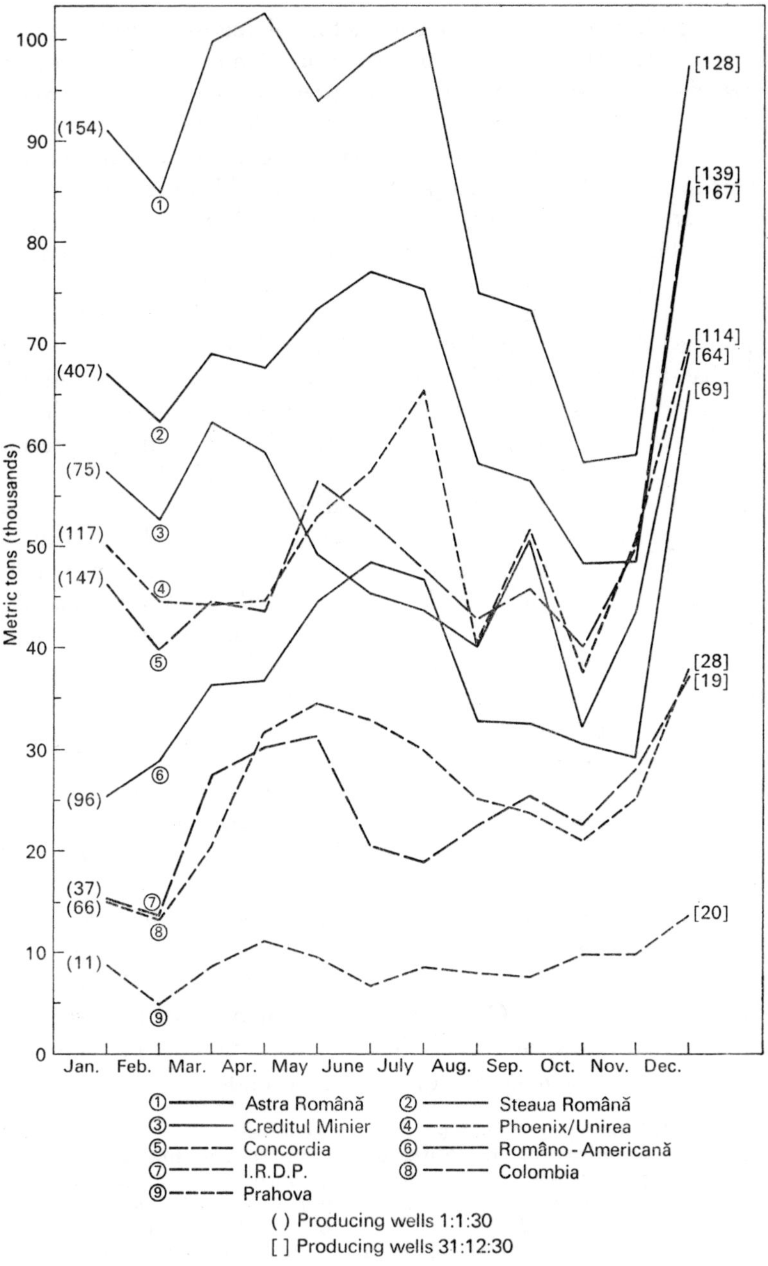

FIG. 7. Production of major Romanian companies, 1930.

was negotiating in Paris for a contract involving the sale of 400,000 tons of products to the Spanish state oil monopoly. The contract was signed three days later. The precise relationship between Steaua and Pallas in this business was not immediately apparent. Osiceanu told Stern[1] that Pallas had concluded the contract on their own initiative but that possibly they would cover part of their requirements with Steaua Română. On 31 January, however, it was publicly announced that Steaua had taken over the entire contract, against which they began to make deliveries.

To the proponents of restriction, the venue of the negotiations appeared significant: in their view Steaua Română had responded ultimately to French initiatives. The Romanian company adhered to the original Conservation Convention only at the last minute "owing to opposition from the French side".[2] Meny of Steaua Française mentioned, as early as 28 July, the difficulties of Steaua Română and indicated a certain restlessness on the part of his company as shareholders.[3] Otto Stern recalled[4] that Steaua's notice of withdrawal "was due to telegraphed instructions which they had received only on the morning of that day from Paris", and went on to give his "impression that Osiceanu is much less responsible than Mr. Meny, who has pulled all the strings as regards Steaua's attitude on conservative matters, the Spanish Monopoly negotiations, and I have no doubt that he is the author of the whole Franco-Romanian game". Whatever Meny's role, it remained correct that a government-guaranteed outlet, whether in Spain or France, negated any idea of restriction: the course of events in this respect demonstrated a change in the attitude of the Romanian Government.

From the outset the Romanian Government had been kept informed, and initially had adopted an attitude of "benevolent neutrality" towards the scheme.[5] Owing to the State's need for funds, however, the Government continued to issue leases on state-owned lands for exploration purposes, in spite of the over-

[1] 71. [2] According to Steaua Română (British): 72. [3] 73.
[4] 74.
[5] The Liberal opposition denounced it as a betrayal of the "national" companies, resulting from the National Peasants' softness towards the world petroleum trusts, whose sinister activity was aimed at a complete monopoly of Romania's oil resources. Nevertheless, Stern wrote to Kessler on 16 July 1930 that he had been informed that Vintila Brătianu had "seen the proposals" and "given instructions" to the Liberal oil companies to sign the Convention—which they did. 75.

production.[1] Madgearu, as Minister of Industry and Commerce, approved the Conservation Convention, subject to reservation of the State's right to intervene in order to produce the maximum profit to the national economy. The State had a direct financial interest in the scheme as, under conservation, receipts from royalties and from railway freights would necessarily be contracted.[2] Madgearu's view was that these would have to be accepted:

the national economic interest compels us to recognize that the continuation of crude-oil production on the present scale would have implied bad administration and waste of one of the most important sources of natural wealth. . . . Moreover, at the present time and in the midst of international circumstances, where Romania sells its cereals at half price, its timber at three-quarters of its value, it would have been a capital error to continue to export our crude oil and products at less than half price.[3]

It appears, however, that Madgearu was ahead of his colleagues in his understanding of the situation. As Otto Stern commented:

the only person in the Government who fully understands the possibility and necessity of our conservation convention is Madgearu, whereas the other members of the Government are more or less against it, the more so as they sufficiently understand neither the situation in the petroleum industry nor our restriction scheme. This was clearly evident when Manoilescu, the present Minister of Communications, who in Madgearu's absence is representing him at the Ministry of Industry, asked us for a meeting two days ago and wanted to know from us to what extent Romanian petroleum production would now be increased as a result of the potentiality test. It took us some time to make him understand—he understood it in the end—that Romanian potential production has nothing to do with the monthly maximum, but still he continues to think, whether, if not for the industry, then for the State, it might not be better to produce at full potential with low prices instead of producing restricted with higher prices.

The weakness indicated by Otto Stern, viz. the isolation within the Government on the issue of the responsible minister, was

[1] Decree No. 7360, 26 Jan. 1931: see *MPR*, no. 3 (1931), pp. 125–6.

[2] The State's straitened financial circumstances induced Madgearu to request that, under the conservation scheme, I.R.D.P. be allowed an exceptional position to enable that company to pay off a debt to the State of 60 million lei. The request was refused. 76.

[3] See *MPR*, no. 15 (1930), p. 943.

emphasized by the policy of the Minister of Foreign Affairs in regard to commercial contacts in both France and Spain. The weakness became more severe in consequence of changes in administration. In October 1930 the National Peasant Government was reconstructed[1] under Mironescu, who combined the office of President of the Council with responsibility for foreign affairs. Madgearu was moved to the Ministry of Agriculture and was replaced at the Ministry of Industry and Commerce by Manoilescu. From the standpoint of the proponents of restriction, these changes meant the highest office was filled by the negotiator of the Franco-Romanian treaty and the most important ministry was occupied by an associate of Osiceanu.

Manoilescu, apparently, was not convinced by Stern's arguments, reported above, since on 11 November 1930, at the annual meeting of the Union of Chambers of Commerce and Industry, he welcomed the end of restriction "thanks to a very audacious gesture on the part of your own President" (Osiceanu) and espoused the idea of maximum production at low prices as state policy. Furthermore, at the same time, he began to implement the Franco-Romanian oil agreement, opening negotiations for the establishment of a distributing company in France by those Romanian interests with French shareholders on export outlets located in France. Româno-Americană and Astra Română were to be excluded from this arrangement.[2]

Notwithstanding the political argument, the position of Romania within the world industry continued to deteriorate, owing to the persistently low price of crude oil, which nullified the efforts of those Romanian producers who sought to compensate for the fall in the value of products by increasing their output of crude and maximizing refinery runs. In consequence of the failure of conservation the Romanian daily average production increased to 18,000 tons, and prices fell to even lower levels than before.[3] The consequences to the larger concerns were serious but were mitigated by the continued existence of marketing outlets: smaller companies lost their outlets and called on the Government for help. From July to December 1931 the official response was to levy taxes on internal sales, as a means of raising a subsidy for small producers—a method which was later abandoned in favour of

[1] See below, p. 181 n. 4. [2] 77.
[3] See editorial in *MPR*, no. 1 (1931), pp. 5–8.

the removal of the embargo on exports of crude oil.[1] Neither measure was adequate, however, in view of the relation of Romania to United States production and prices. During 1931 the increase in production was accompanied by a decrease in domestic consumption and the continuing fall in prices so that stocks, which in January amounted to 673,000 tons, in December totalled 960,000 tons. The composition of these stocks created further difficulties in that the fuel-oil element increased from 266,700 tons to 523,700 tons during the year and compelled some refineries to overcome the resultant disposal problem simply by burning the oil.[2]

From the middle of 1931 the problem of conservation in Romania, which had been unsolved in a purely domestic context, was considered within the framework of a plan for conservation on a world-wide scale, put forward by J. B. A. Kessler of Royal Dutch–Shell for discussion by the international oil industry. He sought to promote the conservation of crude oil "on the principle that the oil industries of all oil-exporting countries should co-operate in retaining the balance between the exports of crude oil and its products and the demand in the world's import markets instead of balancing supply and demand by the more direct method of a regulating production". The mechanism whereby this was to be achieved was a financial inducement to keep net exports below a declared quarterly maximum by the prospect of payment from a central fund *pro rata* to the difference between this national figure and actual exports: on the other hand, exceeding declared maxima would involve penalties. Control of exports in this way would necessarily imply control of production, but this was a matter for national, not international, consideration.[3]

[1] See below, p. 186. In Mar. 1931 the Government banned imports of Russian products into Romania.

[2] According to Minister of Finance Argetoianu, in answer to a parliamentary question: see *MPR*, no. 24 (1931), p. 1172.

[3] It was fundamental to the scheme that it was designed to operate until for a period of six months certain acknowledged standard prices had been ruling for the four main items in the international oil trade, viz. crude oil, gasoline, fuel oil, and gas oil, and would thereupon cease, with the exporter regaining complete freedom of action. Conversely, if prices fell again below the acknowledged standards, then the scheme would automatically be reinvoked. This feature was designed to protect the consumer against excessive price increases through prolonged artificial restriction of output: for details see *PT* (10 Oct. 1931), pp. 475–8; (12 Mar. 1932), pp. 279–83; see also criticism of the proposals from a Romanian standpoint by Mihai Constantinescu (Creditul Minier) in *Le Courrier des pétroles* (Dec. 1931).

After its initial publication, in October 1931, the plan was discussed in a specially convened meeting of the A.I.P.R.,[1] which expressed willingness to send a delegation to an international meeting "to establish a world agreement between oil producers on the basis of the general principles of Mr. Kessler". These were also favourably regarded by the Romanian Government.[2] A series of conferences took place between May and December 1932 between the "Romanian Group"[3] and the "International Group".[4] The first meeting established "Heads of Agreement"[5] whereby Romania's contribution to the international oil trade was to be stabilized at the levels of actual exports during the period from 1 July 1931 to 30 June 1932: exports to this level would be guaranteed, in terms of both individual products and positions in the markets, by the International Group. In return the Romanian Group undertook that its external affiliates would support quota and price agreements to be established in those markets where Romanian oil appeared during the qualifying year, viz. Europe, the Levant, and North Africa.

These terms were initialled by the representatives "exclusively in a personal capacity" but with an undertaking to try to get ratification from their constituent companies and other producers.[6]

[1] On 30 Oct. 1931: see summarized report in *MPR*, no. 21 (1931), p. 1037.

[2] See statement by Vasilescu-Karpen (Minister of Industry and Commerce): *MPR*, no. 22 (1931), pp. 1049–51; and Otto Stern's accounts of discussions with Lugosianu (Minister of Industry and Commerce) and Madgearu: 78.

[3] Represented by Osiceanu, Otto Stern, Masterson (Unirea), Marinescu (Concordia), Constantinescu (Creditul Minier), G. Moriatti (Petrolul Bucureşti —secretary of delegation).

[4] Headed by Arnott (Socony) and including Kessler, Seidell, Fraser (Anglo-Persian). See *MPR*, no. 14 (1932), p. 569. The July conference was preceded by discussions in New York, under Arnott's chairmanship, with representatives of the Soviet Union, in an attempt to bring the Russians into some general accord on export restriction. This failed, and the Russians, though invited to the July conference in Paris, did not attend. On New York discussions, see *MPR*, no. 70 (1932), p. 508; and editorial comment in no. 11 (1932), pp. 433–4.

[5] This summary refers only to the main features of the agreement, which, in total, comprised eighteen articles, relating to (i) proposed action if government policy in a market country gave preference to crude oil, thus displacing products; (ii) the adjustment of quotas with increases in demand; (iii) the status of existing contracts during the initial period of the arrangement; (iv) various administrative procedures. For full text see 79: for results of the conference, as reported by Osiceanu to the A.I.P.R., 3 Aug. 1932, see *MPR*, no. 16 (1932), pp. 665–8; and editorial comment in *MPR*, no. 17 (1932), pp. 695–7.

[6] On ratification discussions see summary of A.I.P.R. meetings of 11 Aug. 1932 (*MPR*, no. 16 (1932), pp. 685–7; no. 17 (1932), p. 721); and of 10 Sept. 1932 (*MPR*, no. 18 (1932), pp. 731 f.). See also 80.

In terms of product, under the "Heads of Agreement", Romania had a right to an annual export of 4,667,860 tons, comprising 1,477,058 motor gasoline, 936,910 kerosine, 809,890 gas oil, 52,257 lubricants, 1,276,531 fuel oil, 6,218 paraffin wax, 76,610 crude oil, and 34,386 asphalt. The question of regulating production was not discussed at Paris, since that was regarded "as a matter to be decided exclusively among the various Romanian producers"; it was, nevertheless, implicit in the regulation of exports and the proposals for regulating production that were introduced by Otto Stern at the beginning of September 1932.[1]

The A.I.P.R. on 10 September 1932 resolved to ratify the "Heads of Agreement", subject to certain adjustments to be discussed at a meeting of the two groups in Paris on 22 September 1932.[2] Nevertheless, in Bucharest, ratification was protracted by reason of the difficulty of arriving at an export, and therefore production, arrangement among the constituent companies and the general misgivings of Steaua Română.[3] The point on which these issues came to depend was whether the agreement should be ratified first, leaving the production and other internal allocations for subsequent discussion, or whether the internal machinery should be agreed before ratification. The first position was held by Osiceanu, the second by Otto Stern, on the grounds that formal ratification might prove valueless owing to insoluble dissensions over the machinery of the agreement. In the end the consensus of feeling within the A.I.P.R., and the possibility of an export war[4] which Romania could only lose, forced a meeting of the participants in Paris, beginning on 29 November 1932, during which the outstanding questions were resolved. By the memorandum

[1] It was envisaged that individual producers should declare monthly their maximum production, on the basis of which they undertook to contribute to a conservation fund. This declared production was to be offset against an estimate of Romanian crude-oil requirements, calculated as the total of requirements for export inland and bunker trade minus stocks plus permanent stocks. The excess of declared maximum production over required production, multiplied by the market value of an average Romanian crude oil, would give the total amount required in the conservation fund—which figure divided by the total quantity declared would give the rate of contribution to the fund by individual producers. The forward estimates of output were to be compared subsequently with actual output, and compensation payments made accordingly: see 81.

[2] See *MPR*, no. 19 (1932), pp. 773–4; no. 20 (1932), pp. 813–15: also 82.

[3] In consequence of these disputes the original ratification date was at Romania's request altered from 15 Aug. to 15 Sept.: 83.

[4] See 84.

of 6 December 1932[1] the Romanian Group undertook to reduce production to 18,500 tons a day, including domestic consumption: this total was to rule for the first three months (1 January to 31 March 1933), "it being understood that for the following three-monthly periods a definite plan on the lines of the Kessler Plan or some similar plan is to be adopted". The provisions for crude-oil restriction were complemented on 12 December 1932 by a draft agreement[2] embodying the general principles of the "Heads of Agreement" of July 1932 on the export aspects of the scheme as a whole: it thus covered the respective export rights and purchase obligations of the participants, the areas in which these rights were operative, and the prices to be considered. The minutes of the meeting at which the draft agreement was initialled specifically tied the latter to the memorandum of 6 December 1932 and outlined specific permitted variations from the general arrangements (including one relating to France).[3]

The draft agreement came into operation on 1 January 1933.[4] Romania's production for January–March 1933, 1,665,000 tons, or 18,500 tons a day, represented a reduction of 5,150 tons a day on the average daily output for the last quarter of 1932. The inducement to maintain this level during the first quarter of 1933 was an average premium of 30 per cent on Constanţa prices ruling at 21 September 1932, i.e. on the date of the ratification of the Agreement in principle by the Romanians.

Adherence to the Paris Convention was voluntary—although the A.I.P.R. urged the widest possible support: this appears, however, to have been given *de facto* except by the companies whose officers had represented the Romanian Group in Paris

[1] For text see 85: the figure of 18,500 tons a day represented a concession of 100 tons a day on the part of the International Group to the Romanian Group on account of "the impossibility of ascertaining at an early date with any degree of accuracy the commitments of the Romanian Group, their claims to gain of trade from the Russians, and the decrease in consumption in the countries named in the Agreement".

[2] Text; see 86.

[3] "5. As regards the French market, the market percentage of the Roumanian Group will not be established in accordance with the qualifying period but shall be based on the year 1932. This special provision is subject to the approval of the Chambre Syndicale. If such approval is withheld, none of the stipulations of the Draft Agreement of this date concerning 'as is' market percentage will apply to the Roumanian Group in France." The altered base period gave Romania an export right of 714,000 tons (including 375,000 tons gasoline), instead of 395,000 tons (including 251,000 tons gasoline).

[4] 87.

and were therefore bound by signature.[1] The effect of the restriction was, however, quickly apparent, and, although some difficulty was experienced in keeping to the daily level, production fell from 2,200,356 (October–December 1932 inclusive) to 1,667,000 tons (January–March 1933 inclusive). Even so, after an initial period of apparent success, U.S. exports proved even more difficult to control and unorganized shipments continued to drive prices down to low levels. Romanian interests sought to remedy this by a higher production quota and by demanding a share in the exploitation of increased market possibilities resulting from the decline in Russian exports in Europe. These issues were thrashed out at meetings in London (February 1933)[2] and Paris (27 March–7 April 1933)[3] in consequence of which the daily Romanian production remained fixed at 18,500 tons for the period 1 April 1933 to 30 June 1933 but restrictions on markets were removed. Romanian concern over unrestrained American production was reflected in the rights to cancellation conceded by the International Group to the Romanians "without previous advice, if the world programme of production does not appear to them to be susceptible of an eventual improvement in world petroleum markets and if the conference now being held in Washington does not result in a satisfactory reduction of the petroleum production of the United States". The Paris meeting also provided for "a new Conference in June to establish a plan to regulate the relations between the two groups for a future period and to study the putting into effect of the Kessler Plan or other plan of similar intent".

This meeting never took place. In Romania the criticism began to be expressed that the control in the United States was inadequate to decrease output or put an end to "bootleg" production,[4] with the result that crude oil prices kept falling and deprived Romania of the expected benefit of increased world market prices. These views were put by Osiceanu to Arnott by telegram

[1] 88.

[2] See 89; summary of consequent A.I.P.R. discussion 14 Mar. 1933 in *MPR*, no. 6 (1933), pp. 317–18.

[3] See 90.

[4] Production 1933:

	January	February	March	April	May	June
U.S.	8,767,726	8,360,973	10,316,374	8,947,881	11,610,339	11,349,217
Romania	578,735	523,107	572,807	562,392	586,499	641,460

MPR, no. 3 (1934), p. 184.

on 29 April[1] and on 24 May[2] threatening Romania's withdrawal unless U.S. output was immediately and permanently reduced to 2 million barrels a day (274,000 tons).[3] Six days later the A.I.P.R. meeting[4] decided not to observe the Agreement but to maintain relationships so as to resume collaboration with the International Group when U.S. production was stabilized at 2 million barrels a day and in the meantime provide for its own agreements on the scale of oil activity. Arnott's reply[5] to Osiceanu's telegrams referred to current U.S. legislative proposals for controlling production and stabilizing the market and asked the Romanians to stand by while they were enacted.[6] This plea was ignored.

There appears to have been no formal renunciation of the Agreement. According to Otto Stern,[7] reporting a conversation of 26 May, Osiceanu considered a conference of oil producers useless so long as the American situation was not settled and intended no official rupture but to leave unofficially to Romanian producers to produce as they wished. He refused to support the idea of maintaining restriction till the end of July. In consequence, Romanian production and export were subject to no further limitations on the part of the world oil industry[8] from the beginning of June 1933.

In neither Romania nor the United States was the oil industry able to regulate itself; in the latter country the formula for failure was compounded of existing legislation and excessive self-interest but in Romania action on the part of industry was frustrated by the positive action of the State. For the State, restriction entailed loss of income from the oil trade both internal and external, and a hindrance to its foreign policy in relation to France: it could therefore never entirely support a limitation of output—except as a temporary measure in an extreme situation and, alternatively, could not stand aside and let the industry work out its own solution.

Within Romania the application of the two restriction schemes demonstrated the impossibility of organizing a united front of oil producers under Romanian conditions and emphasized the isolation of Astra Română and Româno-Americană in the industry.

[1] *MPR*, no. 10 (1933), p. 551.

[2] *MPR*, no. 11 (1933), p. 609.

[3] This was the figure arrived at by the Washington Conference of U.S. oil producers held under Secretary of State Ickes.

[4] 91.

[5] 92.

[6] Code enacted 19 Aug. 1933.

[7] 93.

[8] For the production of the major companies under the 1933 restriction, see Figure 8.

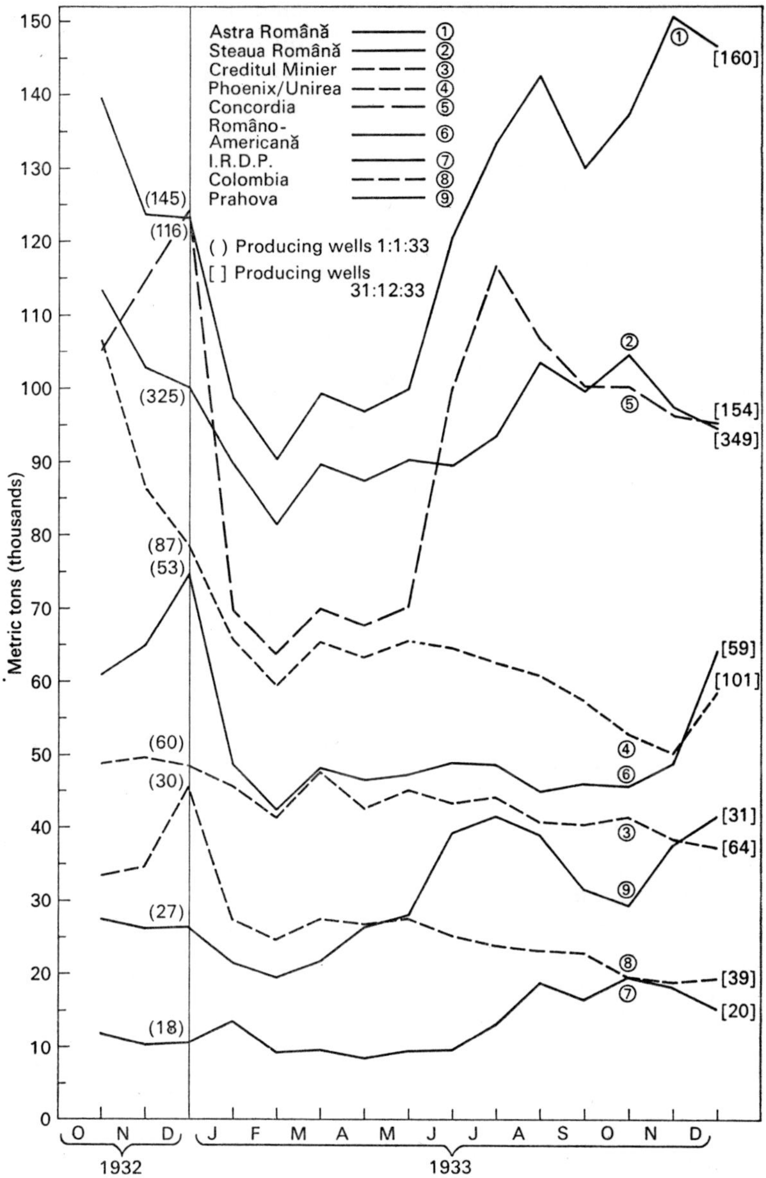

FIG. 8. Production of major Romanian companies, 1933.

In both attempts, the role of Steaua Română as a producer and of its Managing Director, Osiceanu, within the A.I.P.R., was decisive.[1] In contrast, however, during January–May 1933, Astra Română and Româno-Americană took no leading part, confining themselves to expressing their views—naturally in favour of restriction—within the Romanian Group: there was no initiative comparable with that of 1930.[2]

On both occasions conservation affected production: between July and November 1930 Romanian output was held at a monthly average of 266,108 tons as against 362,696 tons in January/June 1930 and between January and May 1933 at an average of 565,151 tons as against 702,137 for July/December 1932.[3] In the first period of restriction, however, control measures, while in force, led to an increase in price of Romanian crude oil: in the second period, not, since the U.S. output never sufficiently came under control to affect the level of Romanian prices. On neither occasion, however, were the periods of overall control adequate to affect export movements: they merely ran down stocks.

After the breakdown of restriction, both "national" and "foreign-capital" companies reverted to planning their activity in line with their expectations from the market, but increases in drilling were offset by a decline in yields. The total output of crude oil for the year, 7,387,000 tons, was only 37,000 tons above that for 1932 and represented a net fall of six tons per metre drilled. From 1934 onwards the position changed, by reason of successful exploration drilling in the Harsa field by Unirea and in Viforata by Creditul Minier. Both discoveries took place on the extensions of existing fields (respectively Boldești and Gura Ocniței). In July the following year the first major discovery away from the Carpathian foot-hills and their associated structures was made by

[1] In 1933, as in 1931 (p. 164 above), Osiceanu appeared to the proponents of restriction to react to initiatives from Steaua Française. On 24 May Stern reported to Kessler his "personal impression" that by request of the French Steaua, Osiceanu was managing so that (conservation) was eventually terminated by the desire of others, in order to keep him and his friends free from blame (94). On 30 May Kessler reported his having been advised by Stern of a "meeting of Roumanian Petroleum Association convened for this afternoon at which he fears breakdown of conservation owing to attitude of French interests in Steaua Română" (95).

[2] 96.

[3] 1930 production figures: *MPR*, no. 6 (1931), p. 267; 1932 and 1933 production figures: *MPR*, no. 6 (1934), p. 329.

Steaua Română at Bucşani,[1] and by the end of 1935 this field contributed 25 per cent of the country's total output of crude oil. In 1936 a rapid decline set in, which, combined with losses through degassing of strata at Viforata and Rasvad, checked the over-all increase in the output of crude oil: this, reaching its peak in July 1936, thereafter began to fall off significantly.

Annual totals		Monthly totals 1936			
1933	7,387,000	January	720,760	July	769,116
1934	8,473,000	February	669,169	August	766,906
1935	8,394,000	March	716,146	September	721,509
1936	8,704,000	April	717,377	October	717,565
1937	7,153,000	May	764,322	November	684,781
1938	6,160,000	June	764,378	December	691,971

The causes of this decline lay in the limited geographical and geological extent of the industry's extractive operations. Increases in production had derived largely from the Meotic layers, which were comparatively thin, containing on average only about fifteen metres of exploitable oil-sand: the "flush" production, i.e. that obtained immediately on opening the strata, was therefore exhausted rapidly: from the late twenties onwards, approximately 30 per cent of total crude-oil output was obtained as flush production.[2] Romania's progress therefore lacked a secure geological basis.[3] The remedy lay in the discovery of new oil reserves, outside the traditional areas, but the legal discouragements to wildcatting mentioned above[4] were made worse by the over-production crisis of 1930 to 1933 and the financial insecurity consequent on the Depression. From 1936 onwards the question of reserves became the central problem for both the industry and the State.

Romania in the Depression, 1929–1934; currency controls

The falling prices and diminished returns which dictated the attempts at conservation were not merely a feature of the oil industry. During this period the general fall in world

[1] On Bucşani developments, see *MPR*, no. 8 (1935), p. 590; V. Zotinca, "Considérations sur le chantier Bucşani", *MPR*, no. 1 (1937), pp. 31–43; E. C. Masterson, "Developments at Bucşani in 1935", *PT* (4 Jan. 1926), p. 3; and article by correspondent of the same journal, "Wildcat Drilling Campaign in Roumanian Plains Area", ibid. (5 Sept. 1936), pp. 309–10.

[2] 97. [3] See Figure 9. [4] See above, p. 118.

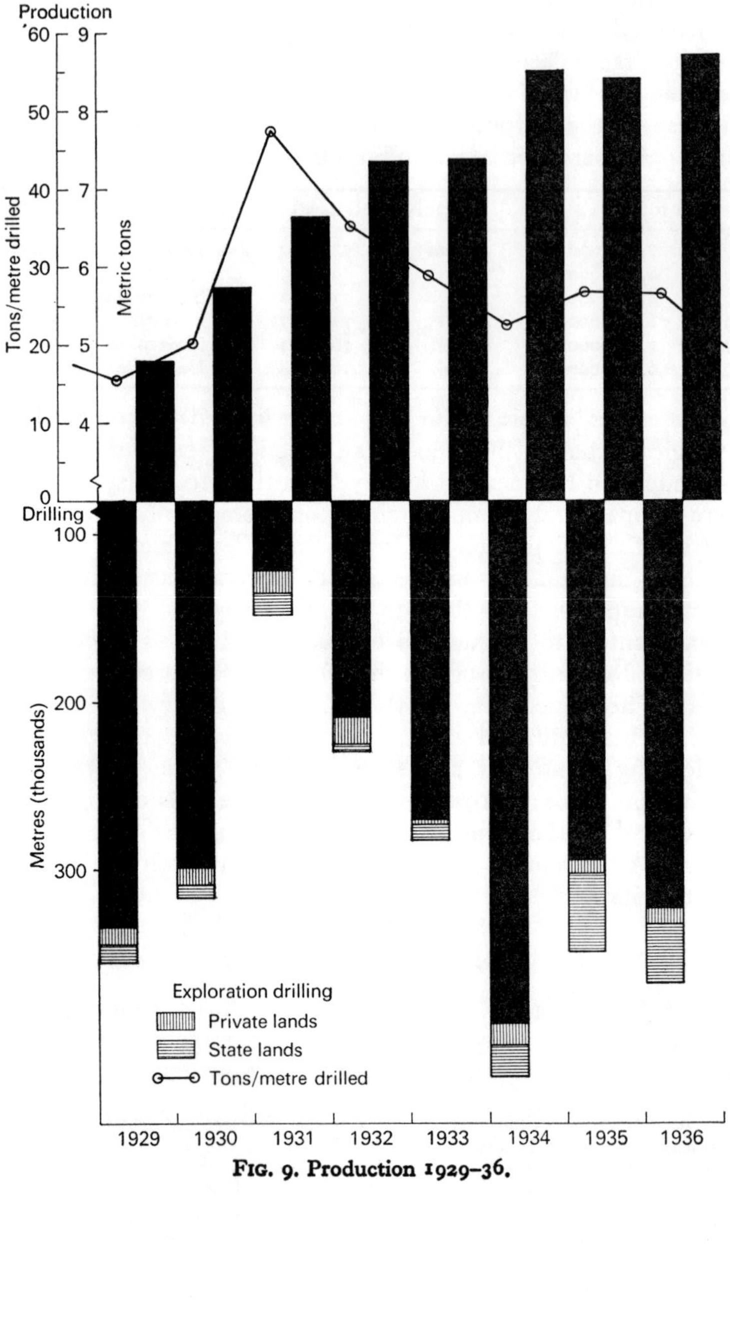

Fig. 9. Production 1929–36.

commodity prices drastically reduced Romania's income from exports. The fall in agricultural prices, particularly, had severe financial consequences in Romania, not only by undermining the taxable capacity of domestic production but also by bringing about a moratorium on agricultural debts and the failure of various banks involved in agricultural credit. At the same time, external supplies of credit dried up in consequence of the crisis in Central European banking after the failure of the Creditanstalt in May 1931.[1] These general causes of disruption were augmented by purely Romanian ones deriving from inadequate methods of state financing.[2] The effect of the total crisis on Romania was demonstrated in the submissions to the Stresa Conference of 5–20 September 1932,[3] which showed that, although by reason of a drastic curtailment of imports, Romania had achieved a foreign-trade surplus from 1930 to 1932, it was insufficient to maintain the balance of payments, burdened with higher service charges in foreign debts and capital withdrawals following the banking crisis. In consequence of the international disruption of trade and finance, in Romania a net capital import of 6,700 million lei in 1929 had turned into a net capital loss of 7,011 million lei in 1931. These factors helped to explain a decline in net national revenue from 173,274 million lei in 1929 to 98,755 million in 1931.

In this situation successive Romanian governments imposed foreign-exchange restrictions[4] and imports controls, declared a moratorium on foreign debts,[5] and turned to clearing-agreements

[1] See League of Nations *Report on Commercial Banks 1931–1932*; H. V. Hodson, *Slump and Recovery*, Chap. III.

[2] See Roger Auboin, *Rapport annuel sur la situation financière et monétaire de la Roumanie*, Section II. The Commercial Secretary to the British Legation, Bucharest, in his report on *Economic Conditions in Roumania. 1931* (p. 37), distinguished 18 (Romanian) "causes" which "tend to accentuate the depression".

[3] Convened, under the auspices of the League of Nations, to concert "measures required for the restoration of the countries of Central and Eastern Europe", particularly referring to transfer difficulties and agriculture. On the conference see discussion by H. V. Hodson in *Survey of International Affairs 1932*, pp. 19–27, 87–95. For Romanian submissions, see *La Roumanie à la Conférence de Stresa*.

[4] Loi des Devises: *Monitor Oficial*, no. 230 (1 Oct. 1932); *CER*, no. 4 (1932), pp. 61–4.

[5] 12 Aug. 1933; subsequently the Romanian State made an agreement with holders of Romanian external state loans for temporary alleviation of the service charges on them. The terms of this agreement were applied unilaterally by the Romanian Government to the war-damage bonds, despite the objection of the annuitants that the sum due to them represented a return of capital for

and bilateral trade in an endeavour to revive Romania's exports. After 1 October 1932 the control of all foreign trade was closely supervised by the State through the agency of the National Bank, which issued licences and rationed foreign exchange.

Romanian exchange-control practice distinguished between countries with "strong", i.e. "free" or "convertible", currency and those with "weak", i.e. "restricted" or "non-convertible", currency:[1] the Romanian Government took all the proceeds of foreign sales and converted them into lei at rates fixed by the National Bank, subsequently giving to the exporter his return: for exports to restricted-currency destinations the exporter received payment entirely in lei: for exports involving free currency, partly in lei and partly in the free currency, in proportions laid down in the relevant agreement.[2]

The system of currency control was complemented by the introduction of bilateral clearing-arrangements[3] establishing cumulative accounts to provide a mechanism for balancing foreign-trade accounts without direct transfer of payment. The establishment of a clearing- and exchange-control regime in Romania accorded with a pattern of relationships issuing from the general financial chaos in Europe and in which all the Danube states,

material damage, whereas the bond-holders had subscribed to loans of their own volition. The coupons falling due in June 1932 and Dec. 1933 were met only in part. This marked the beginnings of a process whereby the service on the war-damage bonds became increasingly difficult, and in 1934, 1937, and 1938 successive agreements were reached between the parties concerned, whereby the values in the original agreements were written down and payment of coupons partly or wholly suspended. In Apr. 1941 all service on foreign debts, including war-damage bonds, was suspended.

[1] Free currencies: £ sterling, Egyptian, Palestinian; Fr. French, Belgian, Swiss; Kr. Czech, Swedish, Danish, Norwegian; Peseta; Turkish lira; U.S. dollar. Restricted Currencies: Reichsmark, Austrian Schilling, Pengő, Drachma, Lev, Dinar.

[2] This arrangement held good also for countries with whom Romania had no clearing arrangement, e.g. Sweden.

[3] The cost of imports was met by payment in the importer's currency into an account maintained in his own national bank for the exporting country, whose exporters drew in payment an agreed counter-value in their own currency from their own national bank. There was thus no loss of currency by exchange, and the sales between the two national currencies were negotiated under the clearing agreements concluded between either governments or national banks: it was assumed that each of the two partners to any particular clearing would have an interest in keeping it approximately in balance over any given period. On the origins of the clearing system, see League of Nations' *Enquiry into Clearing Agreements*; and A. Basch. *The Danube Basin and the German Economic Sphere*, pp. 82–8.

i.e. Romania's immediate trading partners, joined. In the restricted-currency clearing area the predominant state was Germany, whose policy towards the countries of south-east Europe between 1934 and 1939 came to set the scope and limits of their foreign-trade operations. In addition, to provide a means for settlement of outstanding debts, particularly in France and the United Kingdom, Romania also concluded payments agreements with free-currency countries, and thus was subject to some restriction as to the disposal of its free-currency earnings.

From 1932 onwards the overriding financial problem of Romanian governments was to earn as much free currency as possible and to find an efficient use for the restricted currencies they necessarily had to acquire, particularly from their Central European trading partners. From the Romanian Government's viewpoint, oil exports were the most suitable instrument for this foreign-exchange-earning policy, since significant quantities of oil regularly went to free-currency destinations in the periods before the crisis.[1] There were, however, two limiting factors: the first that in the existing world market the price commanded by Romanian oil was related to conditions in the U.S. Gulf—over which the Romanian Government had no control, and the second, that intending buyers holding free currency were in a position to choose whether they would spend it in Romania or not. This decision turned largely on price. In practice, therefore, the relationship between the f.o.b. Constanța price and the Gulf price became crucial to maximizing Romania's free-currency earnings from oil exports.

The institution of exchange and clearing controls defeated the efforts of the oil-exporters for a free regime and entailed a reversion to something like "Birex" conditions. The activities of export companies within Romania were henceforward determined by

[1] Oil Exports (tons)

	1929 Free currency	All destinations	1930 Free currency	All destinations	1931 Free currency	All destinations
Sterling	591,390	..	1,052,228	..	1,584,095	..
Franc	280,728	..	411,331	..	662,802	..
Others	207,765	..	242,612	..	482,498	..
	1,079,883	2,847,398	1,706,171	3,867,459	2,729,395	4,662,789

These figures exclude ships bunker deliveries and "spot" export sales which may or may not have been made against free currency.

the need to balance their lei and free-currency returns to provide respectively for operating expenses and for goods and services received from abroad, including royalties: these calculations necessarily depended on the clearing relations between Romania and the intended destination.

In view of the price difference between Constanța and Gulf, the vital factor became the exchange rate at which the sterling or dollar earnings of Romania's oil exports were converted into lei: as official rates were always low,[1] exporters' quotations, being related to what they needed to earn in lei in terms of these official rates, were correspondingly increased to levels above the world market price. Equally, the artificiality of the official rate of exchange induced exporters to restrict their free-currency earnings to a minimum, and, since they *had* to dispose of their output, enhanced the attractions of restricted-currency destinations. The effect of the Government's regulations on Romania's most important export item was thus to undermine the policy of maximizing free-currency returns.[2]

In addition to financial considerations, Romanian export markets became more difficult of access by the adoption of policies of national self-sufficiency, especially with regard to fuels. As explained above,[3] such policies antedated the onset of the Depression, but the necessity to save convertible currency provided an additional incentive for exploiting national fuel resources. Financial reasons were reinforced by the inauguration of rearmament programmes, with the entailed problem of fuel supply in war. Now that price was not the only factor and the currency in which the price had to be paid had also to be considered, the main objection to synthesized fuel, i.e. its cost, was weakened,

[1] See Madgearu, *et al.*, *Mémoire sur le contrôle des changes en Roumanie*.

[2] For *all* exports, the division between free and restricted currency destinations was, by value, respectively

	m. lei	%	m. lei	%
1932	9,028	54	7,693	46
1933	8,575	60	5,596	40
1934	7,033	52	6,623	48
1935	6,381	37	10,375	63
1936	11,363	52	10,340	48
1937	16,096	51	15,472	49

Ibid., Annexe 1 (2).

[3] p. 152.

and strategic considerations caused experiments to be continued and output to be expanded. These developments reacted on Romania's gasoline exports, particularly to Germany.

In summary, the effect of the Depression on the Romanian oil industry was to restrict markets, and hence revenue-earning possibilities, at a time when external sources of private investment were curtailed. Furthermore, the measures taken by the Romanian State to combat the Depression necessarily brought under its control that part of the oil industry's activity which had—with the partial exception of the "Birex" experiment—hitherto eluded it. Since oil exports remained under exchange control till 1939, and thereafter under comprehensive state regulation, the developments of 1932 were decisive for the oil industry.[1]

Changes in the Romanian State, 1929–1934

At the time when the financial basis of the Romanian State was undermined, the government of the country was disorganized by a constitutional crisis centred on the monarchy and by the fragmentation of political life,[2] partly as a result of that crisis, and partly as a general response to the conditions of the Depression.

In July 1930 Carol returned from his self-imposed exile[3] and was proclaimed King. This volte-face[4] cut across the established political connections, creating public feuds between the party leaders[5] and opening the way for royal manipulation.

The confusions of organized political life contributed to the

[1] For Romanian industry as a whole the regime of payments and currency restrictions, import taxes, and quotas reinforced the degree of protection given by the traditional tariffs and intensified the industrialization of the country. After 1932 imports of finished and semi-finished goods declined in favour of imports of raw materials. Total invested capital in manufacturing increased from 39,984 m. lei in 1932 to 50,069 m. lei in 1938, and the number of enterprises from 3,557 to 3,767. This growth worked to the benefit of Germany as a source of capital equipment. See V. Madgearu, *Evoluţia*, pp. 251 et seq.; and Roberts, Chap. IX.

[2] See H. Prost, *Destin de la Roumanie*, Chap. V.

[3] See above, p. 131. Parliament annulled the Law of 4 Jan. 1926.

[4] Carol returned on the basis of an undertaking with Maniu, then Prime Minister. In Oct. Maniu resigned, claiming that Carol had not observed the conditions. He was replaced by Mironescu, who led the National Peasant Government till Apr. 1931, when it was dismissed by the King. It was in the Maniu–Mironescu reshuffle that Madgearu lost his position as Minister of Industry and Commerce (see above, p. 166).

[5] Gheorghe Brătianu and Vaida Voevod led pro-Carol, Vintila Brătianu and Maniu anti-Carol, factions in their respective parties.

decline in the authority of Parliament, which was also marked by the emergence of various anti-parliamentary organizations. All professed a nationalism extreme even by Balkan standards. Unemployment and the collapse of agriculture were blamed on foreigners, both within Romania and outside, whose activities if tolerated in a purified Romanian society were to be curbed in favour of ethnic Romanians. These common aspirations achieved the greatest coherence in the programme of the most successful of these organizations "The Legion of the Archangel Michael" (also known subsequently by the name of its military wing "The Iron Guard") founded in 1928 by Mihai Stelescu and Corneliu Codreanu.[1] Its composition has been aptly compared with the following of Catiline.[2] The Legion's extremism and violence brought about its official dissolution on four occasions; March 1930 (by Maniu), January 1931 (by Mihalache), March 1932 (by Argetoianu), and December 1932 (by Duca), but the decrees could not be enforced and the Legion continued to function as the Corneliu Codreanu group or the *Totul pentru Țară* (All for the Fatherland) movement. Its survival indicates its divisive effect on the traditional parties. During its formative years, Goga, as Minister of the Interior in Liberal governments, and Vaida Voevod, as his National Peasant counterpart till 1930, gave sympathy and encouragement. Manoilescu was also a supporter. On the other hand, Maniu and Madgearu, among the National Peasant leaders, and Calinescu and Duca, of the Liberals, were its most determined opponents—in consequence of which Duca (December 1933), Calinescu (September 1939), and Madgearu (November 1940) died through assassination by Legionnaires. The relations between the King and the Legion remain unclear, but the known course of events suggests that Carol attempted with some success to use the movement to help destroy the established parties and then to cripple it. In November 1938 Codreanu was "shot while trying to escape", along with his followers who had been responsible for the murder of his former associate Stelescu in a Bucharest hospital on 16 July 1936.

During the Depression the traditional forms of political authority

[1] On Codreanu's background and early activities, see Roberts, pp. 223–33; Prost, Chap. IV; E. Weber, *The Men of the Archangel.*

[2] Prost, pp. 107–8: see also R. G. Waldeck, *Athene Palace: Bucharest*, pp. 17–23.

in Romania began to break down: governments were short-lived[1] and disrupted by feuds within the leadership. The existing parties retained sufficient strength to block an attempt at explicit royal dictatorship in May 1934, but nevertheless in this period it became clear that the political structure of Romania was likely to be drastically changed were either the King or the Iron Guard to get a chance to change it.

Notwithstanding domestic political instability, the external attitudes of Romania remained fundamentally unmodified: the country's international position was formally consolidated by two diplomatic steps taken in February 1934. On the 9th Romania joined with Yugoslavia, Greece, and Turkey in the Balkan Entente. This, from the Romanian point of view, was essentially a "Little Entente" arrangement, involving her partners in the maintenance of Balkan frontiers. Eight days later Romania ratified the convention regarding the Definition of Aggression of 4 July 1933.[2] The Soviet Union did likewise. Official Romanian opinion (optimistically) held this to have implied recognition by the U.S.S.R. of the Dniester frontier, in that her signature had been given to a pact of mutual assistance covering territory in *de facto* occupation.

Both were the work of Nicolae Titulescu, who became Minister of Foreign Affairs in 1932 and conducted Romanian foreign policy, largely from Geneva and his villa in the south of France, till August 1936. Titulescu's diplomacy aimed at guaranteeing Romania's post-war frontiers, using the resources of the League of Nations to this end, and involving also a cautious *rapprochement* with the Soviet Union—a course to which the presence at Geneva of Litvinov gave an appearance of realism.

Romanian governments, of whatever combination of politicians, still contained Francophils who effectively maintained the traditional orientation towards France, and who also encouraged the strengthening of ties with Romania's partners in the Little Entente

[1]

Mironescu	Oct. 1930–Apr. 1931	National Peasants
Professor Iorga	18 Apr. 1931–21 May 1931	"National Union"
Vaida Voevod	6 June 1931–Oct. 1932	National Peasants
Maniu	20 Oct. 1932–5 Jan. 1933	National Peasants
Vaida Voevod	15 Jan. 1933–7 Oct. 1933	National Peasants
Duca	14 Nov. 1933–29 Dec. 1933	Liberal

[2] For text see Shapiro, vol. ii, 1929–32, no. 400. Full diplomatic relations between Romania and the U.S.S.R. were not established until 9 June 1934: see Shapiro, no. 437.

as co-beneficiaries of the post-war settlement. Additionally, the attempted solution of the economic problems of the Danube area, in which Romania was necessarily concerned, proceeded under French auspices. Political understanding between Romania and Germany was hampered by the diametrically opposed views of the two governments on revision of the peace treaties and by Titulescu's policy of dealing with the Bessarabian problem by wary negotiation with the Russians.

CHAPTER VIII

Intensified Nationalism in Industry and State

The effect of the Depression on the oil industry

THE events and developments discussed in Chapter VII left the Romanian State weaker both politically and economically, but stronger in regard to the oil industry, which was wholly subject to government financial policy and was demonstrably disunited within itself. Furthermore, its geographical advantage as a source of supply was offset by the flow of oil from Iraq from July 1934 onwards,[1] and its contribution to the European oil trade was hampered by monetary barriers and experiments with substitute fuels in its customary markets.

Both the oil-conservation crisis and the general economic crisis tended to emphasize disparities within the industry. Large integrated concerns withstood prevailing conditions and declining crude-oil prices better than the small producers, with solely Romanian capital, who succumbed through loss of income from crude oil or through being caught up in bank failures. They were hampered by high production costs deriving from their having to buy power, and from less efficient exploitation methods: insufficient storage capacity compelled them to forgo whatever advantage might have accrued from the quality of their oil. They were therefore more quickly vulnerable to the fall in crude-oil prices.[2] In consequence the producers without refineries favoured restriction, and after the breakdown of the Conservation Convention demanded its renewal. In April 1931 they attempted action within the A.I.P.R., and, on failing to gain their demands, petitioned for direct government intervention—which appeared on 13–14 July in the form of a law for the encouragement of crude-oil production[3] creating *ad hoc* taxes levied ex-refinery on all

[1] See below, p. 197.

[2] See *MPR* editorial "La Variation des prix du pétrole et des produits en 1933" (no. 2 (1934)), which summarizes movements from 1930 to 1933.

[3] See draft law; *MPR*, no. 13 (1931), pp. 646–8; A.I.P.R. discussion: ibid., p. 686; A.I.P.R. memorandum to Minister of Industry and Commerce:

products in the internal oil trade. These encouragement taxes were enforced only till January 1933, when the law was abrogated at the time of the announcement of a new tax structure for fiscal purposes,[1] but meanwhile the situation of the beneficiaries of these taxes was mitigated by the modification by the law of 20 April 1932[2] of Article 198 of the Mining Law to allow the export of crude oil.

The Depression emphasized divisions within the industry as regards not only size and technique but also nationality. Foreign-capital companies—all large producers and/or refiners—were able to withstand prevailing conditions by reason not only of their size but also of their possession of established marketing outlets. This, indeed, was the basis both for the joint offer of Romăno-Americană and Astra Română and for Steaua's action in breaking up the conservation scheme. The demonstrable utility of foreign connections prompted a scheme for the amalgamation of "national" companies for export purposes, evolved by Elie Blank of the banking firm Marmarosch Blank, which was under heavy pressure through the general run on deposits and, after the moratorium of December 1931, the cessation of income from agricultural credit institutions.

M. Blank hoped to combine his bank's interests in Petrol Govora and Sondajul with Creditul Minier, I.R.D.P., Petrol Block, and Petrol Românesc, into an organization which would enable these "national" companies[3] to increase their return from

MPR, no. 14 (1931), pp. 738–40; text of Law: ibid., pp. 701–4 (amendment to Article 1: *MPR*, no. 16 (1931), p. 822); *exposé des motifs*: ibid., pp. 704–6; speech of Eng. Ghica, *rapporteur* in Senate (ex-President, Association of Mining Industry Engineers and Technicians): ibid., pp. 823–4; regulations: *MPR*, no. 18 (1931), pp. 877–8; no. 19, pp. 929–30; no. 20, pp. 969–70; no. 21, pp. 1011–12; results of applications: *MPR*, no. 19 (1931), p. 939; no. 21, pp. 1030–1; no. 22, pp. 1074–5: see also comments of Manoilescu, Minister of Industry and Commerce: *MPR*, no. 12 (1931), pp. 627–8.

[1] See *MPR*, no. 1 (1933), pp. 33–6; no. 2 (1933), pp. 79–82.

[2] Proposal: *MPR*, no. 23 (1931), p. 1124; A.I.P.R. comments: *MPR*, no. 24, p. 1167; memorandum to Government: *MPR*, no. 2 (1939), pp. 88–9; draft law: *MPR*, no. 4 (1932), pp. 153–4; text of law: *MPR*, no. 9 (1932), p. 377.

[3] The potential strength of the combination can be seen from the following:

	Production m. tons	% Romanian total
1929	750,000	15·5
1930	1,085,985	18·9
1931	1,027,412	15·4

See *MPR*, no. 6 (1932), p. 243; no. 8 (1932), pp. 331–3: 99.

overseas markets. The Ministry of Industry and Commerce carried out an inspection of these companies, on the basis of which their participation quotas in the new joint company were worked out. As its contribution to the undertaking, the Romanian State was to supply 400 hectares of land in Gura Ocniței and outstanding royalty crude oil owned by the companies, together with part of the royalty expectations. A Bill was laid before Parliament on 31 March 1932 which made provision for the amalgamation of these companies into the "Romanian Petroleum Company" with a capital of 1,000 million lei, in which the State's participation was to be 30 per cent in cash or kind. Parliament rose on 15 April 1932 without the Bill's having been debated and the scheme was abandoned,[1] presumably because Marmarosch Blank was unable to meet its general business obligations in April 1932 and temporarily ceased payment.

The failure of Elie Blank's scheme again emphasized the relatively strong position of the foreign-capital concerns. They were still more markedly isolated within Romania at a period when the Depression stimulated an agitation for the preservation of the national welfare in all spheres. In the oil industry this agitation demanded restrictions on foreign capital and the extrusion or limitation of foreigners.

The employment of foreigners was governed by the Mining Law of 1929, which permitted a maximum of 25 per cent foreign nationals in each of four determined categories of staff. Management was excluded and subjected to no nationality qualifications, except in respect of those who were also board members. The issue of foreign staff was revived in the Association of Romanian Mining Engineers and Technologists, at whose meetings the strong feelings of its members on the unemployment issue were made clear. A delegate to the annual general meeting in Ploești on 26 June 1932[2] protested against the ease with which residential and working rights accorded to foreigners were prolonged, on the grounds of their being specialists, when qualified Romanian personnel were unemployed. At the next annual general meeting, 4 June 1933, the assembly heard a documented report on unemployment in mining engineering,[3] and the following year

[1] See *MPR*, no. 9 (1932), p. 371; and annual report to shareholders of Creditul Minier for 1932–3: *MPR*, no. 12 (1933), pp. 641–2.

[2] See *MPR*, no. 13 (1932), p. 557.

[3] See *MPR*, no. 12 (1933), pp. 683–4.

demanded unanimously that the authorities should apply existing laws strictly in order to allow the national element its natural rights, one of which was defined as the right to be wholly employed in the mining industry.[1] The natural rights of Romanian technicians conflicted with the legal rights of the foreign owners of Romanian companies to safeguard their technical interests as they thought fit and to benefit from interchangeability of staff—both of which postulated a minimum of foreign nationals.

In the spring of 1934 the Government published a draft law for the employment of Romanian personnel, obliging Romanian companies to employ Romanian nationals to at least 80 per cent of six categories of staff (Articles 1 and 3) and restricting also the sums payable as salaries, bonuses, and other allowances to foreign staff (Article 1): failure to comply within six months involved withdrawal of advantages given by the State, cancellation of state contracts, and suspension or closing down (Article 6).

The draft law prompted diplomatic representations in Bucharest but was passed by Parliament on 10 July 1934, on terms imposing not only the 80 per cent Romanian personnel in each category but also stipulating that existing contracts for foreign nationals must be honoured for a maximum of one year only from the date of the Law.[2] The *exposé des motifs*[3] made it clear that the Law was designed to remedy the unemployment problem, particularly among "intellectuals", defined (curiously) as "engineers, chemists, accountants, architects, etc., completely trained, waiting for years for the moment to profit from their knowledge".

Discrepancies between the draft regulations[4] issued under the Law and the Law itself prompted the "foreign-capital" companies to intervene on 28 January 1935 with the Minister of Industry and Commerce on those grounds and to argue for a more equitable distribution of categories. This modification was not accepted—the regulations came into force,[5] and in a public statement Manoilescu repeated the argument about unemployment of

[1] See *MPR*, no. 12 (1934), p. 803.

[2] "Law for the Protection of National Labour": the preceding sentences summarize Articles 1–6; for full text, see *MPR*, no. 14 (1934), pp. 942–3; no. 15 (1934), p. 997. The Law also provided (Article 1) that at least 50 per cent of the board of directors, the executive committee, and the censors should be Romanian nationals, and that the president of the board had also to be Romanian.

[3] *MPR*, no. 12 (1934), pp. 941–2.

[4] 100.

[5] Text of regulations of 31 Jan. 1935 and "return" form: *MPR*, no. 3 (1935), pp. 202–3.

intellectuals and insisted on compliance with the Law as it stood. What this entailed may be seen from the experience of Astra Română.

The company had had a long sequence of disputes with the Ministry of Industry and Commerce on the employment of foreigners that arose from divergent interpretations of the category "superior technical staff" (comprising various kinds of engineer) and had been subjected to public criticism by the Association of Romanian Mining Engineers and Technologists. It had taken part in the intervention of 28 January and had been officially listed as non-complying.[1] On 14 May 1935 Astra's officials heard over the radio that the company had been fined 10 million lei (representing 0·5 per cent of the capital) for infringing the regulations: the decision was officially confirmed on 30 May 1935,[2] with the added proviso that the higher administrative and technical categories of staff should be brought into conformity with the Law within three months. On 18 June 1935[3] Astra replied, rebutting the specific charges and requesting cancellation of the fine, modification of the Law in line with the recommendations of 28 January 1935, and a period of one year in which to comply with the Law of 16 July 1934. This request was unavailing—as were representations in London to Titulescu and Leon.[4] Astra Română therefore challenged the Government's decision in the High Court, which on 17 June 1936 returned a verdict in its favour.[5] The fine was rescinded on a technicality, i.e. that the decision to impose it was *ultra vires*: the animosity remained.[6]

(*a*) *The revival of nationalist policies*

The frustrated ambitions of the Romanian technical intelligentsia were one facet of the resurgence of nationalism in Romanian oil affairs, and were supported by the State, in pursuance of its employment policies. Budgetary policies at this time were also

[1] See *Monitor Oficial*, no. 78 (2 Apr. 1935); reprinted in *MPR*, no. 8 (1935), p. 587. [2] Ministerial Decision No. 21359. [3] Text: 10.

[4] The signatory of the decision. 102. [5] 103.

[6] The Association of Romanian Mining Engineers and Technologists continued its campaign, demanding the expulsion of all foreign workers; in consequence of which Osiceanu (Steaua Română and President of the Association) and Constantinescu (Creditul Minier and Vice-President) resigned in protest. See *MPR*, no. 13 (1935), p. 997.

directed towards nationalist ends—which dictated a revival of the idea of a state-controlled monopoly for the distribution of products to the inland market and a drastically revised and extended tax and rail-freight policy.

The failure of conservation in 1930 implied the weakening of inland prices to the point where in January 1931 the low levels "much alarmed" the Minister of Industry and Commerce (Manoilescu).[1] The remedy, in his view, was an industry arrangement for the inland trade, failing which the State was prepared to act[2] (thus controverting the provisions of Article 200 of the 1929 Mining Law). The Minister's demand was echoed by the Association of Producers without Refineries in a memorandum addressed to the A.I.P.R. on 14 February 1931.[3] In the A.I.P.R. meeting ten days later the Distribuția partners countered with a proposal designed to guarantee the small producers an outlet for their production[4] and prorate the domestic market. By implication the State's interest in inland trade was confined to the revenue it created. Their scheme did not win acceptance—the small producers were succoured by other means[5] and the domestic market remained unorganized, the only change in the structure of distribution being Concordia's[6] participation as a supplier to Distribuția from 1 April 1934 onwards.

The attempt by Government to control inland distribution was renewed in January 1935 as part of a scheme to create monopolies in basic commodities to increase state revenue.[7] For the oil

[1] The low prices offered to the Government by the companies for royalty crude oil contributed to this feeling: see 104.

[2] 105. The opportunity for government action was provided by the breakdown in Sept. 1930 of the inland cartel agreement of 20 Nov. 1929 by reason of the withdrawal of seven participants, representing 7·6 per cent of the total crude oil treated and 11·6 per cent of the products: *MPR*, no. 20 (1930), p. 1203.

[3] The reformation of the cartel was the second point in the programme, of which the first was a return to conservation, "as the sole remedy to improve the situation immediately". 106.

[4] This was to be achieved by the seven largest companies taking over ex-refinery the output of the small producers—rated at their average monthly production 1929–30—at current (1931) prices less a charge for the expenses of sales: domestic requirements were to be prorated according to refiners' capacity on a given basis. See *MPR*, no. 5 (1931), pp. 254–6; no. 6 (1931), pp. 307–8; no. 8 (1931), pp. 409–10; no. 9 (1931), pp. 467–71; no. 10 (1931), pp. 523–4.

[5] See above, p. 185.

[6] For contract and other details see 107.

[7] e.g. sugar, coffee.

industry, the increase was to result from a larger consumption (which was alleged to be restricted by the inadequacy of existing installations) and from a cessation of tax frauds. These views were put (18 January 1935) to the representative of the A.I.P.R. by the Finance Minister, Victor Slavescu,[1] as grounds for the creation of one large company with state participation: private concerns wishing to remain outside the new organization would be barred from sharing in the inland trade.

The idea of a state monopoly (though not on the grounds alleged) divided the principal members of the A.I.P.R.[2] Astra Română, Steaua Română, Romàno-Americană, and Prahova advocated outright rejection of Slavescu's scheme, basing their arguments on the Mining Law and refusing to admit state interference in their internal business. Concordia and Unirea suggested a compromise as an alternative to a state monopoly's being forced on the industry, and the small refiners pleaded for the scheme to be seriously examined, since any new regime for the domestic market could not for them produce a worse state of affairs than prevailed at the time. The Association formally advised Slavescu that it could not recommend this proposed scheme to its members, and subsequently rebutted the allegations in a memorandum[3] blaming the tax system for the admitted shortcomings of the industry as supplier to the inland markets. The memorandum argued that the plethora of inspectors made frauds by oil companies impossible: if the State suffered losses, these arose through the effect of taxation on the retail trade.[4] Additionally the tax system was equally responsible for the slow rate of growth, since it hampered the profitable construction and operation of depots and increased the cost of distribution (which in respect of the C.F.R. represented an accrual to the State), the result being to leave the company concerned only 10–30 per cent of the sales price per unit outlet according to product. On the retail market the taxes were supplemented by such costs as financing

[1] 108. [2] 109. [3] 110.

[4] In this sector the prevalence of small dealers made control difficult. The fraud operated as follows: standard gasoline (with a density of 0·7 at 15 °C.) would be mixed with kerosine (with a density of 0·815 at 15 °C.) in a ratio of 55:45 to give a product retailable as gasoline, with a density of 0·75 at 15 °C. The tax on gasoline was 6·50 lei/kgm, and that on kerosine 1·50. The admixture diminished the State's return *pro tanto*. The A.I.P.R. argued that a state monopoly was irrelevant to the solution of this problem, which could be achieved only by equalizing the tax—a course which it had advocated for some years.

the construction of public works, demanded by local authorities as a *quid pro quo* for permission to build depots and installations.[1]

The memorandum also set out the special position of the small refiners, demanding, in view of their precarious situation, prorationing of internal-consumption business between all refiners, and an assured rate of profit. Gartner (Astra Română) also stated privately one other ground for refusal[2] which presumably was also in the minds of representatives at the A.I.P.R. meetings—viz. that an internal-distribution monopoly would be in a position to compel its suppliers to deliver at fixed maximum prices.

On 13 February Slavescu went out of office,[3] and his successor Victor Antonescu abandoned this scheme—though the principle of a state monopoly for inland distribution was resurrected the following year in the proposals for revising the Mining Law.[4] The episode, however, illustrates the dilemmas for both Government and industry, raised more acutely by the Depression. The needs of the State demanded increased revenue from the inland market: this comprised additional taxation in all phases of activity related to that market and the improvement of taxable capacity by increased domestic consumption. The interests of the industry lay in improving its returns from the same market: in the absence of a quota or restriction arrangement this involved a reduction in costs—in which the most onerous element was represented by taxes—and the speedier collection of sums outstanding, particularly from state enterprises. Financial policy and practice therefore became the subject of further continuous dispute between the industry, represented by the A.I.P.R., and the State—i.e. the Ministry of Communications (rail freights and taxes), the Ministry of Finance (fiscal taxes), and the Ministry of Industry and Commerce (oil production and refining taxes and royalties).

The taxable capacity of the industry in the inland market was limited by its size and rate of growth. It represented only about 19 per cent of total output, and the trend of consumption of individual products was, from the viewpoint of the fiscal authorities, unsatisfactory. The consumption of motor gasoline—which bore the highest rate of tax—decreased from 1930 to 1934 and

[1] On the problems of obtaining permits, see comments by Zentler (Distribuţia) in lecture to Bucharest Rotary Club, 30 Oct. 1931: 111.

[2] 112.

[3] In the aftermath of the publication of the Aubin report on state finances: see Prost, p. 73.

[4] See below, p. 206.

improved on 1930 levels only in 1935; that of kerosine—the principal product in the rural market—increased by only 7 per cent; and that of fuel oil by 9 per cent over the same period.[1] The industry attributed primary responsibility for the slow rate of increase to the Government's fiscal policy. Other factors adduced were the slow growth of motor transport—itself due to the taxation and import controls—and the restriction on distribution provided by the state of the roads.[2]

In their representations[3] the oil companies drew a distinction between fiscal taxes, representing the State's right to satisfy its needs for money, and transport charges, which represented a charge (exorbitant) for service (inefficiently) rendered. Total revenue continued to increase—which illustrates the futility of the industry's protests and underlines the weakness of its position against that of the State. The latter in its turn ascribed to the oil companies the slow progress in the inland trade and used its criticism in part as a justification for monopoly.

(*b*) *Changes in export markets*

In the same period (1930–5) exports, which increased from 3,867,208 to 6,611,491 tons, offered more promising revenue-earning possibilities, even though prices were lower and the business of export entailed administrative decisions about the relative merits of "free" and "restricted" currencies. Inasmuch as oil still commanded a more favourable price in the world market than cereals, the Romanian Government started from the assumption

1

Inland consumption	Motor gasoline	Kerosine	Fuel oil	Total all products
1930	74,615	155,919	866,340	1,272,875
1931	73,694	168,118	796,420	1,184,373
1932	63,425	158,142	850,719	1,228,282
1933	65,021	160,887	884,846	1,332,776
1934	73,126	165,855	903,978	1,410,604
1935	80,795	165,274	945,366	1,479,685

MPR, no. 3 (1936), p. 179.

2 See R. J. E. Humphreys, *Economic Conditions in Roumania* (dated May 1931), D.O.T., p. 66.

3 See, e.g., memoranda discussions reported in *MPR*, no. 17 (1934), pp. 1133–6; no. 19, pp. 1263–7; no. 22, pp. 1469–70.

that it was the primary earner of free currency,[1] but the policy of maximizing sales to free-currency destinations failed by reason of the freedom of choice conferred on holders of hard currency in the major purchasing countries (as described above, p. 179) and the clearing policies of the two most important restricted-currency states, Italy and Germany.

A clearing agreement with Germany was first arranged between the central banks of the two countries in March 1933. This arrangement, a development in the reorientation of German trade under Schacht's "New Plan" was extended by inter-governmental agreement on 24 May 1935[2] (amended September 1935): under these measures, payment in clearing marks instead of sterling for oil supplies brought about an increase in Romanian exports to Germany, from a total of 171,390 tons in 1932 to 849,382 in 1935. In consequence the Romanian authorities had to find a use for large quantities of Reichsmarks. At first this problem was dealt with by the Romanian National Bank's leaving a part at the disposal of the oil-exporting companies for sale on the free market. This had adverse repercussions on the lei/Reichsmark rate of exchange, and in October 1936, to maintain the old rate, the Romanian National Bank ordered that the petroleum exports to Germany should not exceed 25 per cent of the value of the total of Romanian exports to Germany during the preceding quarter. Exporting companies had to revise their programmes accordingly, and deliveries to Germany fell during 1937.

[1] Oil-importing states did not always agree—relations with Austria providing a case in point. In Dec. 1936 the Romanian and Austrian Governments failed to agree on the terms of financing oil imports, and in consequence the trade and payments arrangement originally signed on 8 May 1935 was allowed to lapse. The Romanian Government stipulated that 30 per cent of all oil purchases by value should be paid for in free currency and the remainder in non-clearing schillings. Romania also stipulated for an increase in the agricultural quota. After the breakdown of negotiations, the Romanian National Bank insisted on payment for all exports to Austria in free currency. The Austrian Government retaliated by ordering payment for oil imports to be made into a blocked account in the Austrian National Bank. In consequence Romania's oil exports to Austria could be made only against free-currency payments or through compensation deals, and fell from 341,804 tons in 1936 to 91,700 tons in 1937. The dispute was resolved after the Anschluss (see below, p. 213 n. 2). The Romanian Government was more successful towards Greece and Bulgaria, which after 1937 agreed to exempt oil imports from their respective clearings and paid for them in free currency.

[2] See U.S. Tariff Commission, *Foreign Trade and Exchange Controls in Germany*, Washington, 1942, Chap. 12. The agreements provided a framework of reference for agreements-in-detail, negotiated annually.

The operation of the clearing with Italy involved a similar problem. An inter-governmental agreement was concluded in September 1934, but its course was affected from the beginning by Italian stockpiling of products for the war against Ethiopia. By the time sanctions were imposed, in October 1935, Romania had an unusable accumulation of lire, which jammed the clearing and impeded its renegotiation. The new agreement of 1937 followed the precedent established regarding Germany in restricting oil exports to Italy to an agreed quota in order to prevent the clearing from becoming choked with lire.

The accumulation of Reichsmarks and lire in the restricted-currency clearings with Romania reflected not merely the commercial but also the military policies of Germany and Italy. For Italy this was simply a matter of purchases of fuel and lubricants for short-term use, namely the operations against Ethiopia from October 1935 to May 1936, which was made easier through the existence of a small but direct stake in Romanian oil.[1] The acquisition of a similar stake became an object of German policy. The chosen instrument was Creditul Minier, on which the German Legation in Bucharest reported at length in February 1935. This initiative was followed up by Schacht on 1 April 1935 in a letter to Manoilescu[2] written in connection with the German-Romanian Treaty that was to be concluded in the following month. Schacht proposed that, in order to overcome Romania's inherent disadvantage in having its most important raw material exploited by foreigners, Germany should undertake the discovery of new oil sources, and accept payment not in free currency but in oil. This suggestion had been broached to Constantinescu of Creditul Minier, but really called for settlement by governments. The Romanian Government for its part should nominate a Romanian oil company ("possibly Creditul Minier") to contract in Germany for oil equipment—to be delivered on credit. The credit arrangement would be backed by the Romanian Government, and would ultimately be taken up in settlement of outstanding debts owed to Germany on reparations account. This settlement would be in terms of oil, not cash. The size of the credit would be governed by the need to put Creditul Minier on its

[1] viz. Prahova and its associates.

[2] Text on AA file *Austauschgeschäfte, Kompensationsabkommen u.s.w. zwischen Deutschland u. Rumänien*, Aug. 1932–May 1936.

feet: as soon as the operation was completed, all installations and equipment delivered would revert to the company.

These proposals were not adopted. At this period the Romanian Government, although wanting to maintain exports to Germany,[1] had no wish to extend them unduly, on currency grounds. Additionally it was trying to foster the production of oil equipment within Romania, principally at the Malaxa plant. The recreation of the German oil interest in Romania had to wait on political changes. For the time being German policy concentrated on the furtherance of "economic ties [which] create at the same time the basis for the interest, which Romania must feel more and more, in also aligning herself with us politically".[2] This was hampered by the feeling in the Romanian National Bank[3] that the country was becoming too dependent on Germany, and that Romania was earning insufficient convertible currency—both of which dictated the clearing restrictions mentioned above.[4] There was, however, a countervailing factor in Romania's need for armaments[5] —which, as before the First World War, involved her in relationships with both France and Germany. Trade and payments negotiations with Germany in December 1937, within the framework of the 1935 Agreement, established a more favourable Reichsmark/lei clearing-rate against an improved position for Romanian agricultural products and the delivery of armaments, and a consolidation of both procedures and transactions.[6]

In contrast with exports to Germany and Italy, Romania's oil trade with France was encumbered with difficulties arising

[1] See the eloquent representations, set out by the Romanian minister in Berlin in a *note verbale* of 11 May 1934, to the German Foreign Office for the maintenance of Romania's oil-import position in the contemplated reorganization of the raw material market in Germany: AA file *Petroleum und andeer Mineralöle.*

[2] Fabricius to Foreign Ministry, 18 Nov. 1937: *DGFP*/D/V, No. 147.

[3] This became the more effective when Constantinescu, the Governor of the National Bank, became Minister of Industry and Commerce.

[4] p. 194.

[5] In one of their routine reports, 22 July 1936, the German Legation stated, that the rearmament programme was the central question of Romanian internal politics, to which all other questions—financial, budgetary, foreign policy, and currency—were subordinate. The King himself took rearmament as his personal task. Titulescu was indispensable so long as the armament programme depended on the credit willingness of France and Czechoslovakia. See AA file *Petroleum und andere Mineralöle.*

[6] Wohlthat to Goering, 13 Dec. 1937 and 14 Dec. 1937, *DGFP*/D/V, Nos. 154 and 155.

from the ease of supply from competing sources in the U.S. Gulf and from the oil policies pursued by the French Government, which through C.F.P. was interested in expanding Iraq production, and which had brought about the creation of a French refining industry to deal with that production.[1]

Negotiations to implement the protocol to the Commercial Agreement of August 1930[2] resulted in an agreement in detail on 5 January 1932. At this stage, however, the designated oil companies, on being approached by the two governments, declined to take part in the scheme, and the agreement lapsed. The principle remained and was resurrected in January 1934 by Slavescu, in an attempt to find a guaranteed outlet for Romanian crude oil and products, the proceeds of which would finance outstanding loan service charges and commercial debts and the purchase of military aircraft. On this occasion the oil was to be supplied not by individual companies grouped *ad hoc* but direct by the Romanian Government using the crude oil it received as state royalties. This, however, implied acceptance of less than the optimum amount of foreign exchange which such oil could earn, since the current return from crude oil was less than that from finished products; further, the crude oil itself would create a refining problem in France, partly by reason of the displacement of existing supplies, and partly by the nature of the crude oil, which was more paraffinous than the Iraq crude it would replace. On the other hand, the alternative of shipping products would require some modification by the French Government of its import policy. For these reasons negotiations were complicated and protracted; the immediate foreign-exchange question was dealt with in the Franco-Romanian Commercial Treaty of August 1934. The solution of the problem arising from the use of "royalty" crude oil, however, took rather longer and was eventually considered in 1936 on the basis of a scheme worked out by M. Wenger of Petrofina S.A.

From the Romanian standpoint the aim of the negotiations was to provide funds for payment for armaments to the value of 725 million Francs, loan interest, and overdue commercial debts.

[1] The position of Iraq was strengthened by the opening of the Tripoli pipeline in July 1934, in consequence of which imports from Iraq into France increased from 279,053 tons (4·5 per cent total oil imports) in 1934 to 2,589,759 (38·2 per cent) in 1935.

[2] See above, p. 161.

The oil element in Romanian sales in France was 25,000 tons aviation gasoline (value—13 million Frs. p.a.) and three-quarters of the royalty crude oil (value—90–100 million Frs. p.a.) for a period of twelve years. These sales represented a guaranteed market outside the clearing. The crude oil was to be handled by a group of French interests headed by Petrofina. A treaty embodying these terms was concluded on 11 February 1936,[1] but its entry into force was made conditional on the conclusion of a contract of sale for the crude oil. Negotiations on the latter emphasize the inherent difficulties in the arrangement. The export of a maximum 750,000 tons crude oil per annum which would otherwise have returned to the Romanian refining industry would increase refining costs within Romania, while the alternative supply of products could only be made at higher prices than those at which the limited import of products was permitted by the French Government. Internally, the payment, instead of being merely a lei exchange between the Romanian Government and the Romanian oil companies, would involve the latter, under the foreign-exchange regulations, in converting part of their receipts into Francs and therefore incurring a currency loss. The contract of sale was ultimately concluded on 5 May 1936: its terms were complicated, reflecting the implications mentioned above,[2] but it ensured a guaranteed outlet for Romanian oil against payment in free currency. Its working was beset by difficulties, deriving largely from the movements in Gulf prices. Its basic drawback, however, was that it tried to establish a workable relationship between two opposed interests—on the Romanian side, the exchange control regime, and, on the French, that policy of self-sufficiency which in agriculture as well as oil undermined the Government's political arrangements in Eastern Europe and enhanced the attractions of Germany and Italy.

Romania's relationships with Great Britain, unlike those with France, were confined virtually to trade and finance. The British Government recognized Romania's acquisition of Bessarabia but, in accordance with its general attitude on territorial readjustments arising out of the peace treaties, had refrained from interesting itself in Romania's other frontiers. Exchanges between the two governments, as evidenced by the public treaties, were concerned

[1] *CER*, no. 2 (1936), pp. 92–3.

[2] Ibid., pp. 99–124.

with the protection of British investments and with overcoming the trade and payments difficulties resulting from foreign-exchange restrictions. Romania's increasing sterling indebtedness after 1932 dictated the terms of the various agreements,[1] which impinged on Romanian oil-exporters only to the extent of helping them to meet their sterling obligations. Until May 1936 exporters were permitted by Romanian exchange-control practice to recover only 10 per cent of the sterling balance of their exports—the rest being paid over to the National Bank. After that date, in addition to the 10 per cent, they were able conditionally to draw on a "special oil account", representing 40 per cent of Romania's oil earnings in sterling, for payment of such items as insurance, interest, and dividends. These measures checked but failed to arrest the decline in U.K. purchases of Romania's oil which set in during 1932 and in which the determining factor was the relationship between Constanța and Gulf prices.

Consideration of the four major European markets for Romanian oil demonstrates that in the absence of a realistic lei/£ exchange rate, which would reduce exporters selling quotations, or of a reduction in the tax and transport elements in the Constanța price—both of which required action by the Romanian Government—the tendency was for Romanian oil to go to restricted-currency destinations regardless of the policies of the authorities in those countries. When these policies—as in Germany—offered the highest return to the exporter in the given circumstances, the tendency hardened into a trend, which the Romanian Government attempted, in June 1935, to offset by introducing export premiums varying according to destination. These failed to arrest the move towards restricted-currency countries, and in November the same year the differential premiums were abolished in favour of a uniform premium of 38 per cent—roughly equal to the extent of lei devaluation. The benefit to oil-exporters was diminished by the imposition of a 12 per cent *ad valorem* tax on the export of petroleum products to cover the premium paid by the National Bank to exporters of *all* goods. This action remedied

[1] 8 Feb. 1935, Cmd. 4802; 3 Aug. 1935, Cmd. 4976; 2 May 1936, Cmd. 5174; 28 May 1936, Cmd. 5187; 25 May 1937, Cmd. 5471. On the course of Anglo-Romanian trade, see reports for 1932, 1933–4, and Mar. 1937 in R. J. E. Humphreys, *Economic and Commercial Conditions in Roumania*; on payments negotiations, see also *Memorandum on the Economic Situation of Roumania,* Sept. 1935 (C4579): all published by the Department of Overseas Trade.

the situation to the extent that the operation of clearings allowed. Nevertheless specific administrative measures by the Romanian National Bank were necessary to stop the "disproportionate" flow of oil to Germany, the principal restricted-currency market.

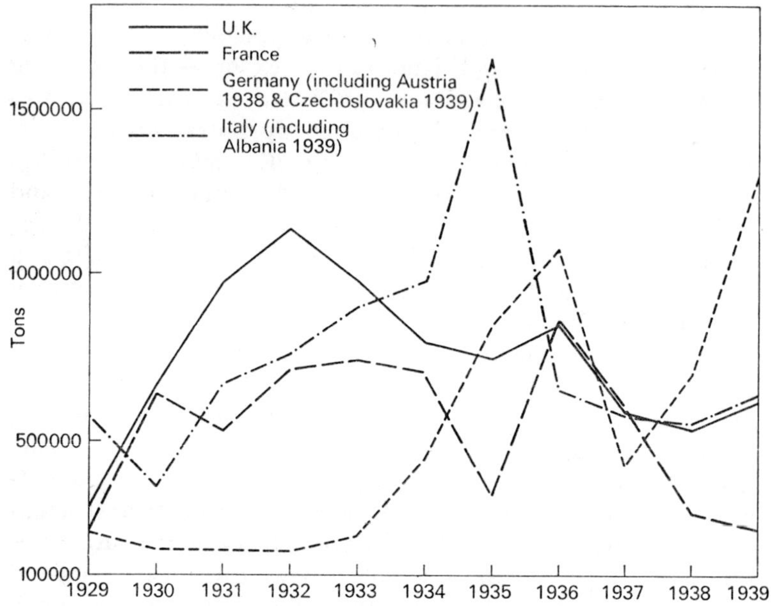

FIG. 10. Romanian oil exports, 1930–9.

The course of Romanian oil exports during the Depression is shown in Figure 10: their value may be seen from the following table:

	Total exports receipts (lei)	Oil exports[1] receipts (lei)	%
1930	28,522,028	10,408,594	36·5
1931	22,196,918	6,901,090	31·1
1932	16,721,593	7,158,850	42·9
1933	14,170,828	7,879,100	55·6
1934	13,612,988	7,213,626	53·0
1935	16,710,730	8,454,009	50·6

[1] Free currencies converted into lei at official rates.

The emphasis on maximizing exports, particularly of increasing quantities of gasoline, raised again the problems of improving rail transport and building a second pipeline to Constanţa, both of which had been discussed at length before the Depression.[1] Since 1929 the beneficial effects of the application of the Stabilization and Development Loan were partly offset by the decline in passenger traffic (38 per cent over 1929–33) and the continuing indebtedness of state institutions to the C.F.R.[2] In these circumstances income from freights became the more important, and, as during the Depression the contribution of oil transport to freight receipts increased from 17·7 per cent (1924) to 36·6 per cent (1934), the C.F.R. was most unwilling to contemplate any loss by diversion of oil products to pipeline transportation (even though such a line would be a source of income to the State).[3]

Nationalist solutions for the problem of reserves

The revival of industrial activity and improvement in agriculture in 1936 mark the emergence of world trade from the Depression. These changes were bound up with a strengthening of the world oil industry. In America the "hot" oil, excessive stocks, and declining demand which had brought about the crisis of 1931–3 disappeared under Federal regulation and the revival of the U.S. economy. In Europe the increase in oil consumption, which had been irregular throughout 1935, became general, stimulated by a revival in shipping, the increase in output and use of automotive vehicles, and rearmament and military expenditure.

The Romanian economy benefited from improved agricultural prices, but its oil industry was not well placed to take advantage of the improved circumstances. Astra Română and Steaua Română[4] publicly drew attention to the problem of reserves in 1935. This theme was also discussed by Professor Macovei,

[1] See above, pp. 149–51.

[2] See reports of R. J. E. Humphreys on economic conditions in Romania: 1931 (published May 1932), p. 51; 1932 (May 1933), p. 53; 1933–4 (1934), pp. 49–50.

[3] On problems arising between the C.F.R. and the oil industry, see Pizanty, *Le Pétrole en Roumanie* (1933 edition), p. 247; and two articles by Ion Miclescu (Commercial Director of Railways), viz. "Les Chemins de fer et l'industrie du pétrole", *MPR*, no. 4 (1936), pp. 267–73; "L'Industrie du pétrole et le transport des produits en Roumanie", *MPR Supplement for 1937 World Petroleum Congress*, pp. 78–82.

[4] Annual reports to shareholders, 1935.

Director of the Romanian Geological Institute, in December of that year. Macovei, reporting the conclusions of a study by his Institute, divided reserves into three categories—proven, i.e. those in territories already in exploitation; probable, i.e. those in neighbouring areas and geological extensions of them; and possible, i.e. where geological conditions seem favourable without the areas having been proved by exploration or adjacent exploitation. The first two categories comprised 6,100 hectares, with an average output of 10,000 tons per hectare. At the 1934 rate of production, this gave the Romanian oilfields a "life" of seven and a half years. As to the third category, conditions indicated 10,050 hectares in Wallachia and 200,000 in Moldavia as possibly oil-bearing. Macovei pointed out that most of these reserves were already in the hands of the State, which should undertake exploration, first, to create new proven reserves[1] to ensure continuity of supply, and, secondly, to apply a programme of conservation, the absence of which had been responsible for the rapid depletion of proven reserves in Prahova and Dâmboviţa.

The implications of this diagnosis of the long-term insecurity of the industry were emphasized when in mid 1936 production began to fall off.[2] In the short term the decline in output meant a decline in quantities available for export at a time when European demand was recovering. During 1936, owing to more stable conditions in the U.S. Gulf, American export prices approached Constanţa prices, and restored the latter's premium advantage on the Danube and Mediterranean oil markets. In the next year Romanian exports fell by 1,216,000 tons and the price disparity reasserted itself:

Exports (m. tons)	1936	1937	Decrease 1937 over 1936 (%)
Crude oil	628,585	472,402	24·8
Gasoline	2,096,671	1,829,865	12·6
Kerosine	1,146,815	952,116	17·0
Lubricating oil	37,208	34,841	6·4
Gas oil	1,181,748	897,339	24·1
Fuel oil	1,764,167	1,454,086	17·6

[1] The necessity to prove new reserves prompted proposals made by Astra Română to the Ministry of Industry and Commerce on 2 Nov. 1934 for a new regime to encourage prospecting and exploration in regions remote from the known oil-producing area: these proposals were not accepted but ultimately became part of the Astra proposals for the revision of the 1937 Mining Law: see 113.

[2] See above, p. 175.

The long-term improvement of the situation raised fundamental issues of state policy governing exploration and exploitation, which had to be taken account of in the plans already under discussion to correct the shortcomings disclosed by the operation of the 1929 Mining Law. The crisis in Romanian oil—in both its production and financial aspects—negated the more liberal prospects offered by the Mining Law of 1929 towards "foreign-capital" companies in that the latter did not take up concessions to explore for new sources or make the long-term investments which their superior financial resources would permit.

Additionally, the return to office of the Liberals in 1933 made probable a modification of the Law in a nationalist sense. Ideological considerations were reinforced by financial ones: I.R.D.P. and Creditul Minier were doing badly, and the smaller companies which were hard-hit or which were forced into bankruptcy were all Romanian-owned. A statement of Liberal policy towards the industry emphasized that production should be realized in conformity with the national interest, for the prosperity, above all, of Romanian companies concerned solely with the exploitation of the country's oil resources.[1]

The Mining Law of 1937

The problems arising from the 1929 Mining Law which practice showed to impede the industry, viz. the grouping of perimeters, validation, and the recognition of exploration rights, came up for discussion with officials[2]—notably Professor Leon[3]—who were politically opposed to the "foreign-capital" companies, and at a time of acute nationalist unrest. Negotiations on technical issues by representatives of government and industry[4] preceded the introduction of a draft "law for the modification of some Articles

[1] Ing. Teodoreanu in symposium *The Industrial Policy of the National Liberal Party*; summarized in *MPR*, no. 1 (1934), p. 12.

[2] On the influence of officials in interpreting oil legislation, see case histories in 114.

[3] Professor of Economics, Under-Secretary of Ministry of Industry and Commerce in the Tătărescu Government of Oct. 1934, author of a nationalist commentary on mining legislation, *Roumanian Oil and the New Draft Mining Law*, Bucharest, 1937: see 115.

[4] *MPR*, no. 21 (1935), p. 1580: the two "industry" participants are not identified, and from the editorial notes under "General Situation" (*MPR*, no. 5 (1936), p. 381, and *MPR*, no. 7 (1936), p. 529) it can be inferred that "foreign-capital" companies were not represented.

of the Mining Law of 1929" in February 1936. The draft, however, attempted to combine technical solutions with the restoration of the privileged position of Romanian capital.[1] It met with opposition in the Chamber,[2] and was criticized extensively on 12 March in a meeting of the A.I.P.R., which submitted its own recommendations to the Ministry of Industry and Commerce at the end of that month. The Bill was in consequence referred to the Mines Committee of the legislature for further consideration. The technical and procedural aspects of revisions, however, were overtaken by a dispute over the scope of acquired rights, which was the kernel of the parliamentary opposition, led by Madgearu. The Government therefore abandoned the rest of the Bill and attempted to deal with acquired rights by decree, published on 10 November 1936.[3] The bid failed, since Parliament refused ratification. The Government thereupon announced that acquired rights would be a feature of a new Mining Law. Meanwhile the decline in production which set in during the year intensified discussion in both the industry and the Government about reserves and the entailed problems such as land acquisition, the regulations for size of perimeters, and the rewards permitted to successful prospectors.

The proposals for a new law were introduced again in January 1937, and acquired rights monopolized debate, in consequence of which the Bill was again referred back, but not withdrawn. At this stage fourteen Romanian national companies, headed by Creditul Minier, intervened with the Ministry of Industry and Commerce. They urged a preferential position for national capital through the award *à l'amiable* of concessions for exploration and exploitation priority in the granting of exploration perimeters, reservation of an important proportion of the inland trade, and preferential rights to supply public authorities.[4] Further nationalist preference was urged by the Association of Mining Industry Engineers and Technicians.[5]

The draft was circulated to deputies by Valer Pop, Minister

[1] Leon's *exposé des motifs*: *MPR*, no. 3 (1936), pp. 234–6.

[2] *MPR*, no. 6 (1916), pp. 469–70; comment in *MPR*, no. 7 (1936), pp. 345–6.

[3] *MPR*, no. 22 (1936), pp. 1697–9: see also criticism by Gh. Dumitrescu in *MPR*, no. 3 (1937), pp. 153–4.

[4] Summarized in article "The Petroleum Industry and the modification of the Mining Law", *MPR*, no. 4 (1937), pp. 237–8.

[5] 116.

of Industry and Commerce, on 13 March 1937: the Bill was unexpectedly introduced on the penultimate day of the session (19 March) and passed through the Chamber of Deputies in four hours—the Opposition abstaining in protest against the Government's treatment of Parliament. The Senate passed the Bill the next day, without debate: it was promulgated by royal decree on 23 March, and came into force the next day.[1]

The Law aimed explicitly at modifying certain articles of the Mining Act of 28 March 1929 in order to strengthen the position of the "national" companies against those owned by foreign capital, and at the same time to increase the area of known petroliferous land and encourage prospecting. Accordingly, the main changes between the two laws related to those clauses governing ownership and the rights of exploration and exploitation flowing from it. The 1937 Law reintroduced the distinction between "national" and "foreign-capital" companies, which had been given legal expression in Article 32 of the 1924 Mining Act. The regime for the exploitation of mineral substances belonging to the State (First Title Chapter I, Articles 4–92) reaffirmed the provisions of the 1924 Law—which had remained unvaried in 1929—regarding the rights of the State over exploration and exploitation, but added (Article 7) that exploration permits would be issued only to companies of proved technical and financial capacity and if such grants were in the national interest. As regards permits for liquid bitumens, consideration would be given to

> Companies with technical and financial means, whose capital is nominative, held in majority by Romanian citizens and whose board of directors is composed entirely of Romanian citizens. For every 4 exploration perimeters assigned to any joint-stock company complying with the requirements of the law, one perimeter shall be reserved for these companies.

The practical relevance of this distinction became clearer in the chapters dealing with concessioning. For the purpose of exploration of liquid bitumens, three categories of area were created (Article 178):

> Category I includes the fields surrounding existing petroleum exploitations, for a distance of minimum 2 kilometres and maximum 20

[1] Text in *MPR*, no. 7 (1937), pp. 471–514 (no *exposé des motifs* printed); also editorial "Considerations on the new Mining Law", ibid., pp. 451–4.

kilometers, where the tectonic faults suspected of being oil-bearing may be ascertained by geological observations on surface.

Category II includes the fields situated between 2 and 20 kilometers round existing petroleum exploitations and where the supposed tectonic faults cannot be ascertained by geological observations on surface, as well as the fields on which exploration perimeters were granted up to the date of publication of this law.

Category III includes the lands from unknown regions where no petroleum exploitations exist.

The different regulations governing each category were set out respectively in Article 179 (Category I), Articles 180–4 (Category II), and Articles 185–9 (Category III), and their political purpose disclosed in Article 187, which set out the distribution of land declared exploitable by reason of successful exploration.

The leasing exploitation perimeters from the reserves of the State not considered as duration reserves shall be made by public adjudication. . . .

The Council of Ministers may decide, following the proposal of the Ministry of Industry and Commerce, on the advice of the Superior Council of Mines, the distribution of concession perimeters as follows:

(a) 60% of the reserve due to the State following the works stipulated by Art. 189; half of the reserve deriving from the works stipulated by Arts. 180 and 185, will be allotted by mutual agreement to those companies having nominative capital held in majority by Romanians, and the board of directors of which is composed entirely of Romanians;

(b) The remainder of perimeters are granted by public adjudication, which may be attended by all companies possessing technical and financial capacity.

The new law also laid down new definitions on those problems—grouping (Articles 193–217) and the validation of acquired rights (Articles 277–81)—which the 1929 Law had left outstanding: the new clauses were restrictive and retroactive, which invalidated areas recognized as having acquired-rights protection until the passage of the Law.

Further, the State assumed rights over gas erupting from oil-wells and unused by the producer—which it took over without payment for marketing through "a state enterprise . . . or a mixed administration with the participation of exclusively Romanian capital" (Article 114 (*a*)). The State also reserved the right to order inland distribution through controlling storage and distribution facilities (Article 207) and through creating a monopoly company with state participation. It also stipulated the basis for maximum internal prices, as having not to exceed export prices at export points, plus state and communal taxes (Article 208).

The 1937 Law did not arouse the intense opposition from representatives of states whose nationals had investments in Romanian oil that had been occasioned by the Act of 1924. As before, the diplomatic representatives had been misled on issues of legitimate concern and were taken by surprise by the parliamentary proceedings. The British ambassador, Sir Reginald Hoare, reported on 26 March 1937 that, in face of rumours that the modification of the petroleum law would entail discrimination against foreign capital, he and his French, Belgian, Dutch, and American colleagues had sounded officials and elicited official assurances that nothing adverse to the interests of their respective nationals would be done. He himself had been assured by Tătărescu on 12 March that foreign interests would be safeguarded. The ambassador also reported that on the day after the Law had been passed he had been visited by a "somewhat embarrassed" Minister of Industry and Commerce, Valer Pop, who sought to explain that the discrimination was more apparent than real, designed to throw dust in the eyes of the Government's more nationalistic supporters. The explanation given by Pop to the Netherlands minister on the day the Bill was passed[1] stressed that the Romanian Government did not intend to make life impossible for the "foreign-capital" companies—the aim was to establish available reserves, to promote their exploitation, and, by rationalizing the industry, to increase gradually the share of national companies in exploitation. Despite the minister's initial disclaimer, the 1937 Law, as passed, destroyed the long-term basis of the "foreign-capital" companies in Romania: concessions with acquired rights were nearly all taken up, and expansion on state lands was made unattractive through the compulsion to share results with competitors. From this point of view, the 1937 Law is the last flourish of National-Liberal policy towards the oil industry, and the long-term future of the two "Liberal" companies, I.R.D.P. (which in 1937 paid its first dividend for ten years) and Creditul Minier (which had not paid more than 5 per cent since 1930), was assured.

Nevertheless, all the oil companies, irrespective of ownership,

[1] 117.

found the Law unsatisfactory on technical grounds, and the cause of the foreign shareholders was necessarily involved in the opposition which arose generally from the industry within Romania. This opposition was expressed through the A.I.P.R. and was specifically concerned with prospecting and exploration, the state royalty provisions, the treatment of acquired rights, the provision for construction of pipelines, and the State monopoly of inland trade. The result was acceptance by the Government of the necessity for a revision of the Law within three months of its enactment.

The work of revision was referred to various commissions, of which one, under Professor Mrazec, was charged by Argetoianu, Minister of Industry and Commerce, with preparing a "five-year plan" to change the serious position of the oil industry and to stimulate prospection and exploration for oil.[1] The preparation of detailed proposals was referred by Mrazec to a sub-committee consisting of Marinescu (Concordia), Ficșinescu (Colombia), and Otto Stern (Astra Română). In mid July the head of the new Ministry of National Economy, Mitita Constantinescu, called for the submission of proposals by 10 August. In the absence of a prepared statement Astra Română's proposals were handed in on that date,[2] and were followed by dissenting opinions on specific points from Marinescu[3] and Ficșinescu:[3] copies were sent to the members of the "five-year plan" commission and to managers of oil concerns.[4] The basis of a revised mining law in regard to exploration and production was therefore the Astra Română draft—on which the other companies were invited by Constantinescu to comment. This evoked on 18 August a memorandum from the Union of National Oil Companies,[5] putting the case for its constituent "national" companies.[6]

Comparison of the submissions of 10 August 1938 and those of the 18th August demonstrates the preoccupation of both "foreign-capital" and "national" companies with the *technical* shortcomings of the 1937 Law. The "foreign-capital" company memoranda restricted themselves entirely to questions of exploration and

[1] See Argetoianu's statement of 14 Apr. 1938: *MPR*, no. 9 (1938), p. 733.

[2] 118. [3] 119. [4] 120.

[5] Uniunea Naţionala a Întreprinderilor de Petrol—founded 11 Apr. 1938 by Creditul Minier, Redevenţa, and other national concerns to represent their interests; see 128.

[6] 122.

production and aimed at making the subsoil again commercially profitable without encroaching on the rights and interests of the owner, i.e. of the State, as created by the 1923 Constitution. This postulated for explorers: (i) rewards commensurate with work done, and therefore risks assumed; (ii) security, so that, when successful, they would receive the reward, however determined, in full and without administrative hindrance; (iii) a free hand in regard to methods employed to carry out working programmes: on the other hand, the State should be guaranteed against the blocking of state lands through slow or inefficient working, while the excessive tax and royalty burdens should be reduced for the industry as a whole. The technique by which these postulates were to be carried out involved dividing oil territories into two categories—viz. I, the known area and its immediate surroundings (i.e. Categories I and II in the 1937 Law), and II, the more remote areas (i.e. Category III in the 1937 Law). In the new Category I the discoverer of an oil-bearing structure was to receive preference in its exploitation—by grant of an exclusive permit "to carry out exploration over the entire area of the tectonic unit",[1] up to a total area of 4,000 hectares. The granting of such a permit was made conditional on the fulfilment of specific obligations, but involved no consideration of the "national" character of the prospecting company (the latter point was made by omission). Similarly, on both the new categories of land, exploration and concession rights were to be granted without "national" distinction: the companies were to be free to apply their own programmes, royalties were to be revised.

The "national" companies criticized the proposals in matters of detail, but did not controvert them, since they subscribed to the view that the object of a new law was to encourage exploration by giving incentives and reducing state charges. They added, however, a second criterion—"the assurance that the State could pursue its own petroleum policy", by, for example, creating reserves, preventing monopolization of oil lands, and protecting national capital and labour by preferential rights. Nevertheless, Constantinescu, having studied these memoranda, insisted in October 1938 on receiving proposals representing the agreed

[1] Ficşinescu and the U.N.I.P. companies dissented from this basis—which disappeared from the discussion of joint proposals, only to be reinstated in the Petroleum Law of 1942; see below, p. 239.

view of the entire industry. These were worked out between November 1938 and April 1939,[1] when the finally agreed proposals[2] were submitted. They comprised a rational scheme for exploration and production, and the improvement of administration—both aimed at increasing the output of crude oil as quickly as possible. The respective roles of State and industry were defined—it being presupposed that the former required a guaranteed return from work done and the latter an opportunity of uninterrupted operation, unburdened by excessive state charges. Exploration was to take place in two categories of area. Each was to have its rule for exploration, on a basis of three years' and nine years' authorization respectively, to give explorers an exclusive right for a limited time over a specific area, by which to follow up the results of preliminary work. Companies would no longer be obliged to publish their findings without some hope of obtaining permits to continue working in the same area. Successful applicants would be obliged to assume certain drilling obligations; an area, once proven, should be opened immediately to concessions—the operator being given as a reward 50 per cent or not less than 100 hectares. The State's share of proved oil-bearing land would form its reserve. Perimeters in the second category could extend up to 100,000 hectares to enable the explorer to spread his costs over as large an area as possible, and thereby provide an incentive to wildcatting. Joint operations should be permitted for the same reason. A minimum annual investment should be stipulated, with financial penalties for non-compliance, to ensure that only those capable of undertaking the work did so. The duration of each permit should be subdivided into three phases: at the end of the first the area was to be reduced by 50 per cent, and at the end of the second by a further 50 per cent leaving a quarter of the original area. The choice of areas to be given up was left to the operator, but the necessity to do so would prevent lands from being uselessly blocked. In consequence of the areas becoming proven, the State would be entitled to 50 per cent as reserves, and the successful operator of up to 50 per cent for further working.

Exploitation concessions should be independent of exploration concessions and allow the operator to proceed as he wished, with

[1] 123.

[2] See *MPR*, no. 9 (1939), pp. 606–7; and extensive summary in article by Prof. V. Patriciu in *MPR*, no. 14 (1939), pp. 923–35.

due regard to health and safety regulations, etc. Concessions on state reserve lands should be negotiated separately. While work was in progress, the operator was to enjoy, as acquired rights, the exclusive and irrevocable right to all crude oil and associated gas deposits, under a given concentration or gas–oil ratio. Operators would be bound, however, to assure the maximum recovery by the most rational methods.

Ideas were also put forward to improve the administration of grouping and validation procedures and the basis of levying royalties. These proposals were overtaken by the Second World War, during which the 1937 Mining Law as regards petroleum was superseded by the Petroleum Law of 1942.[1] They therefore remained academic but are of interest in that they *were* a united approach by the industry and provided a way out of the impasse between it and the State in their mutual and fundamental concern of exploration and production. How the "national" companies were brought to abate their claims to preferential treatment, even though as an association they entered a separate letter to the Minister of National Economy on 19 April 1939, is not clear. The representatives of the "national" concerns individually signed the A.I.P.R. letter of transmittal and, by doing so, subscribed to the proposals which did not admit nationality as a criterion of participation.

While the different viewpoints were being elaborated and reconciled, the 1937 Law remained in force, and did nothing to check the growing exhaustion of the fields, where increased drilling during 1937 was accompanied by a decrease in output—implying both the absence of new discoveries and a declining rate of flow. In 1938 drilling itself generally fell off—only Concordia, Colombia, and I.R.D.P. increased their metreages—and output continued to decline. Permits were taken out for Mining Districts under the 1937 Law, but no wildcat drilling resulted.

The political and economic reorganization of the State, 1937–1939

The fundamental changes in the basis of the oil industry took place during a period of change in the structure of Romanian politics. The increase in the power of the Iron Guard provided

[1] See below, p. 238.

the most blatant evidence of a general sharpening of nationalist feeling throughout Romanian society and the ineffectiveness of the existing political-party system in competition with two external sources of pressure, the King and the *Capitan*.

In November 1937 the constitutional time-limit expired and general elections became due: the breakdown in the political structure is illustrated by the formation of the main contending groups—on the one side the "official" Liberals under Tătărescu and the followers of Vaida Voevod and Iorga, with the German minority organizations opposed by a group comprising the National Peasants under Maniu defensively allied for electoral purposes with the Iron Guard (under the guise of the All for the Fatherland League), to which the Liberals under Gheorghe Brătianu and the Agrarians under Argetoianu adhered. In the voting on 27 December 1937 the Liberals and their allies as the outgoing governing party failed to get the 40 per cent of the total vote necessary to give them the right to 70 per cent of the seats in the Chamber—a situation without precedent for a party in office. The King took advantage of the resultant confusion to dismiss Tătărescu and replace him by an administration under Goga and his National Christian Party, which had polled 9 per cent of the vote, aided by dissident National Peasants.[1] Goga's policy was simply and rabidly "Romania for the Romanians": its implementation involved drastic anti-Semitic measures backed by martial law, which paralysed business, causing a run on the banks and the collapse of the stock exchange. After six weeks, economic dislocation was terminated by the abrupt dismissal of Goga on 10 February 1938 by the King, who now openly dispensed with the parties and set up a "Government of National Union". The new structure of political authority was set out on 27 February 1938 in a new constitution,[2] replacing that of 27 March 1923: all existing political parties were dissolved by law on 14 April 1938.

This change in the balance of internal political forces assisted a reorientation of Romania's external alignments. The failure of collective attempts to solve post-Depression problems, particularly in agriculture, provided the opportunity for the German trade drive in south-east Europe which began in 1936.[3] The growth of

[1] Including Armand Calinescu, who became Minister of the Interior.

[2] Text: *BFSP*, vol. 142, 1938, pp. 783–802: see also below, p. 214.

[3] See below, pp. 218–21.

Romanian-German trade and the demonstration of Germany's increasing international strength given by the remilitarization of the Rhineland in March 1936[1] encouraged a readjustment of Romanian foreign policy towards Germany. Initially the readjustment was seen as an economic matter involving the annually negotiated trade agreements. A closer political alignment was inhibited by Romania's pro-Czech policy during the dispute over the Sudetenland and German support for the Iron Guard within Romania. During 1938, however, the Anschluss[2] (March) and the Munich Agreement (September) successively established German power in Central Europe: from the Romanian side the checking of the Iron Guard without excessive German interference (March), and the continuing need for armaments, promoted a closer alignment with Germany, as evidenced by the armament–gasoline treaty in August and the approach by Carol to Fabricius, German minister in Bucharest, at the end of September while the Munich Conference was in session. The upshot of these manœuvres was Carol's visit to Germany at the end of November 1938.[3]

Furthermore, the divergences of interest within the Little Entente became evident. Czechoslovakia relied on the Soviet Union as an ally against Germany while her Entente partner, Romania, regarded the Soviet Union as the principal external menace. The technical issue on which these opposed views turned was whether or not, and, if so, in what circumstances, Titulescu's diplomacy envisaged or allowed the passage of Russian troops by either ground or air transport across Romanian territory to assist Czechoslovakia in case of attack. Romania and Czechoslovakia had similarly opposite policies towards Poland, both on long-term

[1] Which, *inter alia*, destroyed the technical basis of the Franco-Romanian alliance by making impossible direct access by French forces to Central Europe.

[2] The Anschluss provided Germany with a strong lever in the economies of south-east Europe through the Viennese banks' interests in industry throughout the region. It also brought immediate changes as far as Romania was concerned. After the Anschluss the Romanian Government found itself having to deal with Germany to resolve its clearing dispute with Austria (see above, p. 194 n. 1). It had to abandon its free-currency claim and accept payment for oil supplied entirely through the German-Romanian clearing—the old accounts being liquidated at RM 46·5: Au. Sch. 100 as from 1 June 1938.

[3] See memoranda of Carol's conversations with Hitler, 24 Nov. 1938, and Goering, 26 Nov. 1938: *DGFP* D/V, nos. 254 and 257 respectively. For an analysis of the changes in German-Romanian relationships brought about by the events of 1938, see memorandum by Clodius of 13 Dec. 1938, *DGFP* D/V, no. 264.

foreign-policy grounds and on the transit question. Poland was the first and enduring ally of post-war Romania. At the same time Polish-Czech relationships had not fundamentally recovered from the immediate post-war dispute over Teschen and were continually jeopardized by opposed attitudes towards the Soviet Union.

By 1939 both the nature of the regime in Romania and its deduction of what was necessary to serve its foreign-policy interests had undergone a significant change. Reliance on the "French" system was replaced by reliance on Germany—although the Romanian Government appeared to recognize the dangers involved.[1]

(*a*) *The industry and the Constitution of 1938*

The Constitution of 20 February 1938 reaffirmed the rights of the State to mineral deposits under Article 17 as follows:

> Mining resources as well as all subsoil riches are the property of the State.
>
> A Mining law will lay down the regulations and conditions for the exploitation of these riches, will fix the rights of the owner, which shall be a minimum 50% of the royalty and of the price per hectare of the concession, and will likewise specify the manner and the extent to which such owners may participate in the exploitation of these riches. Rights acquired by the State in virtue of concessions concluded up to the present time will be respected.

The omission of any reference to private acquired rights alarmed the major oil companies, whether "national" or "foreign-capital" in ownership in that, literally interpreted, the clause as it stood would entail either that concessions would no longer be valid or that concessions could no longer be taken out for private subsoil deposits, because they ceased to exist in law. Accordingly the "foreign-capital" companies petitioned the Government, which on 2 March 1938 issued a decree law on the interpretation of the clause.[2]

Although this was the only specific reference bearing on the oil industry, the industry was necessarily concerned in the programme of reorganized state authority, the basis of which was expressed in the Constitution. In its economic aspects the

[1] See King Carol's conversations with Lord Halifax, 17 Nov. 1938: *DBFP*.
[2] 124.

reorganization issued in the creation of a Ministry of National Economy announced on 8 April 1938:[1] it incorporated the Ministry of Industry and Commerce (which ceased to exist) and certain functions hitherto exercised by the Ministries of Agriculture, Health, and Social Welfare, and the Cereals Export Bureau. This concentration was designed to promote the co-ordination of all economic activity in the national interest. The National Bank retained its separate character and functions, but possible conflicts on policy were ruled out by the appointment of the Governor, Mitita Constantinescu, to be the first Minister of National Economy.

At the same time as the formation of the Ministry of National Economy was announced, the "Supreme Economic Council", was also reorganized. This was composed of leaders of Government and industry[2] and had already enjoyed two years' existence during which time its activities were consultative and exhortatory.[3] By a decree of 8 April 1938 its functions were changed to making recommendations to the economic departments of state, and it was specifically charged to examine and report to the Government on the measures necessary to reorganize the Romanian economy. Of necessity, the Council discussed the oil industry, generally in considering questions such as the roles of "national" and "foreign-capital" and fiscal reform, and explicitly in considering the role of the industry in Romanian society.[4] As to the first, the Council stated that (IV) foreign capital was necessary to preserve the national production and that being so it was entitled to the same protection by the State and the same rights to legitimate remuneration as national capital. Taxation should be simplified and its incidence made more equitable so as to promote national well-being by stimulating trade and ending the possibilities of fraud (VII). The explicit discussion of oil problems (X and XI) was extensive: it started from the premisses that the oil industry was the best source of free currency but that its special situation

[1] See *MPR*, no. 8 (1938), pp. 643–5: the Ministry was not in being until January 1939.

[2] See ibid., p. 665.

[3] See speech by Argetoianu, President of the Council, 16 Apr. 1938: *MPR*, no. 9 (1939), pp. 733–4.

[4] This outline considers only the salient features of the proposals relating to oil (to which the roman numerals refer). The commission also proposed a drastic reorganization of agriculture, simplification of legal and civil service procedures, tax revision, and a programme of public works: see 125.

of declining production necessitated a regime designed to hasten the opening up of new areas; this entailed a revision of the law to allow concessions to be awarded of a sufficient size to cover the risks involved in exploration. Exploitation should involve a precise and rational programme of strict drilling obligations; refining operations should be conducted to maximize high-value products; a monopoly should be introduced into the domestic market to eliminate tax evasion. The Council also advocated a reduction in tariffs and the financing of railway development from the state budgets.

This programme, which went far towards meeting the needs of the industry as a whole, was not acted upon: taken into conjunction with the proposals for mining-law revision,[1] it represented the nearest any Romanian government came towards defining and harmonizing the roles of the industry and the State during the period of this study. A month later, on the 16 September 1938, all depots and industrial business concerned with materials for national defence were put under military control,[2] and the reorientation of Romanian foreign policy denoted by King Carol's German conversations in November ensured that the reconstruction of the Romanian oil industry would proceed in a different manner and under German auspices.[3] By the time the Ministry of National Economy came to be established, in January 1939, its declared purpose to be the "generator and organizer of productive activity" ignored the basic recommendation of the Superior Economic Council regarding the scope of private initiative, and its attributes demonstrated the Government's intentions to effect detailed control of the oil industry. Article 18 of the Law called for a state mining-exploration company, the "Administrație Comercială pentru Prospecțiuni și Explorări Miniere" (A.C.E.X.),[4] the organization of which was announced contemporaneously with that of the Ministry of National Economy. The aim of A.C.E.X. was described comprehensively in Article II,

> to undertake field studies and mining research, to prospect, explore, and conduct other mining operations for the discovery and realization of mining rights and interests belonging to the State, as defined by Article 17 of the Constitution. . . . The Administration is able to take over any

[1] See above, pp. 208–11.

[2] 126.

[3] See below, p. 220.

[4] *MPR*, no. 2 (1939), pp. 94–101.

concessions abandoned by the concessionnaire or withdrawn by the State and work it until it is reconcessioned.

For this purpose A.C.E.X. was to work with the Romanian Geological Institute[1] and enjoy various exemptions from the requirements of the 1937 Mining Law. It was to be financed out of the Prospecting Fund[2] referred to in Article 73 of that law, but its initial costs were to be recovered by a grant from the Ministry of Economy of 30 million lei. In A.C.E.X. the State had created an important instrument of policy, the lack of which had vitiated at a technical level the policy of "nationalization" in the twenties. Given the policy towards "national" concerns of the 1937 Mining Law, the days of the "foreign-capital" companies were numbered; fortunately from their point of view the 1937 Law was (as shown above) objectionable to both "national" and "foreign-capital" companies alike. Accordingly, the latter's policy towards this new assertion of state authority was to press on with "all-industry" proposals for an equitable and workable regime.

(*b*) *The industry and the realignments in foreign policy*

Measures to reorganize the legal and operational basis of the oil industry during 1937–9 were paralleled by changes in Romania's "oil" relationships with its main European customers. These changes reflected partly the new political factors already discussed, and partly the State's continuing needs for free currency and for capital.

From 1934 Romanian-French trade declined, leaving Romania with a series of adverse balances. At the end of 1938 the French Government sent a mission to examine Franco-Romanian trade with a view to increasing exchanges. Necessarily the problem of oil supply entered into the discussions—which revolved around the perennial contradiction between the French refiners' need to import crude oil and the Romanian producers' need to export products.[3] The agreement eventually concluded on 21 April 1939 between the Romanian Minister of National Economy,

[1] Professor Macovei of the Institute was nominated President of the Management Committee of A.C.E.X.: *MPR*, no. 2 (1939), p. 126.

[2] On the antecedants of this fund, see 129.

[3] See A.I.P.R. submission to Minister of National Economy, 15 Nov. 1938: *MPR*, no. 23 (1938), p. 1864.

thirteen Romanian companies, and Wenger, representing thirty-two French companies and their affiliates, provided for contracts for the supply of 373,000 tons of gasoline, 14,000 tons of kerosine, and 30,000 tons of gas oil in approximately equal amounts over one year, on a financial basis designed to yield the Romanian exporter a 10–14 per cent return in lei over normal Constanţa business. As the agreement was put into practice, movements in Gulf prices made supply under its terms less attractive to Romanian exporters than supply on the customary Constanţa basis. The difficulties encountered provoked a number of its signatories, on the initiative of Redevenţa, to raise in the A.I.P.R. the suspension of the agreement; and to cause approaches to the Ministry of National Economy and to Wenger for modification of the terms. Meanwhile shipments lagged behind schedule, and by the end of September 1939, i.e. half-way through the contractual period, only 97,300 tons of gasoline had been shipped against the undertaking of 386,500, and 5,100 tons of gas oil against 15,000.

Romanian oil exchanges with Germany during this period offer a striking contrast. In the German trade drive in south-east Europe[1]—in its Romanian aspect—oil was an exceptional commodity in that, unlike other raw materials and agricultural products, it maintained a price in world markets where it could be used to earn free currency. Equally, from the German standpoint, Romania was only one of several sources of supply supplementing indigenous natural and synthetic production, and one, moreover, whose proximity had to be weighed against the higher cost of the oil, the difficulty and expense of transport, and the vagaries of Romanian foreign-exchange policy. As appears from Figure 11, Romania, until 1931, although useful, was not an indispensable source of current supply:[2] its importance to Germany was that, within the limits imposed by the Romanian

[1] On the trade drive in general see Toynbee, "Germany's Economic *Drang nach Suedosten*", in *Survey of International Affairs*, 1936; A. G. B. Fisher, "German Trade Policy in South-east Europe" and "European Reactions to German Trade Policy", in ibid. 1938, vol. i; H. W. Arndt, *Economic Lessons of the Nineteen-Thirties*, pp. 176–206; Basch, pp. 165–226; Wilhelm Treue, "Das Dritte Reich und die Westmächte auf dem Balkan", *Vierteljahreshefte für Zeitgeschichte*, no. 1 (1953), pp. 45–64.

[2] Except for aviation gasoline, see memorandum by Reich Ministry of Economics, 11 May 1938: *DGFP*/D/V, no. 201; for German imports, see Figure 11.

National Bank, Romanian oil involved no hard-currency expenditure and that, provided that Romania had in other respects been drawn into the Germans' economic orbit, it would accrue to Germany in the event of war. Only during 1938 did the shift in the balance of power in Central Europe and the change in

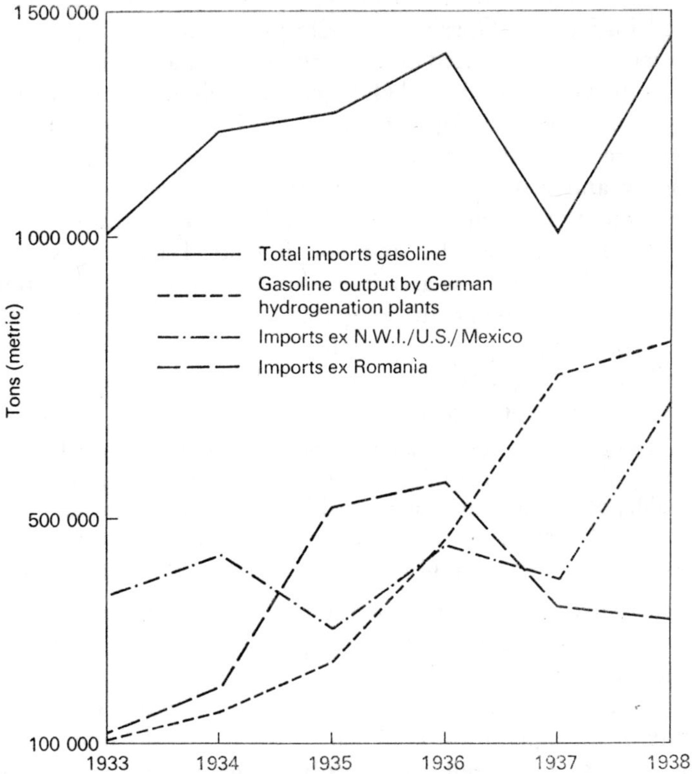

FIG. 11. Gasoline in Germany.

regime in Romania enable Germany to expand its influence over Romanian oil through its relationship with the Romanian Government and through investment in the industry.

The reports of Wohlthat[1] on the negotiations of December 1937 for deliveries during 1938 concern the traditional subject-matter, increases in quotas, exchange-rate parities, and a general "consolidation of exchange of goods and of clearing transactions".

[1] *DGFP*/D/V, nos. 154, 155.

One year later Clodius's reports[1] discussed the political changes facilitating the negotiations by which Germany's share of Romanian trade was increased at more favourable clearing rates. The changes enabled German policy to progress beyond quota and exchange matters to the systematic co-operation for the development of economic relations between the two countries, as discussed initially by Goering and King Carol in Leipzig. Detailed negotiations[2] were carried on in Bucharest by Fabricius and Wohlthat on the German side and Gafencu and Bujoiu on the Romanian, and resulted in the Agreement of 23 March 1939[3] which, in general, recognized the predominance of Germany in the Romanian economy and provided (Article 1) for an economic plan for the development of agriculture, forestry, mining, and communications with German assistance, and for industrial and banking co-operation. Some of these activities were to involve the formation of mixed German-Romanian companies, one of which (Article 1, para. 4) was to prospect for petroleum and carry out a programme of drilling and refining. The terms of this Article were amplified in a confidential protocol:

> As soon as data are available the projected German-Romanian company will draw up a minimum programme of drilling, and, where advisable, of refining, to be approved by the Royal Romanian Government. The Royal Romanian Government will take the necessary steps to facilitate the execution of a large scale petroleum programme.

These clauses met the German objective "to improve and secure deliveries of Romanian petroleum products . . . by participation in the oil industry".[4]

The Agreement has been variously interpreted.[5] The decision of one of its signatories, Gafencu, that "it was definitely favourable to the Reich but not disadvantageous to the Romanian economy",[6] is strictly valid. Germany needed a secure supply of

[1] *DGFP*/D/V, no. 264. Clodius's appearance in Bucharest was accompanied by a naval visit to Romania, a tour of Hitler Youth leaders, and a season by the Frankfurt Opera: see P. E. Mosley, "Hitler and Southeastern Europe", *The Yale Review* (Winter, 1939), p. 249.

[2] *DGFP*/D/V, nos. 293, 294, 297, 298, 306; VI, no. 131.

[3] *DGFP*/D/V, no. 78.

[4] Wohlthat's original instructions as reported back to Goering, 27 Mar. 1939: *DGFP*/D/V, no. 306.

[5] See, e.g., Basch, pp. 210–13; Elizabeth Wiskemann, *Undeclared War*, p. 96.

[6] See Gafencu, *Prelude to the Russian Campaign*, Chap. XI, p. 230.

raw materials, while the Romanian economy needed additional capital and was likely to benefit from German technique. The continuance in balance of such advantages depended, however, on political factors, principally Romania's ability to withstand German and Russian pressure.

As distinct from both France and Germany, Great Britain had more limited objectives in Romania—that is until 13 April 1939, when, in response to the German occupation of post-Munich Czechoslovakia, the British Government gave a unilateral guarantee to Romania.[1] This change in policy, however, did not involve the oil industry in the same way as did the revision of German-Romanian relations. Although the British investment in Romanian oil was much larger than that of Germany, the policy of the British Government in pre-war years conceded economic priority in Romania to Germany on the grounds of its being a "natural" market which in the event of a European war could easily fall under effective German control. Accordingly British strategic planning took account of oil supplied from Romania only in the event of Romanian neutrality in a war between the United Kingdom and Japan, when it was supposed oil supplies would have to be moved from Constanța to India and Ceylon to serve markets normally supplied from the Persian Gulf, whose own production would be needed for naval requirements.[2]

Exchanges between the British and Romanian Governments were confined to the attempt to solve difficulties arising from the settlement of commercial transactions and bonded debt. British-owned oil interests entered into the agreements concluded under these rubrics. Only in May 1939, in the Protocol regarding Commercial and Economic Relations,[3] was particular attention paid (Annex Ia) to "companies in which there is a substantial British financial interest" who were to be accorded "most favoured national treatment . . . in the matter of grants of permits for exploration and exploitation of new lands and in all matters concerned with the grant and operation of concessions and with the production of oil". The remaining paragraphs of the Annex, governing revision of the Mining Law and export controls (Ib) and of freights and

[1] See V. M. Toynbee, "The Apparent Threat to Roumania", and "The Anglo-French Guarantees to Greece and Roumania", in A. and V. M. Toynbee (ed.), *The Eve of War, 1939*.

[2] Information supplied by Mr. Peyton-Smith (Central Office of Information).

[3] 11 May 1939; Cmd. 6018.

export taxes (Ic) imply beneficial treatment as part of a new policy adopted by the Romanian Government towards the industry as a whole rather than towards the specifically British element in it. They also imply the State's acceptance of the case presented by the industry through the A.I.P.R.

A comparison between the terms of the "Wohlthat" Agreement and those of the U.K.-Romanian Agreement signed six weeks later illustrates the difference in the effective influence of the German and British Governments in Bucharest. The difference was in part attributable to the respective aims of the two states but also reflected the impasse resulting from a Romanian foreign policy based on the absolute retention of the frontiers of 1919–20 and the economic under-development, particularly in transport, resulting from rejection of foreign finance in the decade after the war. The most extensive published accounts of the diplomacy of the period 1938 to 1940 from the Romanian side are by surviving pro-Western authors such as Gafencu and Cretzianu: they tend to underestimate those forces in Romanian political and economic policy which, following from the maintenance of the nationalist position in both, made inevitable the decision in favour of Germany in the given situation.

In the short term, however, the internal and external measures taken by the Romanian Government appeared to offer some reasonable prospects for the renovation of the oil industry and its expansion. Requirements of capital and technical skill were to be supplied by Germany. The Romanian State was to promote new mining legislation and examine the lightening of fiscal burdens. By these means, it was intended, the decline in output could be reversed and Romania could begin to share in the revival of the international oil trade. All these were the moderately hopeful signs for the future. As at June 1939, however, in spite of increased drilling, output was continuing the decline of the past three years, and exports of all products (except lubricants) continued to diminish. Refining capacity totalled 11 million tons, but the industry was working to only 60 per cent capacity. Storage in fields, refineries, and export stations, both private and state-owned, totalled 3 million tons.

CHAPTER IX[1]

The Second World War

Romania and the Axis

IN September 1939 the Romanian Government declared its neutrality[2] towards both German-Polish and Russian-Polish hostilities. The rapid defeat and partition of Poland extended the Romanian frontier with the Soviet Union and brought for the first time a direct frontier with territory in German occupation. The new neighbours had already agreed on Bessarabia.[3]

During the same period it was made clear that the British and French guarantee of Romania, while absolute against Germany, was only conditionally operative against the Soviet Union.[4] With the Balkan Entente divided and Italy moving towards belligerence, this left Germany as Romania's only possible external guarantor against Russia, which, on 28 June 1940, occupied Bessarabia and

[1] General surveys of the political events in this chapter, and a study of their relation to the wider context of international politics, are to be found in the articles by Elizabeth Wiskemann in *The Initial Triumph of the Axis*, pp. 318–36, and *Hitler's Europe*, pp. 604–19 (both volumes in the series *Survey of International Affairs 1939–1946*, ed. A. and V. M. Toynbee); and in J. A. Lukacs, *The Great Powers and Eastern Europe.*

[2] For attendant circumstances see Alexandre Cretzianu, *The Lost Opportunity*, pp. 28–32 (the author was Secretary-General, Romanian Ministry of Foreign Affairs, 1939–41, afterwards minister to Turkey, 1943–5).

[3] See Secret Protocol to the German-Russian Non-Aggression Treaty of 23 Aug. 1939; *Nazi-Soviet Relations 1939–1941*, U.S. Department of State, 1948, p. 78, and *DGFP*/D/X, no. 10; A. Cretzianu, *The Lost Opportunity*, pp. 40–6. Subsequent exchanges made it clear that German disinterest did not extend to the Bukovina or to the economy of Romania, which, as in the First World War, represented for Germany principally resources of grain and oil. On 24 June 1940 Weizsaecker (State Secretary, German Ministry of Foreign Affairs, 1938–43) in a report to Ribbentrop identified the German interest as regards oil during the Bessarabian crisis as not permitting (i) a Russian crossing of the River Prut or (ii), in the case of hostilities, Russian air attacks on the oil-fields: see *DGFP*/D/X, no. 8. See also memorandum of Hitler's conversation with Riccardi (Italian Foreign Trade Minister): *DGFP*/D/XI, no. 177.

[4] The British position was further compromised by an outstandingly maladroit attempt, by the British navy to blow up the Iron Gates, at the beginning of Apr. 1940.

northern Bukovina.[1] This was the first adjustment to the frontiers of "Greater Romania". The solution of long-standing disputes with Hungary and Bulgaria was imposed by German arbitration in August and September 1940, in consequence of which Transylvania was partitioned and the Dobrogea frontier existing before the Balkan War of 1912 re-established. The new frontiers of Romania were guaranteed by Germany.[2]

These cessions of territory created an internal political crisis which King Carol attempted to solve by giving General Antonescu[3] virtually complete executive authority: this the General used to force the King's abdication on 6 September 1940. The monarchy was formally preserved in the person of Michael, who became King for the second time, while Antonescu as *Conducător* set up a government with Iron Guard participation.[4] Romania became a

[1] On the transfer of Bessarabia to the Soviet Union, see *MPR*, no. 13 (1940), p. 761 (text of Romanian Government's communiqué); no. 14 (1940), p. 786 (Tătărescu's statement); *DGFP*/D/X, nos. 4, 8, 10, 13, 19, 20, 25, 27, 28, 29, 31, 32, 33, 34, 36, 44, 46, 49, 51, 56, 67; and *Soviet Documents on Foreign Policy 1937–1941*, vol. iii, pp. 43 (Molotov's speech to Supreme Soviet, 29 Mar. 1940), 458 (Tass statement on Russian-Romanian diplomatic notes), 461 (Molotov's speech to Supreme Soviet, 1 Aug. 1940); also Gafencu's account of the events and his subsequent mission to Moscow in *Prelude to the Russian Campaign*, Chap. XIV, and Cretzianu's in *The Lost Opportunity*, Chap. III, and *Journal of Central European Affairs*, (Jan. 1950), pp. 396–403. G. L. Weinberg, *Germany and the Soviet Union 1939–1941*, pp. 99–104 and 128, discusses both the events and their role in German-Russian relations: see also A. Hillgruber, *Hitler, König Carol und Marschall Antonescu*, pp. 70–9.

[2] On these frontier adjustments, see general surveys already cited and relevant documents in *DGFP*/D/X: also C. A. Macartney, *October Fifteenth*, pp. 354–5, 386–92, 404–31.

[3] At that time imprisoned on account of too close an association with the Iron Guard. Antonescu's career had included the posts of Military Attaché in London, Chief of General Staff, and War Minister in Goga's National Christian Government. His political beliefs can best be described in terms of the "fanatical nationalism" which impressed Hitler. Antonescu was pro-French by training and had no ties with Germany, but his hatred of Russia and his appreciation of the balance of military forces in Europe led him to support Germany, even before the war, though the final pro-German policy had already been adopted before he took office: see G. Barbul, *Memorial Antonescu: Le IIIe Homme de l'Axe*; Prost, Chap. XIII. On Antonescu's relationships with the Germans, see Paul Schmidt, *Statist auf diplomatischer Bühne*, pp. 511–12, 538, 548–9, 574; and the revealing language of Antonescu's proclamations on Romania's adherence to the Tripartite Pact (*MPR*, no. 23 (1940), p. 1257) and the anniversaries of his rule (*MPR*, no. 18 (1941), p. 767; no. 18 (1942), p. 559).

[4] Under an amnesty in June 1940 members of the Iron Guard were released from prison and allowed to participate in political life. They promptly demonstrated for a German alliance. Their leader Horia Sima returned from Germany in July and entered the Gigurtu Government.

"National Legionary State" with only one party, devoted to an enthusiastically pro-German foreign policy and, at home, to the cleansing of the country of those anti-national elements which, in the Legion's view, had brought about its downfall.[1] This programme necessarily included a pogrom against Jews[2] and the massacre of political opponents.[3] Its execution began to get out of hand and, in view of the paramount need for order, Antonescu began to disarm the Guard, which retaliated by staging a revolt which lasted from 21 to 24 January 1941. The Germans offered no assistance:[4] the revolt was therefore crushed by the army and the Legion destroyed.[5] Romania thereupon reverted to straight military dictatorship,[6] without ideological trappings, and the structure of political authority, ratified by plebiscites,[7] remained unmodified for three and a half years.

Antonescu continued the *politică de integrare sinceră* adopted by King Carol's last governments and moved into even closer dependence on Germany,[8] joining in the attack on the Soviet Union in June 1941 for the recovery of Bessarabia and northern Bukovina. Successful operations permitted Romania to annex an area of the Ukraine which on 18 October 1941 became "Transnistria".[9]

[1] 128.

[2] A pogrom was also a method of creating vacancies for employment of the Iron Guard or refugees from frontier transfers. See Hillgruber, p. 238.

[3] Including Madgearu and Professor Iorga.

[4] As early as Oct. 1940 Neubacher reported from Bucharest that the economic and political stability of the country was being threatened by the excessive revolutionary zeal of the Iron Guard, who wished to introduce a whole series of new economic measures, which would probably cause serious dislocation within the country if implemented. Some leading members of the Iron Guard were even attempting to force General Antonescu to resign because he was unwilling to try these experiments. Neubacher maintained that the Iron Guard possessed no capable economists and that, therefore, it would be in the interests of neither Germany nor Romania if they managed to force through their programme. On Neubacher, see below, p. 252.

[5] Sima escaped to Germany: on his subsequent career, see Hillgruber, pp. 226–8.

[6] Antonescu's Cabinet of thirteen members contained eight generals and one colonel; only the portfolios of Foreign Affairs, Justice, Labour, and Propaganda were left to civilians. The Minister of Foreign Affairs, Michael Antonescu, was not related to the *Conducător* and, by reason of the latter's preoccupation with strictly military matters, came to exercise great influence throughout other departments of state.

[7] See *MPR*, no. 5 (1941), p. 231; no. 6 (1941), p. 277; no. 22/23 (1941), pp. 895–7.

[8] For Antonescu's ideas on foreign policy, see his conversations with Hitler; *DGFP*/D/XII, nos. 387, 416, 614.

[9] "Transnistria" seems to have been an improvisation for strategic reasons:

Notwithstanding the alignment with Germany and the presence of German forces in the country, Romania remained a sovereign state and—largely through Antonescu's relationships with Hitler—imposed on the Germans the necessity to proceed in accordance with the forms of law. At the same time, the limited freedom of manœuvre thus conferred on Romania gave the Government an opportunity to achieve its ends, which were not necessarily the same as those of its German partners. These factors were of particular importance to the oil industry. As Germany's military position weakened, particularly in consequence of the failure to hold the Caucasus in 1942,[1] so Romania came to play a more important part in enabling the Third Reich to continue the war, and its relative independence from the Germans increased. The most important evidence of this is the attempts from 1943 onwards to arrange a separate peace. From June 1940 to August 1944 Romania was part of the Axis system but remained far from *Gleichschaltung*.

The creation of a German sector in the industry

It is clear that in the Second World War Romanian neutrality was not marked by the nice calculation of the balance of advantage as from August 1914 to August 1916, but was rather a matter of staying out of hostilities if possible, although with external commitments to Germany.[2] The decisive shift towards Germany took place in May 1940[3] in response to increasing Soviet pressure,

the area contained a number of Romanian settlements but they never featured in irredentist agitation (in contrast to, e.g., Romanian populations in Transylvania or the Bánát) and were discovered to be important only at the end of July 1941. See Alexander Dallin, *Odessa 1941–1944: a Case Study of Soviet Territory under Foreign Rule*; Review of Foreign Press, Series B, no. 122 (6 Feb. 1942); Hillgruber, pp. 134–43.

[1] See A. E. Gunther, "The German War for Crude Oil in Europe 1934–1945", *PT* (31 Jan. 1948), pp. 113–21.

[2] Clodius maintained that this line of policy was dictated "by the head and not by the heart. . . . The result is that it will be followed only as long as faith in the victory of German arms, or at least in the maintenance of German hegemony in Europe, endures." See *DGFP*/D/XI, no. 33. See also Clodius's report of discussion with Tătărescu, 30 Mar. 1940: *DGFP*/D/IX, no. 27; and Fabricius's reports: ibid., nos. 117, 252.

[3] The decision was formally taken at a Crown Council meeting on 27 May—the day German forces reached Calais. Belgium capitulated during the night of 27th–28th, and the Dunkirk evacuation began on the 29th. On the process

and at the time of the collapse of French power in the West. From the autumn of 1939 onwards the dominance of Germany was built up on the economic relationship established before the war and on Romania's continuing need for security, particularly as expressed in the tangible form of armaments.[1] Conversely, Romania offered to German military policy[2] a base from which to operate in the Balkans or against the Soviet Union and a major source of oil supplies until the resources of the Caucasus or Iraq became available. The negotiations on the supply of armaments extended from December 1939 to May 1940 and demonstrate Germany's ability to draw political advantage from economic strength. War material was not only a *quid pro quo* for oil supplies but was "the most important factor in our efforts to bring the Romanians more and more into our camp in a political sense also".[3] These efforts, reinforced by the policy of the Soviet Union over Bessarabia and the Danube delta, and extending over a wide range of manufactures needed by the Romanian economy, were ultimately successful. Even before the elimination of Anglo-French military power on the Continent, the representations of Fabricius and Clodius in Bucharest were more telling than those of Sir Reginald Hoare, since they turned on a Romanian requirement which the Germans could, and the Allies could not, fulfil.

Under such stimuli, the provisions of the Agreement of 23 March 1939 took practical shape during the "neutrality" period. In September 1940 a technical-co-operation agreement was concluded between the Hermann Goering Werke and the Malaxa Plant. In November a German War Economy Commission arrived in Bucharest for the purpose of supplying advice and expertise to aid the Romanian economy. Throughout 1941 joint Romanian-German companies were set up in radio manufacture and other industries of immediate use to the German war effort. The Deutsche Bank succeeded to the holdings of the Banque Union Parisienne

of change, see reports by Fabricius on audiences with the King, 16 May 1940: *DGFP*/D/IX, no. 252; and with Tătărescu, 28 May 1940; ibid., no. 345: see also Gafencu, *Prelude to the Russian Campaign*, pp. 282–7.

[1] See Fabricius's report on conversations with Gafencu, 8 Dec. 1939: *DGFP*/D/IX, no. 427.

[2] German bomber units arrived at Braşov in July 1940: closer staff co-operation began in September, and the first army and anti-aircraft troops appeared in October. See Hillgruber, pp. 96–104.

[3] Clodius/Fabricius to Foreign Office, Berlin, 29 Mar. 1940. *DGFP*/D/IX, no. 21.

and the Société Générale Belgique in the Banca Commercială Română, and the Dresdner Bank through its agency, the Societatea Bancară Română, took over the local affiliate of the Prague Kreditbank.[1]

Meanwhile, the inter-governmental basis of the activity, i.e. the Agreement of 23 March 1939, was extended into a plan ratified on 4 December 1940,[2] for reorganizing the entire economy of Romania under German auspices. This agreement promised the full financial and technical support, including long-term credits, which had been recognized after the First World War as a prerequisite of the economic reconstruction of Romania, but which had been frustrated by Romania's refusal, on nationalist grounds, to accept the obligations of international loan procedure, as then construed. In 1940 the rendering of such assistance involved a much closer *de facto* political dependence on an external power, although in wartime conditions Antonescu was able to limit the extent of collaboration. In discussion with Goering in March 1941 the *Conducător* defined three principles governing the country's policy:

(i) that Romania was ready to exploit her mineral resources jointly with Germany, but that there must be no alienation of property in the process;

(ii) that the direction of the enterprises and the exploitation of mineral resources must lie in Romanian hands; and

(iii) that Romania herself would define the extent of foreign penetration of her economy.[3]

The communiqué issued after Antonescu's return from Vienna made it quite clear that German penetration was to take place either at the expense of foreign or Jewish capital or in the form of new capital, but in either case only with the knowledge and consent of the Romanian Government and only to the extent which permitted the State to control and direct its basic economic interests.[4]

[1] See *Dresdner and Deutsche Banks*; a special report of the Military Governor U.S. Zone, June 1947, p. 20; section on Romania in *Conditions in Occupied Territories: The Penetration of German Capital into Europe*, H.M.S.O., 1942 and Hillgruber, p. 156.

[2] See protocol: *MPR* no. 24 (1940), p. 1201; and comment by Cancicov Minister of National Economy: *MPR*, no. 1 (1941), pp. 19–21.

[3] *DGFP*/D/XII, no. 126.

[4] See *MPR*, no. 6 (1941), p. 281.

In practice Romania was able to ensure that these principles were respected, but, nevertheless, through the series of economic arrangements from March 1939 onwards and the political relations established in the autumn of 1940, Germany had achieved direct control over a large segment of Romanian economic life and an indirect control—by reason of a virtual monopoly of purchase and supply—over the economy of the country in general.

Germany was thus well situated to overcome its initial disadvantage in the Romanian oil industry—in contrast to the situation in 1914—of there being under German ownership only four small companies with a total capital of 77 million lei (or 0·5 per cent of the total invested in the industry) and a crude-oil output of 51 million tons (or 0·86 per cent). German policy aimed at enlarging this base through the acquisition of further assets and by bringing pressure to bear on the Romanian Government. The former yielded more satisfactory results than the latter, since the conquest of particular countries—Czechoslovakia in 1939, the Netherlands, Belgium, and France in 1940—gave Germany a command of firms with Romanian interests, a process which could not be blocked by Romanian Government action.[1]

The first German acquisition was the Petrol Block, which had an interest in the I.R.D.P., and among its industrial assets a small field in Băicoi and two refineries with a total capacity of 540,000 tons per annum. The throughput of crude oil had fluctuated in the immediate pre-war years, but the refinery did contain a lubricating-oil plant and had been extensively modernized in 1937 by the erection of a Dubbs cracking unit supplied by a Czech firm which, after the occupation of Bohemia and Moravia in April 1939, came under German control. Petrol Block found itself, however, unable to meet its outstanding obligations to the firm and was reorganized financially in such a way that the debt was paid off in exchange for German capital.[2]

[1] On German policy regarding industrial assets, see memoranda by von Maltzan, 25 May 1940, and by Clodius, 30 May 1940: *DGFP*/D/IX, nos. 313 and 354; Goering's directive of 2 Aug. 1940, in *Nazi Conspiracy and Aggression*, vol. vii, pp. 310–11; report of discussion in Reich Ministry of Economics, 16 Aug. 1940, on acquisition of shares of important foreign enterprises in south-east Europe, ibid., pp. 258–9; and Fisher's note on objectives and procedures, ibid., pp. 255–8.

[2] See report of General Manager 14 Dec. 1939: *MPR*, no. 24, p. 1560. After the company was put under German management, 320,741 tons of crude oil were treated and 86,550 tons of pacura cracked to provide high-octane gasoline,

The Germans attempted, similarly, to fulfil the policy begun in 1935 of establishing a German holding in Creditul Minier, using as a means debts outstanding to a Czech firm, itself taken over by a German bank. This attempt failed. Creditul Minier was the Romanian "national" company *par excellence* and, unlike I.R.D.P., successful, and the Romanian Government was determined that the expansion of German influence to which it had consented under the Agreement of March 1939 should not take place at the expense of "national" capital.[1] This consideration dictated also the acquisition of shares in Petrolul Românesc by Creditul Minier. Nevertheless, the latter's association with Germany was secured by an agreement made in September 1939 under which Elwerath would co-operate in the exploitation of certain Creditul Minier properties in Moreni, and S.O.C.O.P. was to be turned into a German-Romanian company.

The German conquest of Western Europe in the summer of 1940 yielded control of other Romanian companies through control of their parent concerns—Colombia, Concordia, and Astra Română. The first two named were taken over through the work of the Armistice Commission and the negotiation of the sale of shares with shareholders.[2] Astra Română came under German control through the imposition of *de facto* German control on the majority shareholders, BPM, The Hague, which then altered the management of Astra Română accordingly.[3]

The Romanian Government legally recognized these changes in property rights and interests in the Law of 10 June 1941, promulgated at German request, as follows:

> The legal dispositions and measures taken by the authorities who effectively exercise the attributes of public control in countries now at

in 1940. Petrol Block sold in the home market 18,415 tons of products and exported 342,734 tons, virtually all of which went to Germany and Italy.

[1] See 129. Gigurtu suggested to Hitler on 26 July 1940 that Germany "possibly wished to assume the succession of the *British* and *French* shareholders": *DGFP*/D/X, no. 234 (author's italics).

[2] On details of share transfers, see correspondence on AA file *Petroleum und andere Mineralöle* during period May 1940–Mar. 1942, particularly between 22 May and 13 Nov. 1940; and a report of German Armistice Commission to Foreign Office, Berlin, 3 Mar. 1942, on the acquisition of certain Concordia shares and on results achieved in carrying out the policy.

[3] German and Romanian policies clashed over Astra Română, as the leading oil undertaking. Examination of the procedures adopted demonstrates (i) the 'correctness' imposed on the Germans as allies of the Romanians and (ii) the latter's capacity for administrative delay: see 130.

war [and] concerning commercial companies who have their seat in these states on September 1st 1939, will also be applied to those companies throughout the territory of the Roumanian State.[1]

The Germans were thus enabled to appear in Romania as owners of oil undertakings with a total of 47 per cent of the country's crude output. They were and remained, however, foreign nationals, even though the Romanian Government in varying degrees welcomed or tolerated their presence or simply yielded to *force majeure*.

The co-ordination of these German interests was undertaken by the Kontinentale Oel A.G., a company founded, according to Goering in November 1940, "to take over companies belonging to enemy and neutral powers in the countries occupied by Germany".[2] Kontinentale Oel was legally constituted on the 27 March 1941 with a capital of RM. 80,000,000 granted by the German State, leading banks, and German oil enterprises. The voting power of the preference shares, which were exclusively in the hands of the German Government, ensured compliance with that government's wishes.[3] In his letter to Admiral Raeder, Goering was in fact describing the technique by which the Kontinentale Oel built up its position: the published aims of the concern were rather more extensive, namely the centralization of control over the production, refining, and transport of German oil requirements. At the outset the pursuit of this aim was thought to be compatible with observance of the acquired rights of other German companies. Political developments did not force this contradiction into the open, though a trace of what might be expected to have happened can be seen in Kontinentale's taking over Foraky Românească and the Elwerath interests in Romania.

[1] *Monitor Oficial* (11 June 1941); the Law was occasioned by the need of the German-controlled Royal Dutch management to appear in Romania with the legal right to direct Astra Română, without production of the share certificates (which they were otherwise required to do). The Law, however, applied to *all* transfers. 131.

[2] Goering to Raeder, 14 Nov. 1940: AA file *Petroleum und andere Mineralöle*, Nov. 1940–May 1941.

[3] For diagrams showing (i) how the interests were represented at the level of the board of directors and management, and (ii) the place of Kontinentale Oel in the general organization of German oil production and distribution, see Charts E and A respectively on p. 14 and p. 10 in *Notes on the Organisation of the German Petroleum Industry during the War*, BIOS Final Report, no. 513 (1945), item 30. See also discussion by A. E. Gunther in Chap. III of "The German War for Crude Oil in Europe", *PT* (6 Dec. 1947).

Otherwise, Kontinentale was the beneficiary of Germany's acquisition of foreign assets as described above. The company's interests in Romania were conducted through the Bucharest branch of the Kontinentale Oel GmbH, Berlin, the operating company of Kontinentale Oel A.G. founded in August 1941,[1] though the foreign shareholdings and conclusion of major agreements remained with the parent concern.

(*a*) *The growth in German control*

While the Romanian Government admitted the creation of a German enclave in the oil industry as part of the price of external support, it tried to limit the extent of such penetration and did not abate its own efforts to bring the industry under more comprehensive Romanian authority. The situation after September 1939 provided the occasion for extensive supervision of Romania's economic life on grounds of defence and security. The Government took powers of requisitioning and made regulations for increasing national output and controlling prices. The latter particularly referred to the oil industry, which was required to submit a detailed annual working programme to the Ministry of National Economy.[2]

Two days after the Crown Council's decision on definite alignment with Germany, a sweeping "Decree Law for Exceptional Powers for Mobilizing the Economy in Peace Time" was promulgated.[3] It ascribed to the Ministry of National Economy (Article 1) "full powers to take measures necessary concerning the organization, co-ordination, and control of production, distribution, and consumption of materials necessary to the army and to the civil population as regards the internal economy as well as exchanges with abroad": the other articles set out its powers in relation to those of other departments of state, including the General Staff, and the validation of its decisions. Three months later the Antonescu regime introduced in a "new" economic policy a series of measures designed to make government control effective over individual enterprises. The basic instrument was the "Decree

[1] See under "Business Reports" in *Oel und Kohle*, no. 31 (15 Aug. 1941), and under "Trade Register", ibid., no. 32 (22 Aug. 1941).

[2] *MPR*, no. 8 (1940), p. 490.

[3] *Monitor Oficial*, no. 125 (31 May 1940), reprinted in *MPR*, no. 12 (1940), p. 700: see also *exposé* by Mircea Cancicov, Minister of National Economy: ibid., pp. 699–700; and 132.

Law for the Supervision and Protection of the National Economy",[1] whereby the Ministry of National Economy was empowered to appoint, without giving any reason, a special Commissar to any enterprise for the purpose of integrating its activity with general state policy whenever the Ministry considered such action necessary (Article 1). The Commissar,[2] who had to be ethnically Romanian, was empowered to call on any authority, or state institution or the executive power to compel the execution of the ordinances or decisions of the Ministry (Article 4), failure to comply with which carried imprisonment for individuals (Article 5) and sequestration of assets for institutions (Article 6). Articles 8 to 16 introduced the notion of economic sabotage—involving a wide range of acts of omission or commission considered to be "damaging to the interest of the national economy or the general interests of the Romanian community". Separate regulations on commercial and industrial and financial sabotage were issued on 1 November 1940,[3] being modified in February 1941[4] and August 1941.[5]

These laws, which were generally applied to all economic activities, strengthened the power of the State over the oil industry. There were in addition special measures which were designed to control in detail the various phases of its operations, but which were rooted in the general pre-war nationalist policy for the industry. In this way the creation on 16 January 1940 of the *Commisariatul General al Petrolului* (General Oil Commissariat)[6] derived from the same considerations as the setting up of the Ministry of National Economy and the A.C.E.X. (which at this

[1] *Monitor Oficial*, no. 214 (14 Sept. 1940); see *MPR*, no. 18 (1940), p. 1034. See also summary of broadcast by Prof. Leon (Minister of National Economy in the Antonescu Government) on the changes in economic policy, in *MPR*, no. 17 (1940), pp. 977–8 and 133: it is clear that Leon was not expounding a newly embraced totalitarian political ideology but was merely expressing his long-standing nationalist political convictions.

[2] Commissars, all drawn from the Iron Guard, were in fact only appointed to "Allied" companies (including Româno-Americană): see *MPR*, no. 19 (1940), p. 1073; and 134.

[3] See *MPR*, no. 22 (1940), p. 1190.
[4] See *MPR*, no. 4 (1941), p. 162.
[5] See *MPR*, no. 15 (1941), p. 687.
This anti-sabotage legislation became one of the essential elements in the post-war "Communization" of Romania: see below, p. 284.

[6] See *MPR*, no. 3 (1940), pp. 139–40; and note of Angelescu's (M.N.E.) discussion with A.I.P.R.: 135. On activities of C.G.P. see 136 and, in regard to tankage, 137: see also below, p. 259.

juncture became an operational as distinct from a merely planning organization). The task of the C.G.P. was to "supervise, co-ordinate, and direct the production, refining, sale, and export of oil in accordance with the interests of the country": in practice it concerned itself with the level of crude-oil output and refining, the improvement of refinery methods, and the maintenance of refinery out-turns at a sufficient level to ensure supply to domestic markets and the army. The Commissariat also functioned as agent for the General Staff in the imposition of its requirements on the industry. In September 1940 the Commissariat was wound up and these functions devolved on the reorganized Ministry of National Economy. During the period of the C.G.P., exports similarly passed under direct state control by the creation of a Ministry of Foreign Commerce,[1] which, like the Commissariat, disappeared at the outset of the Antonescu regime.

The Romanian authorities complemented this direct supervision of the activity of the industry by imposing a control on the movement of capital through the registration of shares by their owners either at the Ministry of National Economy or at the offices of diplomatic representatives abroad.[2] The same end was served by making the transfer of the *siège sociale centrale* of a company conditional on government approval.[3]

Control through civil authorities was supplemented by that exercised through the military administration, which tended to pursue its own independent policies towards industries considered of strategic importance. Beginning in December 1939 with company responsibility to the military for air-raid and anti-sabotage precautions,[4] the network of ordinances governing the oil industry became detailed and comprehensive, so that, for example, in March 1940 Professor Netta, President of the C.G.P., promised the companies official intervention "against the instructions that permission for every construction must first be asked with the Military Command in Braşov".[5]

From May 1940 military authority over the industry was strengthened by restrictions on movement in the oil zones

[1] See *MPR*, no. 6 (1940), pp. 337–47: also *exposé* by Christu (Minister of Foreign Commerce), ibid., p. 359; and 138.

[2] 11 July 1940: see *MPR*, no. 14 (1940), p. 833; and no. 15 (1940), p. 886.

[3] 27 Aug. 1940: see *MPR*, no. 17 (1940), p. 981; and 138.

[4] See *MPR*, no. 2 (1940), p. 103.

[5] 139.

(including ports)[1] and compulsory building of depots and maintenance of stocks at the industry's expense under the direction of the C.G.P.[2] Further, at this time the sale or transfer of refining assets without permission of the Commissariat was forbidden.[3] In the following months the Ministry of National Economy required the registration of company records,[4] forbidding their removal without permission—for which purpose an Office of Documentation and Co-ordination was set up in the Ministry, with combined civil and military staff.[5] Guards were installed in offices and plants, and special teams of officers made inventories of documents, removing those they required for further study to the Ministry of National Economy; documents left in company archives were placed in the charge of selected, ethnically Romanian, employees.[6] At this time military requisitioning of vehicles, assets, and personnel[7] began, and in December state and private enterprises themselves were "militarized"—a term defined according to the relevant law (Article 2) as "the institution of a regime of control, discipline, and military jurisdiction in state and private institutions and enterprises, including the entire personnel, which will be considered requisitioned".[8] Under the Antonescu regime there was no longer any practical distinction between military and civil authority, and the apparatus of state control was complete and made effective.

The day-to-day control which resulted from these arrangements was exercised ostensibly to preserve the neutrality of Romania in matters of oil and to ensure supplies to the armed forces. In practice, however, two important qualifications applied: firstly that Romania was "neutral" for Germany, and secondly that measures whether taken by civil or military authorities were used to promote nationalist ends. In consequence, the effect of the

[1] *MPR*, no. 11 (1940), p. 655. After the exposure of British sabotage intentions (see above, p. 223 n. 4) the fields were put under guard and access governed by a pass system. The few foreigners left in the area were kept under surveillance by the Siguranţa, till the general expulsion of "Allied" foreign nationals in July. Some were in the meantime picked up and interrogated by the Iron Guard.

[2] 140.

[3] *MPR*, no. 10 (1940), p. 592.

[4] *MPR*, no. 12 (1940), p. 696.

[5] *MPR*, no. 13 (1940), p. 763.

[6] 141: on the use of this authority see 142.

[7] *MPR*, no. 16 (1940), 926.

[8] *MPR*, no. 24 (1940), p. 1295; no. 5 (1941), pp. 235–6. The repair shops of Steaua Română, Petrolifera Română, and Concordia, and the Distribuţia network were all militarized: see *MPR*, no. 4 (1941), p. 191.

Romanian Government's policies was felt by Allied companies.[1] Controllers were successively appointed to Astra Română[2] in July 1940 and to Concordia[3] and Unirea[4] in August. The senior foreign management and technical personnel of these companies were expelled. The appointments were based on the powers conferred on the Government by the Decree Law establishing the C.G.P. (Article 8), and in each instance the ground alleged was non-compliance with regulations concerning company records.[5]

Control through Commissar was in practice not necessarily or wholly detrimental to the interests of the company concerned;[6] Engineer Damaschin in Astra Română used his position to settle current transport and production problems but as Government and General Staff agent in the company—whose approval was necessary to each administrative act or decision—he restricted the company's freedom of action. Further, he received applications

[1] i.e. excluding Româno-Americană, since the Romanian Government did not wish to antagonize the United States. Later the company received a Special Commissar under the Law for the Supervision and Protection of the National Economy. In Oct. 1941 the dilemma of its situation was resolved when it was handed over to the German affiliate, D.A.P.G., which acted as trustee: see 143.

[2] Astra had in the past consistently refused to make records available, but had managed to avoid a direct clash with the authorities on this issue: on the background and circumstances of the appointment see 144.

[3] Alexander Miller of Astra Română commented, "a Commissar has also been appointed to Concordia. The reason given in the 'Official Gazette' for this punishment is that drilling reports were removed from the office of Concordia's Technical Management on 29th June, 1940 and that a case containing all the original concession contracts, validation decisions, etc. had been removed from Concordia's office, in both cases without having been communicated to the mining authorities. The Concordia Commissar is well known as one of the 'club of unemployed engineers' who are so keen to get into the service of the bigger oil companies. . . ." Miller later explained that "the alleged removal of documents had taken place . . . a zealous employee of the company had removed certain documents to safety on the evening when general mobilisation was declared in connection with the events in Bessarabia. Not only is this stretching the law rather wide to treat this as an offence, but moreover the law of 16th January only permits the appointment of a Commissar as a punishment for non-observance of orders given by the General Commissar. It seems obvious therefore that the Government are ready to snatch at the slightest excuse to get their people into the foreign companies." 145.

[4] The appointment of a Commissar to Unirea was preceded by no previous warning: the company received a letter stating it was being penalized on the grounds of intentional deviation of wells without advising the authorities or placing the relevant documents at the disposal of the military. The Commissar arrived an hour later.

[5] See 146.

[6] Oil Commissariat Decision No. 2692/1940: 147.

for jobs and denunciations of the management by dissatisfied shareholders and employees.[1]

(*b*) *The rivalry between German and Romanian authority*

Alliance for reasons of foreign policy as described above and direction of the oil industry's output and trade towards Germany did not rule out a conflict of interest between the Romanian and German Governments over the oil resources and operation of companies within Romania itself. This conflict expressed itself in the Romanian bureaucracy, which remained the repository of economic nationalism and which now employed on the Germans its long and elaborate technique of administrative frustration. Astra Română provides a case in point. As the concern representing between 25 per cent and 30 per cent of the total activity of the industry, with acquired rights on concessions and with three-quarters of its shares in foreign hands, Astra Română was the main object of Romanian nationalist attention: additionally, from the Romanian official standpoint, it was after September 1939, though formally "correct" in its attitudes, less than wholly helpful towards the Government's foreign economic policy. Hence the Commissars, denunciations, inspections of records, and so on. From the German standpoint Astra Română in virtue of its producing interests represented the most important European affiliate of the Royal Dutch–Shell Group of companies in their control. Hence the attempts to secure legal recognition of German authority and alterations in the management of the company, and the removal of Romanian Government "watchdogs".[2]

The pattern was for laws to be decreed or regulations promulgated that were applicable to all foreign nationals by definition,

[1] The precise degree of Commissar interference was obviously a matter of individual temperament and background. Damaschin's position in law was based only on the alleged disobedience of the C.G.P., and, when, in consequence of the change of economic policy under Antonescu, the Commissariat was abolished, he was replaced on 23 Sept. 1940 by Calleya, formerly of Româno-Americană and a leading member of the Iron Guard. Calleya's activities were much more drastic, particularly with respect to the expulsion of Jewish employees, and encompassed changes in the management through the promotion of political extremists. His appointment coincided with the arrival in Romania of a new German and Dutch-Nazi management, under Rost van Tonningen, resulting from the German occupation of the Netherlands, and did not long survive the assertion of German rights in this respect. See 148. Damaschin was reluctant to leave: on the change-over, see 149.

[2] 150.

from which the German interests in Romania then had to claim exemption through their special relationship with the Government. They often ultimately prevailed—though success became noticeably more difficult after El Alamein and Stalingrad. Nevertheless, whatever the outcome on particular issues of day-to-day administration, on the fundamental matter of mining-law revision the Germans' failure to impose their own concepts—at a time when the military situation still favoured the Wehrmacht—was complete. Reform of the Mining Law was of concern to the oil industry, irrespective of the control under which it functioned, and to the German authorities, both to facilitate an increase in production and to lay the long-term foundations for German-Romanian co-operation.

(c) *The Petroleum Law of 1942*

At the outbreak of war the agreed industry proposals[1] for mining-law revision were still under consideration at the Ministry of National Economy. In December 1939 the Ministry announced its disagreement on certain aspects of these proposals, particularly on leasing rights, working obligations, and rewards to successful explorers. The official alterations were submitted to a new commission, on which the A.I.P.R. was represented by its chairman, Zanescu. At this stage the problem of revising the fundamental legislation governing the oil industry in all its phases became involved in the reconstruction of the Romanian economy as part of German-Romanian collaboration. On 8 March 1941, at their Vienna meeting, Goering complained to Antonescu that the mining laws were an obstacle to increased production.[2] More detailed discussions were held in Bucharest a week later between Neumann (State Secretary), Neubacher, and Michael Antonescu, in which a "fundamental reordering of the regime for oil extraction" was agreed.[3] The style of this new ordering of the industry's legal basis became clearer on 11 April 1941, when the Council of Ministers, discussing petroleum policy, announced that the Mining Law would be amended to allow exploration and exploitation of

[1] See above, p. 210.

[2] See *DGFP*/D/XII, no. 126. The point was also put by Otto Gramsch (Kontinentale Oel GmbH) in an article in *MPR* of 15 Sept. 1941: see *MPR*, no. 18 (1941), pp. 771–3.

[3] See file note of discussion, 15 Mar. 1941: AA file *Petroleum und andere Mineralöle*, Nov. 1940–May 1941.

entire geological structures (instead of surface areas under concession)[1] and that the commission working on the new law would take into account "not only the interests of the companies but also the great interests of the State and the national economy". The Government thus demonstrated its determination to maintain political control over the terms on which Romania's oil resources were exploited and to ensure a preferential position for Romanian nationals.

The working out of these implications, in spite of German comment on their urgency, took yet another year. The Germans were aware of the possibility that the Romanian Government's ideas of reform might not tally with their own, and in consequence the officials sent from Berlin to advise on the problem were also charged with watching over Germany's interests.[2]

The Law was eventually promulgated on 17 July 1942. In his *exposé des motifs*,[3] Ion Marinescu, Minister of National Economy, recapitulated the nationalist view of the history of Romanian oil legislation and its effects on exploration and production, concluding that, by basing itself on purely fiscal considerations, government policy had not correctly assessed the problems of finding new reserves or enabling maximum use to be made of existing oil-bearing areas. In order to promote "a vast exploration programme, the State was now willing to encourage and reward the explorers". Marinescu's analysis of the principles on which the new policy was based demonstrated that encouragement was to be both technical and financial. The basis of exploration and exploitation was to be the geological structure (defined in the Law, Article 2, as "any simple tectonic unit . . . in which hydrocarbons can accumulate"): it would obtain both in known areas

[1] See *MPR*, no. 8 (1941), pp. 347–8. This notion had already been canvassed in January by Ing. Gh. Georgescu in an article "The Roumanian Oil Industry in the Framework of the Economic Policy of Roumania": summarized in *MPR*, no. 1 (1941), p. 42.

[2] See letter from Otto Gramsch (Office of Four Year Plan) to Clodius, 17 Apr. 1941, concerning Klewitz and Haymann, whose "presence in Bucharest" to speed up preparations for the new law "is particularly supposed to prevent the Romanian Government from making of its own accord changes in the law which might harm German interests": AA file *Petroleum und andere Mineralöle.* Dr. Fischer (Office of Four Year Plan) in a note to the Reich Ministry of Economics on 3 Dec. 1941 identified Stoicescu of the National Bank and Creditul Minier as the main obstacle to progress. Ion Marinescu, the minister responsible, was also on the board of Creditul Minier.

[3] See *MPR*, no. 15 (1964), pp. 463–80.

and in the Mining Districts of 10,000–20,000 hectares which the State would create in order to facilitate rational prospecting and build up long-term oil reserves.

Similarly, the organizational basis for both these phases of activity was to be no longer the individual company but "associations established between the companies which have adequate technical and financial means and which possess rights or territories on these structures". The inducement to explore would take the form of exploitation rights, royalty and tax reductions, and compensation for losses in the exploration of Mining Districts by the grant of new structures out of the long-term reserves. Marinescu summed up the fundamental idea behind the new law as "that the Romanian State will direct the oil economy in everything that concerns the grant of permits for prospecting, exploration and exploitation, the control of activity, and the treatment and commercialization of products".

The Law itself[1] was the first concerned specifically with oil and natural gas[2] as distinct from other mineral resources, for which the 1937 Law remained valid. It comprised 293 Articles and set out a comprehensive regime for exploration and exploitation on the principles mentioned above. The right to prospect was permitted to physical and corporate legal persons only on the basis of an annual permit issued by the Ministry of Mines and Petroleum: prospectors who proved the existence of a possible oil- or gas-bearing structure, within the period laid down and after reference to the Romanian Geological Institute, had a right to exploit the structure, i.e. "any simple tectonic unit", concerned. Exploration was to be allowed only to companies of recognized technical and financial adequacy, and was to be carried out through an obligatory working programme in three kinds of area: within Mining Districts, on "isolated structures", and on "extensive structures".

The first named were relatively large areas where prospecting was incomplete. Rights were to be given to applicant companies—

[1] German text in *MPR*, no. 15 (1942), pp. 481–98; no. 16 (1942), pp. 517–34: this analysis covers only the political principles on which it was based, and the technical provisions entailed by them.

[2] Gas deposits were defined as (i) those containing more than 95 per cent CH^4; (ii) the deposits with a gas–oil ratio higher than 3,000 $m^3:m^3$. Gas deposits in crude-oil strata or found under oil deposits came under the same exploitation regime as oil. This provision overcame one of the long-standing difficulties facing the oil companies: see 151.

grouped into an association if there were more than one—who were obliged in return to prospect the whole district, beginning exploration work after a maximum of eighteen months on at least two structures, and thereafter operating so that all identified structures would be actually explored. All exploration obligations would be set out in detail for each applicant in a contract to be concluded with the State as a precondition of the granting of exploration rights in a Mining District. Structures proved oil-bearing were to be formed into groups of three in the order in which they were discovered, and were to be granted for immediate exploitation or retained by the State as long-term reserves. "Isolated structures" were also to be explored by compulsory associations of prospecting companies and holders of private concessions with an obligatory minimum schedule of work. "Extension structures" were of two kinds—by area, i.e. extensions to existing oilfields, and by depth, i.e. those below the last known petroliferous horizon. Rights on both were to be granted preferably to companies with existing activity in either exploration or exploitation and in association with holders of private concessions. Compliance with an obligatory work programme was necessary but the rewards of success were slighter than in either "isolated structures" or Mining Districts, since the probabilities could in part be reasonably calculated: again only 50 per cent went to the explorer, the State controlling the disposal of the rest.

This basis for future oil exploration and exploitation had a twofold importance. Technically, it provided for a rapid programme of work in the hitherto unexplored areas outside the traditional districts and ensured that in these old areas the "acquired rights" of individuals would be worked according to state-imposed schedules. Further, the stipulations regarding partnerships implied association between Romanian and "foreign-capital" companies.

Politically the arrangements provided the means of prejudicing the future of "foreign-capital" companies and thus inducing them to convert themselves into Romanian concerns, defined by Article 269 as "those in which 75 per cent of the capital was held by Romanians by ethnic descent and in which the Chairman of the Board of Directors, Committee of Management, and Auditors were also Romanians in the same proportion". Such companies were to enjoy decisive advantages for successful exploration and exploitation. Article 23 accorded to them "the right to participate

25 per cent in the exploration of any isolated and unexplored structure granted by the State over and above the rights due to them for territories held in the same structure" without sharing the costs of prospecting. This meant that Romanian participants in an association would have an immediate advantage over their non-Romanian partners. Further advantages accrued from the distribution of structures discovered within Mining Districts,[1] and on "extension structures", where Romanian companies were to receive (Article 96 (3)) the right to 25 per cent of the state land proved oil-bearing, without prejudicing their right to bid for the 25 per cent of the total which was put up to public auction. They also enjoyed certain financial privileges.[2] In this way the complete ascendancy of Romanian capital was to be assured.

The Law (Section IX) regulated transport and refining activity in detail, ascribing to the State extensive powers over pipelines (Articles 201–3), compelling the refining in Romania of all crude oil (Article 204) and therefore prohibiting its export, with reservation only in favour of the State in time of war (Article 209), asserting priority to the satisfaction of the domestic market, which was to be organized by the State in association with refiners and producers (Article 210). Further, all this activity was underpinned by the creation of a National Petroleum Institute (Articles 215–31) with far-reaching powers over the organization of research and development.

Immediately after the publication of the Law, Marinescu[3] decreed the setting up of a committee under the chairmanship of Preda (Secretary-General, Ministry of National Economy) to work out the necessary regulations.[4] Applications were invited for

[1] From each group of three the first was granted to the explorer; one of the other two was divided 75 per cent/25 per cent between the explorer and Romanian companies (Article 96 (4) (*b*)); and the remaining structure was retained by the State as long-term reserves. When the State decided to exploit these reserves, however, 50 per cent were to accrue to companies suffering losses through exploration in other Mining Districts and 50 per cent to Romanian companies (Article 87 (4)).

[2] viz. a 2 per cent mining tax and a 12 per cent state royalty, instead of 3 per cent and 14 per cent respectively; reduction on taxes levied on sales of shares and the issue of new shares; exemption from dividend taxes and from imposts on mergers.

[3] In their 1942 annual report Creditul Minier appealed to shareholders to prove their ethnic origin to allow the company to benefit from the "Romanian capital" provisions: the result gave them a favourable margin of 4·03 per cent: see *MPR*, no. 4 (1942), p. 109.

[4] 25 July 1942: see *MPR*, no. 18 (1942), p. 586.

the leasing of isolated non-explored structures and were published on 20 May 1943.[1] The rules governing the creation of associations as required by Article 269 were promulgated on 20 October[2] in the same year, and from that time till the end of the Antonescu regime the Ministry of Mines and Petroleum delineated and granted to companies a total of eight Mining Districts, mostly in Moldavia. These were granted only in principle, i.e. without the completion of the new legal formalities, but test work began immediately.[3]

The German reaction to the Law concentrated on its technical possibilities rather than its discriminatory provisions. On 20 August 1942 Neubacher sent to the office of the Four Year Plan the text of both the Law and Marinescu's *exposé des motifs*, with the comment that it was to be welcomed because it provided for the distribution of entire geological structures, though it contained a number of inherent difficulties. These he instanced as likely to arise from the set time-limits which were not always justified, the State's concern to ensure that small operators had the possibility to participate in exploration, the reservation of decisions to the State at numerous stages of exploration and production, and various technical flaws.[4] There appears to have been no diplomatic *démarche* about the subordinate status of non-Romanian nationals. In keeping with this reaction, German policy concentrated on urging the distribution of Mining Districts and maximizing the Romanian export potential by such measures as subsidizing the conversion of C.F.R. locomotives from oil to coal. Diplomatic intervention had regard to practical means of cutting down Romania's domestic consumption rather than to the political principles underlying the Petroleum Law.

In this way, without any of the international complications attendant on the passage of earlier nationalistic mining legislation, the Romanian State had written into the constitution of the country the basis for prolonging the comprehensive detailed control which wartime developments had enabled it to assert. Although not denying the right of foreign nationals to invest in Romanian oil, the State ensured that any such investment could only play a

[1] *MPR*, no. 7–8 (1943), p. 214.

[2] *MPR*, no. 19–20 (1943), p. 553.

[3] 152.

[4] See AA file *Petroleum und andere Mineralöle*, 12 Aug.–20 Oct. 1942.

minor role in any enterprise. On these terms it was unlikely that foreign funds would be forthcoming.

Whether this legal and political situation could be maintained depended on the outcome of hostilities. The Allies served notice that they regarded the property transfers as invalid, in the *Inter-Allied Declaration against Acts of Dispossession Committed in Territories under Enemy Occupation or Control*[1] of 5 January 1943, and the repeal of the Mining Law was put on the agenda for an eventual peace settlement. Similarly, it seems most probable that a Kontinentale Oel organization representing a triumphant Germany would be at the least restive about its inferior foreign-national status. As it was, Germans and Romanians had to be content with the balance arrived at in order to get on with the war, which in October 1942 turned against them. Romanian divisions were among those encircled at Stalingrad, and the clearing of North Africa opened the possibility of bombing attacks on Romania itself. There was every incentive to leave matters as they were. Meanwhile, regardless of the ownership and control of its constituent units, the operations of the oil industry were conducted to further the Axis war effort.

The operations of the industry in the war

While Romania kept out of direct participation in hostilities the distinction between "neutral" and "un-neutral" conduct was defined by practice rather than by reference to the concepts of international law. The result for the oil industry was that its commercial significance faded, and, after the summer of 1940, disappeared, as its objectives were set not by its relation to export markets and the world oil industry as a whole, but by its participating in a European war economy, in which cost considerations were secondary and for which the world oil industry was of no relevance. This broken relationship was not re-established during the period of this study.

(*a*) *Economic warfare for Romanian oil*

On the outbreak of war Romania's export trade became involved in the economic warfare between the Allies[2] and Germany. The

[1] Cmd. 6418, Miscellaneous no. 1, 1963: for the effect of this, see below, pp. 280 f.

[2] On 12 Sept. 1939 the British Cabinet scrapped the pre-war policy and authorized the purchase of oil stocks in Romania. This policy was presented

Allies had a superior position "on the ground" by reason of their established companies. These, as Romanian legal persons, were formally obliged to serve the legitimate interests of the State. In practice, the State was becoming increasingly pro-German. The policy of the British and French Governments concentrated therefore on keeping Romania as "open" as possible and on intervening to protect the position of the companies owned by their nationals in order that such companies should not be forced by the Romanian Government to assist the German war effort.[1] The effect of this diplomatic activity, at the level of business, was that Allied companies minimized the quantities available for export to Germany both by refusal to sell from their own sources,[2] and by making purchases from the free-market surplus.[3] This had to be done in such a way as not to present the Romanian Government, which at the time was increasing its own administrative control over the industry, with an excuse for establishing a quota system[4] or for adopting German suggestions that it should take over and market the output of the entire industry.[5]

The policy of minimizing availability for Germany, however, was followed by two contradictory methods. In accordance with the policy of the British Government, the "British" companies concluded contracts with their parent concerns to cover their entire export surplus—thus enabling them to plead successfully to the Romanian authorities that they had no oil available for Germany[6]—and, by paying higher prices,[7] bought as widely as

to the representatives of British oil interests in Romania, at a meeting in the Ministry of Mines on 15 Sept. 1939, and subsequently developed by that department (as part of the Ministry of Fuel) and the Ministry of European Warfare. On the general policy regarding economic warfare in oil, see W. N. Medlicott, *The Economic Blockade*, vol. 1, Chap. VI: on the meeting of 15 Sept. 1939, see 153.

[1] In terms of personnel, this was carried out by E. A. Berthoud (Petroleum Attaché) and M. Bridgeman (seconded from the Ministry of Economic Warfare).

[2] Though the Romanian Government insisted on the fulfilment of outstanding contracts to Germany after 3 Sept. 1939.

[3] They also attempted to make an artificial shortage of middle distillates by leaving as much of the gas oil fraction as possible in kerosine and fuel oil, and by cracking their surplus gas oil (see 154). Proposals to buy up small Romanian companies were not adopted: the Allies contented themselves with bidding up the price: 155.

[4] From the Allied point of view, any quota system based on past performance, especially if the base year had been 1938 or 1939, would have given the Germans far more than they in fact received.

[5] On trading problems under neutrality, see 156.

[6] 157.

[7] The Germans followed suit, with the result that the price increase created

possible in the free market.[1] French policy was based on the Wenger agreements, the unsatisfactory working of which enforced revisions in October 1939 and March 1940. Part of the difficulty of operating under wartime conditions arose from the free-market purchasing policy of the British companies, which by driving up prices destroyed the price basis established by Wenger and left importers in France with the much cheaper source of supply ex-Gulf. The essence of these Franco-Romanian arrangements was the notion of a government-negotiated market at agreed low prices. In the British view, however useful this policy might have been pre-war, its continuance in wartime merely admitted the right of the Romanian Government to set destinations. Further, the French (and Belgian) parent companies made no arrangements for tying up their affiliates' export surplus, which accordingly continued to appear on the local market.[2] In this way it was possible that oil from French interests could be acquired by the German purchasing missions.[3]

The harmful effect of these divergent approaches became evident before the end of 1939, and Kessler sought to overcome them by a plan, discussed with M. Pineau on 6–7 December,[4] whereby the six major exporters, representing directly about 75 per cent of possible Romanian exports, should be divided into two groups according to the nationality of shareholding, viz. (i) Astra Română, Phoenix, the British holding in Steaua Română, with Româno-Americană,[5] (ii) Concordia, Colombia, and the French holding in

difficulties in the export trade, particularly as at the same time the Romanian Government attempted to use the opportunity created by wartime trading to demand payment in dollars: see pp. 250, 258 n. 2.

[1] In Jan. 1940 such purchases were discontinued on the grounds that, as the available evidence suggested that the Romanian Government had committed itself to supplying oil to Germany in quantities exceeding the whole of the free market surplus (which was, in fact, true) continued operations would merely hasten the State direction of all exports. See 158.

[2] See L. Wenger, "Rapport sur le Protocol Franco-Roumain de Pétrole du 21 avril 1939 et sur son exécution", 15 Mar. 1940: 159. On general Anglo-French differences of approach to the problems of economic warfare, see Medlicott, vol. i, Chap. III.

[3] 160. According to Neubacher, oil from the Colombia company began to go to Germany in Mar. 1940: see his memoirs, *Sonderauftrag Südost*, p. 42.

[4] 161.

[5] The position of Româno-Americană, as a Romanian national affiliated to a (neutral) U.S. parent concern, and therefore subject to U.S. anti-trust legislation, was particularly acute. Though the general sympathies of the Jersey

Steaua Română, to account for respectively 50 per cent[1] and 25 per cent[1] of Romanian exports on long-term contract at fixed prices for disposal through their affiliated companies in the U.K. and France and French North Africa. Of the remaining export possibilities 20 per cent were to be taken up by purchases in the market, and shipments divided equally between France and the U.K.[2] The proposed division of purchases approximated to the ratio of consumption in the British and French markets and ensured that the burden of lifting highly priced oil would be shared *pro rata*. The arrangement on the British side virtually represented the course of action adopted by the companies concerned, with government support. M. Pineau accepted the Plan in principle and arrangements were put in hand.

There were, however, two dangers to the operations of the "British" group, viz. that an excessive export to the U.K. might provoke German demands for parity of treatment, and that Româno-Americană could not request U.S. diplomatic support for any refusal to supply Germany. These two factors dictated a modification in the plans for disposing of the constituent companies' output. The destination was changed to the "Levant" countries (defined *ad hoc* as Egypt, Palestine, Syria, Greece, and Malta). In these areas the U.K. companies had established outlets of their own and Româno-Americană had for some time supplied the Socony Vacuum subsidiaries on contract. Such a solution involved shorter, safer hauls than from the West Indies or the Persian Gulf—thus saving tanker tonnage—and improved the market position for Romanian oil, which had been sagging under high Constanța prices. The given market situation in each country, resulting particularly from government price controls and from the varying qualities of product required, entailed that supplies be laid down on the cheapest basis, i.e. ex U.S. Gulf. This accordingly set the limit on the price which could be charged for

management in New York and London were with the Allies, and they were willing to co-operate in diversion schemes, the attitude of the local management was not so pro-Allied. In the event, commitments to D.A.P.G. and the Socony Vacuum in Czechoslavakia were fulfilled minimally, and arrangements were made for Româno-Americană's surplus for 1940 to be contracted to Anglo-American (now Esso Petroleum): see 162. On the general difficulties of getting Româno-Americană into a "loyal" framework, see 163.

[1] Subsequently amended to 45 per cent and 30 per cent: see 164.

[2] The remaining 5 per cent of availability was represented by Prahova and was committed to Italy.

Romanian oil but the Romanian Government did not permit a cut in the f.o.b. Constanța price. Romanian supplies were therefore uncompetitive unless the price difference could be bridged. The British Government agreed to subsidize the "Levant" operations.[1] Movements to the Atlantic seaboard virtually ceased in June 1940 in consequence of the rapid German military successes in Western Europe. The "Levant" areas continued to be supplied until October, that is until the time of comprehensive German-Romanian collaboration, after which the denial of transport and detailed control of "Allied" companies, with the concomitant withdrawal or expulsion of their expatriate staff, brought economic warfare to an end.[2]

Allied policy aimed at denying to the Axis not only oil but also storage and transport facilities. This was to be done by refusing to lease tankage to German concerns or to third parties for the storage of products thought to be ultimately destined for Germany, and by the chartering of river barges. Allied companies laid up their own vessels and went into the charter market for their own needs. They also sought to restrict the tonnage available by taking over vessels on charter to Germany. The crux of this policy was finding alternative employment for the vessels, since the Romanian Government insisted on the maximum use of all tonnage and supervised the fulfilment of the relevant charter party. This aspect of economic warfare and the difficulties involved are illustrated by the scheme[3] devised by the Anglo-Danubian Shipping Company for the chartering of twenty-nine vessels at that time engaged on shipments to Germany and owned by the two Romanian river companies. The scheme provided for these vessels to be used as part of a river-rail link to Switzerland with trans-shipment at the Yugoslav river port of Vukovar. There soon appeared two principal obstacles: the lack of sufficient rail tank-cars to service the barges at Vukovar, and the restrictions imposed in Switzerland on the import of oil from Romania, which were based on an official comparison of the laying-down costs of supplies from different sources. On this basis the United States was a much cheaper

[1] On all aspects see 165. Consideration of the "Levant" scheme emphasizes that, for the Allies, success in preventing oil from going to Germany was only half the problem; disposal outside Romania presented considerable difficulties and involved government action on oil markets.

[2] The course of the economic war appears in Figure 12.

[3] 166.

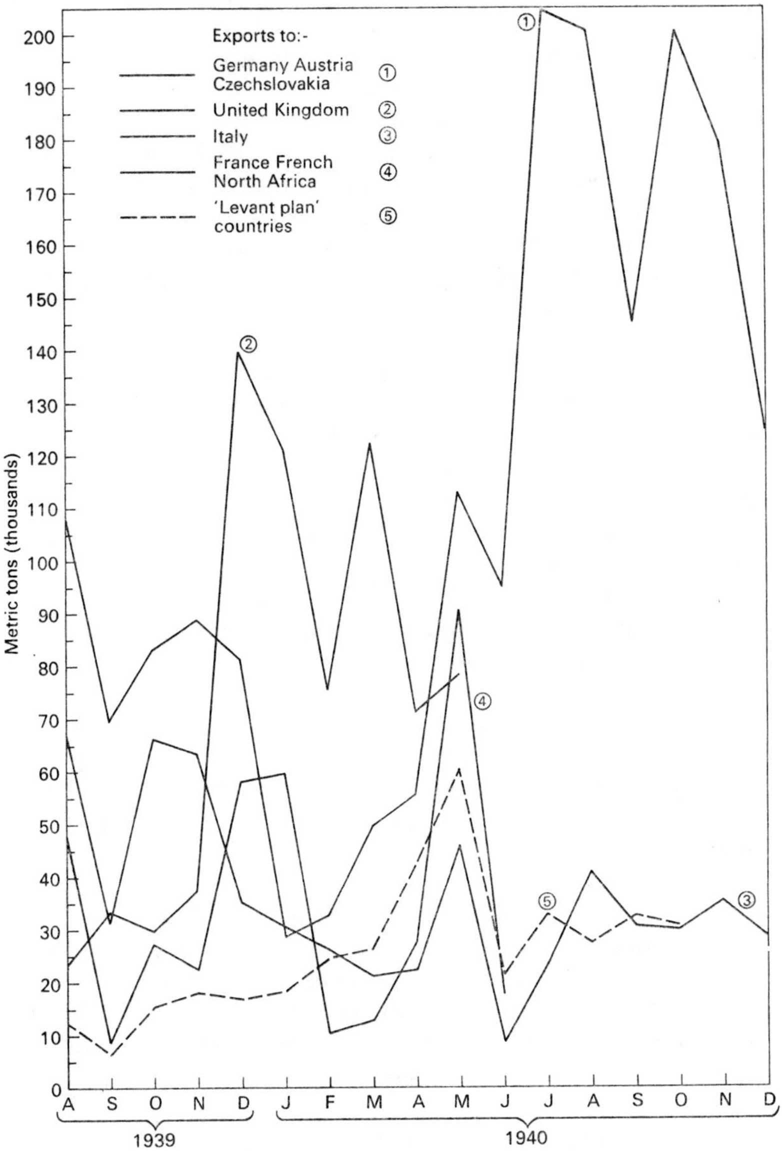

FIG. 12. Economic warfare for Romanian oil.

source than Romania. Costs of supply from Romania by the proposed Danube route and the existing sea-rail route through Vado were compared and found to be only marginally different.[1] But, as in the "Levant" scheme, there remained the gap between these laying-down costs and those from the United States: it was proposed that this gap be covered by a subsidy, for both the tank-car hire charges and the price differential in Switzerland,[2] either by direct payment or by artificially adjusted freights.

Further examination of the proposal revealed a number of more far-reaching problems: Vukovar was a mobilization centre which the Yugoslav military authorities were unwilling to have congested by rail tank-cars: the negotiation of reduced rail freights through Yugoslavia (in order to minimize laying-down costs in Switzerland) had to wait on the satisfaction of the transport needs of the state-owned Yugoslav River Shipping Company: economic relations between Yugoslavia and Romania were impaired through the Yugoslavs' retaliating for the high prices demanded for Romanian oil[3] by counter-demanding complete dollar payments for copper exports: the continuance of even the 15 per cent and 30 per cent quotas for gasoline and kerosine respectively in Switzerland could not be guaranteed, and the outlet was to that extent unstable: both Switzerland and Yugoslavia were vulnerable to the export of products from Italy necessitated by high stocks there created through heavy crude-oil imports after the outbreak of war. There was also the acceptance by other marketing companies in Switzerland of this method of supply, the delimitation of responsibilities at each stage, and the working out of the relevant freights.

These problems were overcome, and the Goeland company, through agents, the Société de Surveillance de Genève, chartered the lighters and collected 50 out of the 250 tank-cars required—the balance being organized from Yugoslav and Hungarian sources by Jugoslavensko Shell d.d. On 29 March the Danube became open for shipping and the movement began. The episode demonstrates the complexity and range of problems entailed by the

[1] 167. [2] 168.

[3] 169: the high prices reflected, in part, Allied buying policy in Romania (see above, p. 245) and the Romanian Government's attempt to capitalize on the oil shortage to maximize its free currency earnings (see below, p. 258 n. 2).

[4] 170.

original idea to deprive Germany of the benefit of only twenty-nine river barges and the difficulties of operating on an ostensibly commercial basis in the crepuscular circumstances of Romanian politics.[1]

During Romanian "neutrality" the Germans based their struggle for oil not on existing assets but on their over-all political relationship with the Romanian Government and their ability to supply arms. The latter, from the German standpoint, strengthened a friendly state and provided a lever to make relations even closer. In the short term they provided a *quid pro quo* for oil. As early as 16 September 1939 Fabricius, minister in Romania, reported a suggestion from Calinescu, the Romanian Minister President, that the Germans should deliver arms against a long-term commitment to supply oil and other materials.[2] The suggestion was adopted on 29 September and terms discussed.[3] The main problem between the two governments was the value to be placed respectively on the oil and on the war material: the Romanians pleading the current (i.e. increasing) prices and the Germans demanding a basis of pre-war prices for both sides of the exchange. This point was eventually conceded and deliveries of war materials began in March 1940 under a preliminary agreement concluded on the 7th of that month.[4] Final agreement was not reached until 27 May, in the "Oil Pact".[5] The predominant share in Romania's oil exports won by these means was secured by the political and military events already discussed.[6]

The "machinery" of German activity followed from its necessarily political basis. At the outbreak of war the minister in

[1] It was alleged at the time that the explanation for the Allies' failure in sabotage and in economic warfare lay in the unpatriotic preference of the "Allied" oil companies for profits—a view repeated by Mr. St. John in his *Foreign Correspondent* (pp. 145–6), published in 1960. Mr. St. John's assessments of this and the other issues he discusses appear to be based neither on proper observation at the time nor on subsequent research.

[2] *DGFP*/D/VIII, no. 74.

[3] *DGFP*/D/VIII, nos. 111 and 166.

[4] See Clodius/Fabricius to Foreign Office, Berlin, reporting conference with Tatarescu: *DGFP*/D/VIII, no. 660; and Ribbentrop's memorandum to Goering, 16 Mar. 1940: ibid., no. 678; also Neubacher's report to Foreign Office, Berlin, 28 May 1940: *DGFP*/D/IX, no. 338; and Hillgruber, pp. 84–7.

[5] See Neubacher's report to Foreign Office, Berlin, 28 May 1940: *DGFP*/D/IX, no. 338; Hillgruber, pp. 84–7.

[6] German successes stimulated discussion of the role of Romanian oil in the new political and economic conditions in Europe: see Hans Bahr, "Neue

Bucharest actively promoted German purchases through discussions with Romanian ministers and civil servants: the actual supply remained in the hands of agents of German oil companies supported by a representative of the *Reichsstelle fuer Mineraloele*.[1] Available transport facilities restricted actual shipments to less than the 100,000 tons a month, which was the level of German requirements from Romania.[2] After the German-Romanian protocol of December 1939 special arrangements were made to supervise its working and to overcome existing deficiencies in organization. Carl Clodius was assigned to Bucharest as "Special Representative for Economic Questions" assisted by Hermann Neubacher, a leading Austrian Nazi who was at that time Bürgermeister of Vienna.[3] Their mandate, as set out by Ribbentrop, involved initiating, effecting, and following through "the necessary purchase contracts, ensuring the required transport facilities, and counteracting attempts at sabotage by enemy powers".[4] Clodius later withdrew to Berlin, leaving Neubacher as sole Bucharest representative. He became the key figure in the execution of German-Romanian oil policy. Under him functioned from March 1940 the local office of the Rumaenien Mineraloel G.m.b.H. (Ruminoel), an organization set up in Berlin to co-ordinate all purchase and transport questions in the Romanian-German oil trade and staffed by personnel from the companies concerned.

Transport was crucial: Constanţa was open only to Allied and neutral shipping by reason of the blockade, the Danube froze in

Erdoel Politik, im Kontinental-Europaeischen Raum", *Wirtschaftsdienst*, no. 29 (19 July 1940); and two articles in a special number of *Oel und Kohle* devoted to south-east Europe, no. 40 (22 Oct. 1940): F. Friedensburg, "Die gegenwaertige und zukuenftige Mineraloelwirtschaft von Sued-Ost Europa" and Dr. L[aufs], "Das Donau-Versorgungsgebiet".

[1] 171: also information supplied by Herr Gensch (formerly Deutsche Shell) to writer.

[2] See *DGFP*/D/VIII, no. 402.

[3] See report by Wiehl (Director, Economic Policy Department, German Foreign Office) on conference on Balkan questions held under Goering's chairmanship at Karinhall on 2 Jan. 1940, and subsequent Ribbentrop/Goering correspondence: *DGFP*/D/VIII, nos. 502 and 508.

[4] Presumably with regard only to raw materials: general anti-sabotage competence had already been assigned to Manfred von Killinger, former S.A. Obergruppenführer, who replaced Fabricius as minister in Romania in Jan. 1941. See instructions to Legations, 30 Dec. 1939: *DGFP*/D/VIII, no. 495, and para. (3) of Ribbentrop/Goering, 4 Jan. 1940, ibid., no. 508; on von Killinger's promotion, see Hillgruber, p. 117.

winter,[1] and the railway was limited both as to choice of routes and because of the lack of tank-cars. Of the four rail routes, those through Curtici and Episcopia Bihor suffered the common drawback of single-track system over the section Predeal–Brașov, and that via Halmei was closed to international traffic by the Hungarian Government in December 1939.[2] The fourth route, Cernăuți (Orașeni)–Lwow–Krakow,[3] ran through the Russian-occupied area of Poland, and, though the Russians agreed not to alter the gauge in this sector, its working presented considerable difficulty[4] and it was eventually closed to through traffic in early 1941.

The restrictions imposed by the network were accompanied by operational difficulties which Otto Stern analysed in the following terms:[5]

> The maximum number of oil-trains per day which can be carried over rail routes is the maximum number of trains which the Romanian authorities will permit, and this is governed by a number of variable factors, the principal of which is the amount of internal traffic that must be carried over the various routes. At present this internal traffic is considerably above normal, due mainly to the preparation of extensive defence works, troop movements, and transportation of military supplies. In these circumstances, it appears that the Romanian authorities have just allowed the Germans to run a maximum of six oil-trains per day in all over the two available routes combined. In actual practice they are running anything from one up to six trains (on rare occasions) already and have averaged two trains per day in January and about two and a half a day in February. But it they fail to make use of their full quota of six trains one day they cannot of course make up the arrears the next day.

Measures for immediate improvement were the doubling of the line Predeal–Brașov and the conversion of all possible tank-cars to the German traffic. Both were undertaken: the first rested on co-operation with the C.F.R. and on finding the requisite capital

[1] In 1940 this was exceptionally prolonged—from 28 Dec. 1939 until 29 Mar. 1940.

[2] 172.

[3] The line was specially mentioned in German-Russian economic negotiations: see *DGFP*/D/VIII, nos. 102 and 237.

[4] 173.

[5] 171. Stern also suggests that the Romanian authorities lengthened permitted delivery schedules to provide officially for delaying the dispatch of oil-trains to Germany. See 174.

which the Romanian Government had for so long failed to make available: this was achieved under a basic transport plan approved in September 1939 and the "bottle-neck" was moved by March 1940.[1] The tank-car problem was rather more intractable, since questions of ownership and use and the Romanian Government's policy were entailed. In January 1940 the total number of tank-cars registered in Romania was 10,369.[2] It had been standard practice before the war for oil companies to hold directly only part of their requirements, hiring the rest from tank-car companies or from the State. The companies were therefore particularly vulnerable to official policy, since, even without exceptional measures, the State had a direct or indirect lien on three-fifths of the total stock. However, until problems of congestion—which themselves limited the effective deployment of tank-cars—were overcome, the Romanian Government had no need of exceptional powers, though it had threatened them in January 1940 for its own purposes.[3] After unreserved co-operation with Germany had been decided on by the Romanian Government in May 1940,[4] the railway authorities attempted to get the companies to surrender their hired tank-cars, and, when the proposals appeared likely to become bogged down in argument, abruptly terminated the contracts.[5] But even before the notice expired they requisitioned all cars into a pool.[6] They were supplemented by German-controlled tank-cars, which were permitted into Romania in increasing numbers. By the autumn of 1940 the traffic was running smoothly, although with delays imposed by troop transports.[7]

In transport via the Danube the key factors were the availability of tugs and barges and delivery and handling facilities.

[1] On wartime C.F.R. developments, see Report of the Superior Council of Transport and Tariffs: *MPR*, no. 13 (1942), pp. 407–9.

[2] Astra Română figures: see 175.

Ownership		Use	
by C.F.R.	3,463	by C.F.R.	1,733
by users—for 25 years	2,953	by oil cos.	6,066
by users	3,953	by other industrial cos.	161
		rented (mostly to oil cos.)	2,409
	10,369		10,369

[3] 176.

[4] See above, p. 226.

[5] This measure affected 3,057 cars: see 177.

[6] 178; *MPR*, no. 15 (1940), p. 813.

[7] 179.

As to shipping, the Germans were in a superior position to the Allies even without Romanian assistance. In 1940 there were under the German flag 57 tugs, 534 barges, and 85 lighters, the two latter giving a combined capacity of 396,178 tons. The British and French commanded 22 tugs, 80 barges, and 49 lighters (capacity 115,767 tons). Additionally, there were under the Romanian flag 127 tugs, 553 barges, and 66 lighters, with a total capacity of 511,149 tons.[1] The German companies had for some time operated their fleets as a pool, with common shipping programmes and freight-sharing arrangements: they were thus already organized for the conditions that their enemies now had to meet by improvization through the established ship-chandling firm of Watson and Youell, Galaţi.

At the outbreak of war both German and Allied agents competed in chartering vessels laid up at Brăila—a measure designed on the Allied side to permit laying up Allied-owned vessels as part of the policy of denial. The change in political relations with Germany, brought all oil-vessels remaining in the Danube into the traffic with Germany—with the exception of vessels under the Yugoslav and Hungarian flags, which did not follow suit till April 1941 and March 1944 respectively. There were, however, more than enough to move the available cargo.

The main river port available for this traffic was Giurgiu, which already had facilities adequate for 166,000 tons a month:[2] the task was to improve and extend them. This was undertaken by a joint company, S.A.R.D.E.P.,[3] formed from an existing trading company "Transat", with 20 million lei and management provided by Creditul Minier (then in close association with Elwerath).[4] The new company acquired a site within the oil zone of Giurgiu to build a tank farm, pipelines, and railway facilities.

[1] See *MPR*, no. 14 (1940), p. 834: the table also includes under "Holland" 4 tugs, 33 barges, and 31 lighters—presumably the fleet belonging to the "Cosmos" company of Vienna (Continentale Motorschiffahrtsgesellschaft), whose Dutch nationality was not recognized by the Netherlands Government: Italy is shown as having 3 tugs and 3 barges. See also *MPR*, no. 20 (1940), pp. 1097–8.

[2] 180.

[3] See *MPR*, no. 12 (1940), p. 111; no. 15 (1940), p. 882. In Oct. 1940 S.A.R.D.E.P., in association with the firm of Meotica Română, bought up the leading broking business of Sanielvici: see AA file *Petroleum und andere Mineralöle.*

[4] 181.

Work was held up by flooding after the unfreezing of the Danube in April 1940 but the installation was completed and in operation by June.[1]

The effectiveness of Giurgiu was enhanced by improvements to the pipeline system, by which the supply through the two existing 5 in. lines from Băicoi was augmented by the construction of two new parallel 10 in. lines from Teleajen. These were in full operation by August 1941, and the total capacity from the fields to Giurgiu was thus raised from 1,000 tons a day of crude oil to about 7,000 tons a day.[2] Further, in March 1944, the 5 in. line supplying crude oil to the Petrol Bucureşti refinery in the capital was relaid over the stretch from Bucharest to Giurgiu and used to supplement the other lines of the same diameter.[3]

Under these conditions, from mid 1940 onwards Giurgiu handled approximately 50 per cent per annum of Romanian oil exports.[4] Other ports used were Olteniţa[5] (connected by rail with Ploeşti but disused since 1934) and Orşova,[5] where tankage from an old disused refinery was, under German auspices, re-erected alongside the Danube. Orşova began working in May 1940, and by reason of its position, upstream of the Iron Gates, permitted better synchronization of trains and barges and the full loading of the latter. It thus tended to become the terminal for packed products. These ports' largest traffic was in 1940: thereafter it declined by about two-thirds to a level of 60,000 tons per annum over the next three years.[4] The improvement in Danube facilities[6] and the streamlining of administrative procedures were responsible for cutting the time for round trips to Vienna from thirty to fifteen days. The immediate success of this programme may be gauged by the increase in quantities shipped to Germany, as shown in Figure 12.

Romania's oil-export trade followed the course of military and political events which put an end to long-established outlets in the east Mediterranean and north-west Europe, and which enabled

[1] 182.

[2] For movements through Giurgiu lines, 1940–1, see *MPR*, no. III–IV (1945), p. 110.

[3] 183.

[4] See *MPR*, no. XI–XII (1944), p. 69.

[5] 184.

[6] Work in Romania was complemented by extensions to other Danube ports, principally Bratislava: see *MPR*, no. 19 (1940), p. 1084.

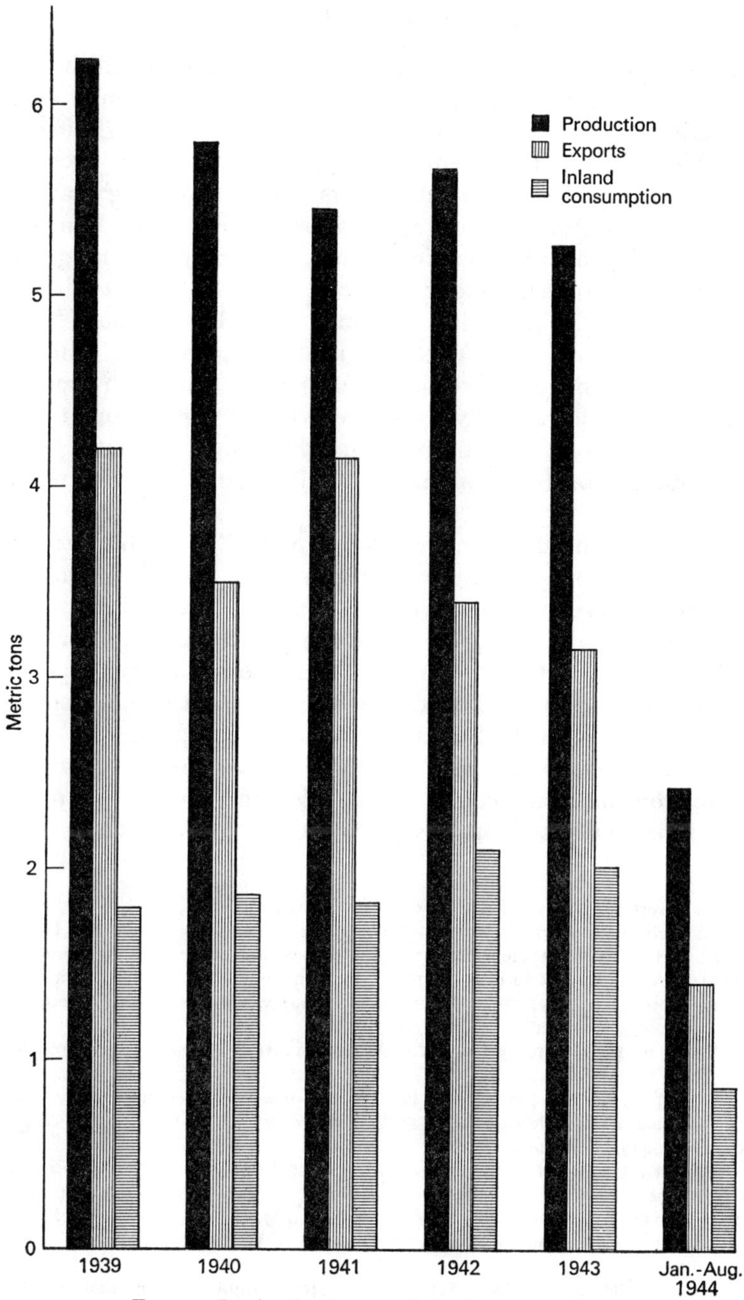

FIG. 13. Production/exports/inland consumption.

Germany to turn to account the economic position established in regard to the industry before the war. From the autumn of 1944 deliveries to Germany (including the Protectorates of Bohemia and Moravia and the Wehrmacht) accounted for, on average, 70 per cent p.a. of total oil exports.[1] The remaining markets comprised the Axis partners and associates with whom Romania retained independent economic relations, Italy, Bulgaria, Hungary,[2] Serbia and Croatia,[3] Vichy France,[4] and the Soviet Union[5] (till June 1941). Trade also continued with the neutrals, Yugoslavia[2] (till March 1941), Switzerland, Sweden, and Turkey. All these destinations (except the last) were reached by rail or river. Constanța virtually closed down for the export trade, becoming a point of storage and, while the two states were co-operating, of trans-shipment for Russian oil en route to Germany.[6]

There was in addition one other "market", viz. the German Military Mission, whose requirements were taken over at the refinery.[7] These were used for the military and airforce units stationed in the country, and to sustain operations, i.e. the attack on Yugoslavia and Greece (for which purpose a temporary pipeline was erected across the Danube between Giurgiu and Ruschuk), and the invasion of the Soviet Union.

The relation of Romania's exports during the war years to production and inland consumption is demonstrated in Figure 13: their value appears from the table[8] on the opposite page.

[1] For German figures of deliveries, see Hillgruber, pp. 249–51.

[2] In early 1940 shipments to both countries were hampered by the break-down in the relevant clearing arrangements consequent on the Romanian Government's attempt to exact higher prices in dollars for oil exports against the iron ore (Hungary) and copper (Yugoslavia) which respectively were the basic exchange product.

[3] Trade and payments agreement, 7 Aug. 1941: *MPR*, no. 19 (1941), pp. 809–10.

[4] In Sept. 1941 the French and Romanian Governments reached agreement on the outstanding balance of the "Wenger contract" deliveries: see *MPR*, no. 1/2 (1942), p. 31; and 185.

[5] Under a trade and payment agreement of 26 Feb. 1941 Romania delivered 46,151 tons of products: for text of agreement see *MPR*, no. 10 (1941), pp. 480–1; on deliveries see *MPR*, no. V–VI (1945), p. 187.

[6] See 186. Varna (Bulgaria) was also used for this purpose.

[7] 187.

[8] See: *MPR*, no. XI–XII (1945), pp. 346–7, quoting Romanian Central Statistical Institute.

Year	Total exports Quantity	Value (million lei)	Oil exports Quantity	Value (million lei)	Oil exports as percentage total
1940	5,374,415	36,780	3,528,868	23,037	62·6
1941	4,820,574	41,286	4,169,721	32,233	78·1
1942	3,975,145	52,816	3,400,847	32,002	60·6
1943	3,652,092	71,132	3,178,607	47,061	66·2
1944 (Jan.-Sept. only)	1,855,447	46,030	1,445,672	26,358	57·3

(*b*) *The reorganization of the industry under German auspices*

As explained above, since 1937 the problem of increasing oil-production occupied both the industry and the State and underlay the Mining Law of that year and the proposals for its revision. In wartime conditions the problem was not so much one of making secure valuable export markets as of serving a political alliance: it therefore assumed even greater importance from the State's point of view. In consequence, increasing output was among the stipulated tasks of the C.G.P. This could only be achieved by the discovery of new wells—which would take time—and the opening up of existing wells, closed in as marginal producers, or by allowing sufficient gas pressure to build up for economical working. The Government's interest was for as rapid an expansion of output as possible.

The C.G.P., in February 1940 circularized oil producers,[1] asking for details of their 1940 programmes with an estimate of what work would be necessary to increase output by "at least 20 per cent" (later increased to 30 per cent).[2] Ways and means were discussed in a series of meetings between the C.G.P. and the industry, from which it appeared the main impediments were the

[1] 188. The Government's demands posed a serious problem of economic warfare for the "Allied" companies, who had no interest in anything more than the minimum possible increase in drilling and production. The difficulty arose from the fact that diplomatic intervention could only be invoked against measures either explicitly aimed at increasing sales to Germany or imposing compulsory drilling programmes in a way that could be construed as an invasion of shareholders' rights. These considerations dictated a policy of ca' canny co-operation by the companies, basing themselves on commercially and technically valid arguments in support of their slow progress. See 190.

[2] By Professor Netta during second general meeting, 26 Feb. 1940: see 191.

Mining Law and the shortcomings of equipment and machinery necessary to maintain drilling and operating schedules. Other factors were administrative, such as the effect of the call-up on specialist personnel. There also emerged a dispute over the available reserves from which increases in production could be expected: these were evaluated during three meetings of state and industry geologists held at Câmpina. Professor Macovei reported that there were 290 proven wells still to be drilled, of which 135 were programmed for 1940: this would, however, only maintain but not increase Romania's production. This conclusion left out of account any increase in production resulting from increased exploration, which could only depend (in the companies' view) on a new law: this was accepted by Professor Netta, but Macovei had already hinted during the first "geological" meeting on 28 February at an alternative policy—i.e. the revival of A.C.E.X. as a state drilling organization.[1] The question of materials necessary to fulfil programmes was settled at the fourth Commissariat meeting (14 March), when Professor Netta confirmed that the necessary import permissions would be given for the line pipe and casing over 6 in. in diameter, but stipulated that requirements under this size would have to be met from Romanian sources.

Although drilling metreages slightly improved, no significant addition in output resulted from these discussions and the consequent action.[2] The change of regime and the arrival of the Germans in the autumn of 1940 presaged tighter government control and greater efforts. The German activity was directed by Kontinentale Oel A.G.,[3] whose position rested on its ability to supply drilling-rigs and other material (as against its total lack of proven oil areas). Accordingly Kontinentale concluded joint-

[1] 192. The structure of the administration was designated on 31 July 1940 (see *MPR*, no. 16 (1940), p. 934) and officials named in September (see *MPR*, no. 18 (1940), p. 1025): see also report on first year's activity (*MPR*, no. 20 (1941), pp. 839–42) and article by Radu Cadere (A.C.E.X.), "The role of the Commercial Administration A.C.E.X. in the national mining economy", *MPR*, no. I–II (1945), pp. 15–19.

[2] For drilling and production statistics 1940, see *MPR*, no. V–VI (1945), p. 175.

[3] Kontinentale's long-term production interests were to be served by the creation of a Geological Research and Advisory Institute set up under Kontinentale Oel G.m.b.H. to undertake a complete geophysical survey of the country in collaboration with the Romanian Geological Institute and the laboratories of the various companies. See Dr. Otto Gramsch (Kontinentale Oel), "German Participation in the Roumanian Oil Industry", *MPR*, no. 18 (1941), pp. 771–3.

operation agreements with Astra Română and A.C.E.X. The agreement with Astra[1] provided for joint exploration of a total of 44,188 hectares in Moldavia, Astra providing the concessions and Kontinentale ten deep-drilling outfits: expenses would be shared and Astra compensated for work already done. The agreement with A.C.E.X. was similar: it provided for Kontinentale to receive a 70 per cent share of expenses and output in a joint venture in the Manești oilfield.

Under these agreements Kontinentale became the main drilling concern in Romania. Drilling programmes concentrated on Dâmbovița, where total metreage increased from 63,258 in 1940 to 140,106 in 1943—the year of maximum effort. During the same period the totals in other traditional areas rose as follows: Prahova by 16,582 metres, Buzău by 9,146 metres, and Bacău by 603 metres. Furthermore, in 1943, 1,344 metres were drilled at Putna and 1,935 metres at Arad in the Bánát.[2] In the next year the drilling-rate fell off through the non-delivery of rigs and spare parts needed to maintain the programme. The activities outlined above, combined with running existing wells at maximum flow, arrested the decline in Romania's oil production,[3] although the total fell far short of the aims of either the Romanian Government or the German authorities.[4]

The refining policy adopted during wartime[5] was simply to maximize runs of fuels and lubricants, particularly for aviation.

[1] The agreement was disliked by the German management of Astra Română, but was imposed on them by the B.P.M., as shareholder, acting on the instructions of the Reich Ministry of Economics: 193.

[2] Total metreages drilled (*MPR*, no. IX–X (1945), p. 297):

1940	235,000 metres	
1941	252,942 ,,	
1942	343,789 ,,	
1943	341,458 ,,	
1944	146,827 ,,	(entire year).

[3] Total production of crude oil (*MPR*, no. IX–X (1945), p. 297):

1940	5,810,000 tons	
1941	5,453,179 ,,	
1942	5,665,357 ,,	
1943	5,273,432 ,,	
1944	3,525,000	(entire year).

[4] 1943 represents only 45 per cent fulfilment of plans: Hillgruber, p. 201.

[5] A description of the refining industry, with flow diagrams of the main units, is given by V. Cerchez in a series of articles in *MPR* (1939): no. 16, pp. 1061–5; no. 17, pp. 1125–30; no. 18, pp. 1189–93; no. 20, pp. 1305–9; no. 21, pp. 1361–6.

The supply of high-octane aviation fuel rested on the availability of asphaltic crude—viz. about 15 per cent of total Romanian output—and on upgrading the gasoline yield from other crudes by blending with alkylates.[1] By reason of the declining supply of asphaltic crude—and the Romanian Government's reluctance to part with it[2]—the German Air Ministry decided to erect an alkylation plant at the Astra Română refinery at Ploești. The plant was completed only in 1944, so that its original purpose of contributing to the supply of high octane gasoline to the Luftwaffe was never served. It was, however, the only technical innovation in the Romanian refining industry during the war years. Otherwise capacity remained as it was, both in quantity available and in pattern of use, and was affected only by bombing attacks. These, beginning with sporadic raids at the time of the attack on Russia, developed into a systematic air offensive after 1 August 1943, with the heaviest bombardments in the spring of 1944,[3] and seriously disabled the refining industry: the lubricating oil plants of Astra Română and Creditul Minier were put completely out of action.

Before the bombing there had been a degree of rationalization of the refining industry in the interests of the war effort. The management of the Socony Vacuum plant at Brașov was instructed to discontinue production of automotive lubricants and concentrate on aviation lubricants, even though this entailed breaking contracts.[4] Similarly the authorities enforced the segregated refining of asphaltic crude in order to maximize the yield of base gasoline of aviation quality (+70 octane) for blending with tetraethyl lead to improve the octane rating to the maximum 87. Shortage of catalysts, normally of U.S. origin, held up the production of iso-octane.

The effect of Government and German policies on gasoline production was to raise the yield proportionately to the quantity

[1] On technical processes, see "Alkylation" in A. E. Dunstan *et al.*, *The Science of Petroleum*, vol. v, part V.

[2] On 21 Nov. 1941 Clodius complained to Neubacher (letter no. 391) that the Romanian Government had refused to permit the export of 7,000 tons of asphaltic crude for aviation gasoline. The Germans had been forced to agree, therefore, to its being topped in Romania, and the residue exported as a lubricating oil distillate; this, however, was held up for lack of available topping plant: see AA file *Petroleum und andere Mineralöle*.

[3] Table of air damage to refineries by value: *MPR*, no. III–IV (1945), p. 107.

[4] 194.

of crude oil submitted for refining from 22·8 per cent in 1939 to 26·7 per cent in 1943: the increased percentage, however, related to declining throughputs, so that for the same period there was a net loss in physical terms of over 16,000 tons. The output of gas oil increased in 1941 but fell off relatively and absolutely in 1943 to below the yield for 1939. Fuel oil decreased slightly in percentage and absolutely by about 380,000 tons. The lubricants programme was successful.[1]

[1] *Refinery out turns.*

	1939	1940	1941	1942	1943
Total excluding refinery use and loss	5,411,619	5,041,193	4,829,069	4,887,702	4,574,777
of which:					
Gasoline	1,331,905	1,381,543	1,306,257	1,359,917	1,315,543
Gas oil	770,221	796,393	897,650	760,114	618,127
Fuel oil	1,953,981	1,807,627	1,545,144	1,626,001	1,573,456
Lubricants	63,471	69,702	103,891	158,783	142,014

MPR, no. I–II (1945), p. 9. Additionally the *MPR*, no. III–IV (1945), shows gasoline manufactured in the oilfields from well-gases, for the same years, as 246,135; 240,000; 198,131; 194,352; and 190,052 m.t.

CHAPTER X

The Armistice Period

The Armistice Agreement and the State

As stressed in the previous chapter, the weakness of Germany's position in Romania lay in the fact that her allied status and the personal resilience of Antonescu limited the scope of effective German authority and permitted the survival of a government and administration not wholly under German control. In consequence, there remained open the possibility of, at the least, non-co-operation on issues in which German and Romanian interests diverged. By mid 1943 the continuance of Romania's participation in the war fell under this rubric; and the Romanian Government, in the person of Michael Antonescu, began to elaborate schemes for withdrawal from the war.

In spring 1944 the advance of Russian forces into Moldavia and Bukovina and the treatment given to Italy after the surrender of September 1943 encouraged negotiations between Romanian representatives and the Allies in Cairo and Stockholm.[1] In consequence of the signing of a draft armistice agreement, the Antonescu regime was overthrown by a royal *coup d'état* on 23 August 1944.[2] Thereupon Romania reverted to the constitutional arrangements of 1923,[3] and a provisional "Government of National Union" was formed. Red Army troops arrived in Bucharest a week later.[4] The country's external position was that of an independent state

[1] On the Armistice negotiations, see A. Cretzianu, "The Roumanian Armistice Negotiations, Cairo 1944", *Journal of Central European Affairs* (Oct. 1951); F. Nano, "The First Soviet Double Cross", ibid. (Oct. 1952); also J. Lukács, *The Great Powers and Eastern Europe*, pp. 487 et seq. The Romanian delegates claimed that the terms agreed in Cairo were not those they were subsequently required to sign in Moscow. According to Cordell Hull the agreement was "arranged by the Russians" (*Memoirs of Cordell Hull*, vol. ii, p. 1461).

[2] For the Communist view of these events, see speeches by Gheorghiu-Dej and Ana Pauker reprinted in *The Planned Development of the Roumanian Peoples' Republic*, Bucharest, 1950.

[3] Decree of 31 Aug. 1944: *Monitor Oficial* (2 Sept. 1944).

[4] Military operations are analysed in *Contribuția româniei la victoria asupra fascismului*; and H. Kissel, *Die Katastrophe in Rumänien 1944*.

waging war against its former allies and associates on the side of its former enemies,[1] with whom its relationships were covered by the Armistice Agreement signed in Moscow on 12 September 1944.[2] This recognized Romania's complete reversal of policy, particularly in its military implications (Articles 1–3), and stipulated certain retributive actions on the part of the Romanian Government (Articles 5–13); Bessarabia and northern Bukovina were to revert to the Soviet Union (Article 4), while the Vienna Award regarding Transylvania was declared null and void, it being provided that Transylvania (or the greater part thereof)[3] should be returned to Romania, subject to confirmation at the peace settlement (Article 19). War criminals were to be apprehended and Fascist-type organizations dissolved (Articles 13 and 14), and information media and civil administration subjected to the supervision of the "Allied (Soviet) High Command". Finally, to regulate and control the Armistice Agreement until the conclusion of peace, an Allied Control Commission was to be established, under the general direction and order of "the Allied (Soviet) High Command, acting on behalf of the Allied Powers" (Article 18).

These articles, in practice, provided the basis for the transformation of Romania into a Communist state. With the entry of the Red Army[4] the country came under Russian control. The Allied authority set up to supervise the execution of the Agreement was, *de facto*, Russian. Soviet policy in Romania[5] was designed to

[1] Romania was never accorded "co-belligerent" status, either during the war or at the Peace Conference of 1946, where a Czechoslovak proposal to this end was defeated.

[2] See *Conditions of an Armistice with Roumania*, Miscellaneous no. 1, 1945 (Cmd. 6585).

[3] The absence of any mention of Bulgarian-frontier adjustments, and the qualification regarding Transylvania, presumably indicate that Soviet policy was not yet clear on these issues: see Lukács, p. 657; R. V. Burks, *The Dynamics of Communism in Eastern Europe*, p. 156.

[4] Soviet units arrived in Bucharest on 30 Aug./1 Sept.

[5] In the debate whether Russian policies in Eastern Europe followed a blueprint thought out in advance or represented a series of *ad hoc* responses to the situation, the evidence of Russian policy in Romania supports the former view. Over and above their retributive and security aims, the Russians appear to have wanted to settle a long-standing feud once and for all, by punishing the guilty and educating the volatile Romanian people in the higher, more serious, realities of Soviet life. On the place of Romania in Soviet policy, see H. Seton-Watson, "Soviet Aims in South-East Europe" (article originally published in *The Times* (11 Nov. 1946), reprinted in *Nationalism and Communism*, pp. 110–14); J. M. Mackintosh, *Strategy and Tactics of Soviet Foreign Policy*, Chaps. I and II.

exact retribution for the Romanian invasion and to provide for permanent military security—a notion implying not merely disarmament and treaty guarantees but also the abrogation of the political power of those who had launched the invasion. The carrying through of a revolution depended internally on the creation of Communist agencies, using the presence of the Red Army as the ultimate deterrent to opposition, and externally on the Soviet Union's imposing its own policies for the settlement of Eastern Europe.

Both necessarily took time: the nullifying of both internal and external opposition was not completed till 1947—the one marked by the trial of opposition political leaders[1] and the currency reform,[2] the other by the conclusion of the Peace Treaty.[3] The rights formally granted to the Allies under the Armistice were defined and enforced by the Russians, favouring the Communist or pro-Communist elements in Romania and discriminating against the opposition. In this the Russians appeared to draw a distinction between the ruling groups and the peasants, but not, except tactically, between the various "historic" Romanian parties: they were portrayed as betrayers of the people and as pro-Fascist regardless of their actual conduct in the past or the surrounding circumstances.

U.S. and U.K. policy recognized the Russian preponderance in Romania—at least until the end of hostilities and the conclusion of a peace treaty. British policy used this recognition[4] as a bargaining counter to restrict Russian activity elsewhere, while the U.S. Government adopted the more general view of joint responsibility for Romania's rehabilitation, pending which the Soviet Union should in some sense act as a trustee for Allied interests.[5] The

[1] See below, p. 302.

[2] See below, p. 312.

[3] See below, pp. 297–301.

[4] This refers to the so-called "percentages agreement" between Stalin and Churchill, discussed in the latter's *Triumph and Tragedy*, p. 198. Its immediate circumstances are still unclear, but Churchill's motive appears to have been to probe Russian intentions in a situation in which the only restriction on Russian influence in the Balkans was the Red Army's own operational problems: see also Cordell Hull, *Memoirs*, vol. ii; Roberts, pp. 265–73; Lukács, pp. 625–6; R. L. Wolff, *The Balkans in Our Time*, pp. 251–64; M. F. Herz, *Beginnings of the Cold War*.

[5] See P. E. Moseley, "Hopes and Failures in American Policy towards East Central Europe, 1941–1947" in W. D. Kertesz (ed.), *The Fate of East Central Europe*.

position of the latter was held to be safeguarded by Soviet declarations that Russia had, apart from Bessarabia and Bukovina, no desire to acquire Romanian territory, to change the existing social structure, or to limit Romania's independence.[1] This general external guarantee was reinforced within Romania by the tripartite nature of the Control Commission set up under Article 18 and by the Romanian Government's undertaking to restore the legal rights and interests of the United Nations' nationals under Article 13 of the Armistice Agreement. In practice these safeguards were feeble: the authorities set up by the Red Army on their passage through the country were Communist; in Bucharest the Control Commission operated under statutes promulgated by the Soviet Government.[2] Until the Potsdam Conference the British and American Missions were not structurally part of the Control Commission but were treated as delegations to it: this situation was rectified after Potsdam, but even then meetings were called only on the invitation of the chairman. As to Article 13, the effective restoration of legal rights and interests depended on a government authority which from 6 March 1945, under Russian pressure, became increasingly hostile to Britain and the United States. In this situation the position of Western-owned companies was weakened at the outset by the reluctance or inability of the governments of their shareholders, particularly while the war was still on, to make an issue of any Romanian problem which might jeopardize relations with the Soviet Union on more far-reaching questions such as the settlement of Germany.

The failure of the British and American Governments during the Potsdam Conference effectively to controvert or modify the position achieved by the Soviet Union in Eastern Europe as a result of military operations, and the absence of any Western

[1] A note in these terms was submitted on 1 Apr. 1944 by the Soviet ambassador, Gromyko, to the U.S. Government, and the declaration was repeated by Molotov the next day to foreign press correspondents. The April declaration was recalled by Moscow radio on 25 Aug. It is still impossible to determine whether, until they arrived in the country itself, the Russians believed they could achieve their general aims in Romania on these terms, and only later found they had miscalculated, or whether the declarations were designed to smooth the path of the Red Army into the Balkans by encouraging disengagement from the Axis, and soothe Russia's allies, with whom there had already been serious rifts over Poland. The allies' general inference *after* the Russians arrived in Romania was that the declarations were merely propagandist.

[2] *The Conferences at Malta and Yalta*, p. 238.

forces in the area, ensured that the exercise of Russian authority in Romania remained virtually unrestricted. This authority was used to reorganize government and administration on Communist lines, which culminated in the transformation of the Romanian monarchy into a "Peoples' Democracy" in January 1948.

The Armistice Agreement and the industry

During this period, therefore, the political and international context of the oil industry in Romania was drastically changed. German authority over the industry had been exercised either through or with the consent, however reluctant, of the Romanian State. The dichotomy which is the central feature in this study still held. With the advent of the Russians, however, it broke down, since, due to the way in which the Russians enforced the Armistice Agreement, there were two rival "state" authorities in Romania—one increasing in effectiveness at the expense of the other—with conflicting policies towards the oil industry. These tensions lasted approximately two years. Further, since the basic instrument defining Romania's status was signed by Russia's allies (and not merely by the Soviet Union), they were concerned in the fulfilment of its terms. The political problems of the oil industry therefore became a factor in East–West relations. British and American involvement was confirmed by the existence of the "Allied" oil companies, whose position the U.K. and U.S. Governments attempted—and failed—to protect. The dilemma of the "Allied" companies is illustrated by the operational outcome: as Romanian nationals they sustained the burden of oil reparations to the Soviet Union from which, as "United Nations" nationals, the Potsdam Agreement, in principle, exempted them.[1] In this sense Romania's oil obligations to Russia were paid by Britain and the United States.

[1] See *Protocol of the Proceedings of the Berlin Conference*, paragraph XX and Annex II. The Protocol was published as Cmd. 7087 in 1947. These items and the paragraph on "Oil equipment in Rumania" (XIII) were omitted from the report on the Conference issued by the Foreign Office on 2 Aug. 1945 and published in *The Times* the following day (p. 8). It is not clear how far the acceptance of the principle was known to interested parties in the oil industry at the time, who in their representations to government tended to rely on the clauses for the restoration of their property, rights, and interests in the Armistice Agreement with Romania. The drafting history of the Protocol (including paragraphs XIII and XX) was published in *FRUS: The Conference of Berlin 1945*, vol. ii, pp. 1477 ff., in 1960.

The fact that the industry as a whole was common to both a protracted internal revolution and an international conflict makes it necessary to examine the political consequences of the Armistice and its fulfilment in some detail.

After the overthrow of the Antonescu regime, parliamentary government was resurrected in Romania but, owing to Russian and Communist action, remained in a state of almost continuous crisis. A provisional "Government of National Union" was set up under General Sanatescu, A.D.C. to the King, with a Cabinet composed predominantly of non-party military officers but including the leaders of the National Peasants (Maniu), the Liberals (Dinu Brătianu), the Socialists (Petrescu), and the Communists (Pătrăscanu). Since their dissolution by law in February 1938 the traditional party organizations had atrophied: the Communists' initial strength was recruited mainly from members and sympathizers released from jail after 23 August. Ana Pauker, Vasile Luca, and Emil Bodnăraș arrived from Moscow with the army of occupation, which sealed off large areas of the country from Bucharest and set up Communist administrations. The traditional parties had no opportunity of resurrecting their provincial support, and the concentration of political life in Bucharest worked to the advantage of the Communists, whose party was styled *Partidul muncitoresc român* (Romanian Workers' Party) and whose policy-statements at this time stressed non-controversial issues such as prosecution of the war, alliance with the Soviet Union, and "government according to the principles of democracy".

On such a basis, and with Red Army assistance, the Communists began to build up their organization through the P.M.R. and a number of "Front" organizations.[1] The trade unions were reformed under Communist auspices, in virtue of an agreement concluded between Petrescu, the Socialist leader, and Pătrăscanu in May 1944 by which all labour leaders who had collaborated with Carol or Antonescu were barred from holding office in the post-war union structure: this ruled out most of the Social Democrats. Apart from subscriptions, the Communists financed themselves

[1] Of these the most important were: (i) *Frontul Plugarilor* (the Ploughmen's Front), originally a radical agrarian party under Petre Groza, which the Communists infiltrated to the point where it became their "rural branch". (ii) The Union of Patriots, designed to enrol "progressive" business men and intellectuals. (iii) ARLUS, a society to foster intellectual and cultural friendship with Russia.

by levying contributions from industrialists like Auşnit and operating the black market. From the outset they enjoyed a privileged position in regard to means of communication—radio time, supplies of newsprint,[1] etc.—and could reckon with the prudential support of those who saw them as the "Russian" party, and therefore the more likely to prevail: support from this source increased as the ineffectiveness of Britain and America became more obvious.[2]

The four political parties represented in the cabinet were grouped together in a "National Democratic Bloc" but this grouping broke down over the policy of purging officials and over the reorganization of local government in the provinces. Street demonstrations, at which the armed squads organized by Bodnăraş made their first appearance, were accompanied by the creation under Communist auspices of the National Democratic Front (F.N.D.), comprising the Communists, Socialists, United Trade Unions, the Ploughmen's Front, and the Union of Patriots. Pătrăscanu demanded that the last three organizations be admitted to a share in the government and, on refusal, withdrew from the Cabinet on 18 October. Petrescu resigned on the same issue. After two weeks of political crisis, a new government was formed on 4 November under Sanatescu, in which seven Cabinet offices, including those of Vice-Premier (Groza), Justice (Pătrăscanu), and Communications (Gheorghiu-Dej), were held by representatives of the F.N.D. Petrescu, Maniu, and Brătianu remained outside the new administration, although their supporters were included.

The immediate political problem was the fulfilment of the Armistice Agreement, particularly Article 12 concerning the restitution of Soviet property, which occasioned discussions with the Soviet Government in the person of Vyshinsky, Vice-Commissar of Foreign Affairs (8 November–8 December). The Russians maintained that the obligation to return property extended not only to Romanian seizures still in existence on 12 September 1944 but also to a quantity of goods equal to those consumed or destroyed during the invasion of the Soviet Union

[1] Their leading journals were *România Libera*, *Scînteia* (dailies), *Contemporanul* (weekly), and *Viaţa Sindicală* (weekly, the trade union paper).

[2] On methods, see R. L. Markham, *Roumania under the Soviet Yoke* (Markham was the Balkan correspondent of the *Christian Science Monitor* and was expelled by the Control Commission in July 1946 for having too close an association with Maniu's party): also P. B. Steanu, "The Nationalisation of Roumanian Industry", *The Journal of Central European Affairs*, vol. xi, no. 1 (Jan.–Apr. 1957).

from June 1941 onwards. Antonescu's communiqués were taken as the basis for quantifying this claim. The Romanian Government countered by referring the Russians to the war-indemnity clauses of the Armistice, which referred explicitly to "losses caused . . . by military operations and by the occupation . . . of Soviet territory". The impasse on this issue remained while the Romanian Government set up a "Commission for Implementing the Armistice" and on 9 November issued peremptory orders for the return of Soviet property within five days. Difficulties arose, in particular over the return of items purchased from the official trading bureau created in Bucharest by the wartime "Government of Transnistria" and also from Bessarabia which, in the Soviet view, had the same status as "Transnistria", i.e. Russian territory occupied by the enemy. Since the chairman of the commission was a close associate of Maniu and its members included the Minister of Foreign Affairs, these disputes were carried into the Cabinet, still divided over the policy regarding "war criminals" and the reconstruction of local government. In consequence a new Cabinet crisis developed at the beginning of December. Sanatescu resigned and was replaced on 6 December by General Rădescu, whose record showed him as anti-Carol and anti-German and whose Cabinet included the same F.N.D. elements as before, strengthened by the presence of an F.N.D. nominee, Teohari Georgescu, in the Ministry of Home Affairs, which controlled the police.

The Rădescu Government promulgated on 21 January 1945 a law for the prosecution of war criminals and a decree on the organization of trade unions.[1] However, during that month the F.N.D. began a campaign to take over the government on the grounds of the need to fulfil the Armistice terms, to carry out drastic purges, and to put through an agrarian reform. Additionally, the F.N.D. represented itself as assured of the return of Northern Transylvania if it gained power. The agitation—the local success of which was assured by Red Army support and the absence of Romanian forces at the front—was marked by street demonstrations and strikes and culminated in a second intervention by Vyshinsky at the end of February 1945[2] resulting in

[1] Law No. 52: *Monitor Oficial*, no. 17 (21 Jan. 1945).

[2] See J. F. Byrnes, *Speaking Frankly*, pp. 50–3; on political events after the Armistice see also article by Vasile Luca in the Cominform journal *For a Lasting Peace, for a People's Democracy* (4 Nov. 1949).

the imposition of an F.N.D. regime under Petre Groza on 6 March.

The Cabinet contained seven F.N.D. members, including the principal officers. Immediately after it was installed, representatives left for Moscow for discussions eventually issuing in the treaties of 8 May 1945.[1] The inclusion of opposition elements and the fact that until 1947 the Communist leadership still included Pătrăscanu, who displayed tendencies later anathematized as "Titoist",[2] entailed that the Government's performance was not always satisfactory to the Russians.

Fulfilment of the Armistice Agreement remained the major problem of Russo-Romanian relations. The Russian attitude was summed up by General Susaikov, who complained on the 13 June that the peasantry, not the bourgeoisie, was bearing the burden of the Armistice; that much of the cereals and livestock collected from the peasants went to the black market, for which the Russians got the blame; that the Romanian Government was purposely evading Article 8 of the Agreement and dissipating enemy assets; that Article 10 deliveries were not being fulfilled and Russian troops in Romania had to be supplied from the Soviet Union; that reparations deliveries were behind schedule due to lack of control of the black market; and that the restoration of property under Article 12 had not been completed. The Russian peasants knew what the Germans and the Romanians had done and were impatient.

The outstanding questions were settled in September 1945, when the Groza Government concluded a series of agreements in which the Romanians accepted the Russian interpretation of Article 12 deliveries and thus incurred an obligation to pay in goods (valued at prices ruling on 1 April 1945) to a total of 3,000 million lei over and above those already shipped or ready for shipment on 11 September.[3] They also conceded the "capture" of the Romanian navy and the appropriation of merchant vessels by the Russians,[4] as well as the locomotive and railway equipment.

[1] See below, pp. 276 and 294.

[2] Pătrăscanu was disgraced in Feb. 1947, tried and executed in Apr. 1954: on his "nationalist" heresies, see discussion in G. Ionescu, *Communism in Roumania 1944–1962*, pp. 151–6.

[3] The Romanians were subsequently obliged to pay freight on these deliveries to destinations in Russia, and hiring charges for rolling-stock ceded under the Armistice but left in Romania.

[4] The Soviet Government returned twenty-three of the vessels taken over,

In return the Soviet Union scaled down its demands for provisioning the Red Army in Romania, accepting a cash settlement in lieu. The terms agreed, however, recognized the right of occupation (that is, as distinct from transit) not granted by the Armistice Agreement. On the operational aspects of the Agreement of 11 September, the Romanian Government made further concessions.[1]

The creation of a Russian sector and its international repercussions

Under the Armistice and these first post-war agreements, the Soviet position in Romania thus became much stronger than that of the Germans during the war. From this date onwards ultimate compliance with the Soviet wishes was assured. The opportunities for administrative dissent diminished as the Groza regime adopted tougher domestic policies; the opposition was neutralized and the Communists created their own institutions to take over the running of the State.[2]

The Armistice Agreement not only altered the context within which the industry had to operate but also bore on it directly, as the supplier of fuel to the Red Army and Romanian forces while the war continued (Article 10),[3] and as a source of reparations (Article 11).[4] Owing to the particular construction put on the

on condition that the Romanian Government brought them into the joint company, Sovromtransport (see below, p. 279 n. 3).

[1] 195.

[2] See below, p. 304.

[3] Article 10: "The Romanian Government must make regular payments in Romanian currency required by the Allied (Soviet) High Command for the fulfilment of its functions, and will in case of need ensure the use of Romanian territory of industrial and transportation enterprises, means of communication, power stations, enterprises, and installations of public utility, stores of fuel, fuel oil, food, and other materials and services in accordance with instructions issued by the Allied (Soviet) High Command. Romanian merchant vessels, whether in Romanian or foreign waters, shall be subject to the operational control of the Allied (Soviet) High Command for use in the general interest of the Allies. (See Annex to Article 10.)"

Annex to Article 10: "The Romanian Government will withdraw and redeem, with such time-limits and on such terms as the Allied (Soviet) High Command may specify, all holdings in Romanian territory of currencies issued by the Allied (Soviet) High Command, and will hand over currency so withdrawn free of cost to the Allied (Soviet) High Command."

[4] Article 11: "Losses caused to the Soviet Union by military operations and by the occupation by Romania of Soviet territory will be made good by Romania to the Soviet Union, but, taking into consideration that Romania has not only withdrawn from the war, but has declared war and in fact is waging war against Germany and Hungary, the parties agree that compensation for the indicated

Agreement by the Soviet Union, Article 7,[1] whereby the Romanian Government undertook to hand over "war material", and the protocol to the Agreement, Article 2[2], defining "war material", both became relevant to the oil industry. Among the territorial provisions, the Russian acquisition of Bukovina and reacquisition, from the Soviet standpoint, of Bessarabia (Article 4) implied the loss of distributing assets, and the cancellation of the Vienna Award (Article 19) implied the return to Romania of the Transylvanian gas-fields.

The Armistice Agreement replaced German control of Romania's oil industry by Russian through both the general supervision of Romania instituted under Article 18 and, more specifically, the obligations assumed by the Romanian Government under Articles 10 and 11. Control over Romanian output to secure these two objects was vested by the Control Commission in the Russian Oil Directorate, which set up offices at Ploești and Câmpina with key personnel at all refineries and export stations. The Directorate's function was to organize the repair of refineries and to supervise the distribution of oil products, the latter in co-operation with the Russian traffic office, which controlled the railways and allocated tank-cars. The Directorate began a systematic inspection of the industry: through its subsidiary offices it initiated a system of regular reports by oil companies concerning their activities. Officials also visited members of the companies' staff to cross-examine them about the companies' business, directors, personnel,

losses will be made by Romania not in full but only in part, namely to the amount of 300 million United States dollars payable over six years in commodities (oil products, grain, timber products, seagoing and river craft, sundry machinery, etc.). Compensation will be paid by Romania for losses caused to the property of other Allied States and their nationals in Romania during the war, the amount of compensation to be fixed at a later date. (See Annex to Article 11.)"

Annex to Article 11: "The basis for settlement of payment of compensation provided for in Article 11 of the present Agreement will be in the American dollar at its gold parity on the day of signing of the Agreement, i.e. 35 dollars for 1 ounce of gold."

A convention setting out the detailed procedure for executing deliveries under Article 11 was concluded on 16 Jan. 1945. See below, p. 292.

[1] "7. The Romanian Government and High Command undertake to hand over as trophies into the hands of the Allied (Soviet) High Command all war material of Germany and her satellites located on Romanian territory, including vessels of the fleet of Germany and her satellites located in Romanian waters."

[2] "2. That the term 'war material' used in Article 7 shall be deemed to include all material or equipment belonging to, used by, or intended for use by, enemy military or para-military formations or members thereof."

and so on. By these means the Directorate acquired a detailed knowledge of the activity of each individual company, which they were able to follow from the drilling stage until products were dispatched from the refineries.[1] Their methods made for extra work, much *paperasserie*, and a great deal of wasted time owing to those interminable discussions which seem so congenial to the Russian mind.[2]

In the oilfields themselves, between September and December 1944, the Russian authorities removed 50,678 tons of equipment.[3] This action vitiated the refinery repair schedules and field drilling programmes, and thus retarded the restoration of the industry after the extensive Allied bombing. The Soviet removals extended to other industrial plant and equipment—including railway rolling-stock and the Malaxa 14 in. steel-rolling mills,[4] both of which affected the current and future performance of the industry. Rail tank-cars became "Russian" by the simple expedient of painting out their registration markings.[5]

During the armistice period the Oil Directorate remained the ultimate authority over the operations in the oil industry.[6] Its activities were directed at maximizing the output of gasoline and other fuels and enforcing delivery schedules—and involved friction both with oil companies and with the Romanian State. Disputes with the former were mainly technical in character, but those with the latter arose from the interpretation of Articles 10 and 11 of the Armistice Agreement[7] and from Soviet requests for the production of company records on, for example, refining-plant design, and geophysical work.[8]

[1] 196.
[2] 197.
[3] Malaxa's subsequent co-operation with the Communists earned him in 1946 compensation in free currency for this and other losses, and, through a Romanian Government decree, the complete restoration of his property, rights, and interests as they were before German control. It also earned him a passport.
[4] Comprising 47,712 tons of tubulars, 2,966 of miscellaneous equipment, spare parts, electric motors, etc. 33,380 tons of tubulars were left in company stocks. The Russians also began dismantling Astra's alkylation plant with unskilled labour, but on representations by the British Mission abandoned work.
[5] In that the Soviet requisitions extended to Allied property, the action also brought about the first breach between the Russians and their Western allies over oil policy: see below, p. 280.
[6] After Apr. 1946 authority was shared with the Oil Co-ordinating Commission of the Romanian Government, but the latter remained subordinate.
[7] See below, p. 293.
[8] 198.

The Oil Directorate ensured Soviet control over the operations of the oil industry as it was, but the Directorate's activity, deriving as it did from the Armistice Agreement, was subject to at least formal limits and did not extend to the ownership of oil rights. The Soviet interest in the latter was secured in consequence of the formation of the Groza Government as part of a general policy of providing for long-term Soviet control of Romania, the basis for which was the Agreement on Economic Collaboration signed in Moscow on 8 May 1945.[1] The protocol to the Agreement posited (under Part I) the creation of a Russian-Romanian company "for the exploration, exploitation, treatment, and commercialization of oil and oil products" and briefly indicated the means whereby this was to be done. Romania was to contribute lands and crude oil (Article 2), and Russia the equipment necessary for the company's development (Article 3); membership of the board of management was to be equally shared (Article 4); and the company was to be accorded preferential rights on non-state oil lands (Article 5). It was specifically stated that "private Romanian companies, individuals, and associations may be called in to take part in this company"—a proviso made necessary by the absence of a Romanian state-owned refinery. Further, Article 1 implied either that the Romanian State would monopolize distribution or that it would expropriate the "foreign" companies hitherto concerned with that function.

The convention giving effect to the protocol was signed on 17 July 1945. The new company was called Sovrompetrol; its capital was established at 5 billion lei, divided between the founders. The Romanian contribution was to consist of shares in a number of Romanian companies (principally Creditul Minier and Redevența) and their assets, together with 50 per cent of state crude-oil royalties valued at 1939 prices. This element was necessary to bring the total Romanian up to the required 2½ billion lei. The Russian share was to comprise oil equipment[2] and goods valued at 1939 prices and shares in oil companies representing (in the Soviet view) former German assets. Sovrompetrol's privileged status was indicated by Article 2, implying that it would not be subject

[1] *BFSP*, vol. 149, 1955, part III. There were in fact two agreements signed on this date, the other being for the exchange of goods (see below, p. 294).

[2] "Oil equipment", in practice, comprised the tubulars and other materials removed from the fields between Sept. and Dec. 1944 and materials rejected by the Russians from reparations on grounds of inferior quality.

to the legal restrictions imposed on other "foreign-capital" companies, and Article 5, setting out the practical benefits it was to enjoy in the day-to-day operations of the industry.

Among these privileges were preferential rights to oil concessions on state lands, the opportunity of obtaining other concessions by mutual agreement with the State—thereby obviating the need for participation in the usual auction for oil concessions—and the right to meet any bid made for any concessions which was put up by the State for such auction. Sovrompetrol was also guaranteed 75 per cent (150,000 tons p.a.) of all crude-oil royalties accruing to the State from the production by all companies from state concessions. The remaining 25 per cent, theoretically available for allocation by the State among refinery-owners with little or no production of their own, in practice often reverted to Sovrompetrol. The company also enjoyed certain fiscal privileges in respect of tax exemptions and state royalties.

The realization of this privileged position was not without difficulty in practice. The original protocol stipulated that the company would be set up by 11 July, but the convention was signed only on the 17th and included two provisions involving possibilities of further delay—that constituent members should establish their legal right to dispose of the shares[1] in the companies they were to contribute to Sovrompetrol, and that the latter was to function under Romanian law.

The first provision occasioned no difficulty on the Russian side, but the Romanian directors of Creditul Minier and Redevenţa refused to allow the Romanian Government to sign away their assets,[2] and eventually, on 28 July 1945, the Romanian Government

[1] The transfer of shares was regulated by Decree Law No. 673, published in the *Monitor Oficial*, no. 161 (19 July 1945). Shares of banks and oil companies (enumerated in schedules) were to be transferred to "Prombank" and "O'bedinenie Ukrneft" respectively, within a period of 15 days after an official intimation to do so: failure to comply rendered the general manager or other stipulated executives of a company or the individual shareholder liable to correction or imprisonment of from one to five years, with no consideration of extenuating circumstances.

[2] The board of Creditul Minier refused to recognize the validity of arrangements made without its consent, though it declared itself willing to co-operate with the new organization. In return the Minister of Mines and Petroleum appointed to the company a "sole administrator", Anton Dumitriu, by profession a professor of philosophy, under the law of 6 Oct. 1944, revised in May 1945, that related to cases where the attitude of a company was deemed detrimental to the political or economic interests of the State. In June 1947 Dumitriu was

issued a special decree empowering it to sign the agreement in the names of the two companies. The recalcitrant officers were later removed. The second provision delayed the legal constitution of the company. Since from August 1945 onwards the King refused to sign the decrees of the Groza Government, Sovrompetrol was the subject of another special decree, on 24 October, whereby the company was "deemed legally constituted . . . without the necessity for authentication of documents or fulfilment of any other formality".[1] The charter and statutes were set out in the same document, and Sovrompetrol representatives were installed in the offices of the constituent companies.

The effect of these enactments was to create a Russian interest in the oil industry extending to 30 per cent (3,800 tons a day) of total current production, with refinery capacity of 6,000 tons a day and cracking of 2,200 tons a day.[2] Until January 1948 the constituent companies retained their separate identities but were managed by Soviet technicians in liaison with a central office under a general manager, Stupin, of Soyuznefteexport. The adhesion of the companies on the Romanian side was piecemeal. The process of welding all the companies together was protracted by resistance on the part of the shareholders and personnel of the companies concerned. Concordia was the last to be absorbed into the new combine, largely due to personal rivalry between its Russian-nominated manager, Ciopashvili, and Stupin.

The privileged position of the company was reinforced by subsequent action by the Romanian Government, particularly in 1947, when it conceded to the concern a guaranteed level of profit, accorded "compensation" for loss of profits due to the State's low-price oil policy, and permitted remittance of profits

dismissed in consequence of charges of embezzlement of funds in Switzerland, and, after a month's interregnum by a—politically unsatisfactory—Social Democrat nominee, was replaced by Lassersons, associate of Stupin in Sovrompetrol, who carried through the fusion of Creditul Minier with that organization in 1948. See 199.

[1] *Monitor Oficial*, no. 243; see also 200.

[2] Other companiea' shares, for comparison:

	Production	Distillation	Cracking
Astra Română	34% 4,500 t/d	4,500 t/d	1,000 t/d
Românо-Americană	11% 1,400 t/d	1,800 t/d	250 t/d
Steaua Română	10% 1,300 t/d	2,000 t/d	. .
Unirea	9% 1,200 t/d	1,200 t/d	. .

in U.S. dollars. Sovrompetrol's working relationships with the Romanian Government were at the outset from its own standpoint not wholly satisfactory. Under the convention of 17 July 1945 the Romanian Government only undertook to ensure the participation of Romanian companies; it was not a direct participant and this permitted it a limited freedom of manœuvre with regard to Sovrompetrol.[1] Nevertheless the final submission of the Groza Government to Soviet policy during 1947 ended any source of friction, and was marked by the granting of the further privileges to Sovrompetrol already referred to.

In establishing itself as an operating company, Sovrompetrol could offer as an inducement to its prospective business associates drilling materials, equipment, and transport—a prerequisite to their meeting the work schedules imposed on them by the Russian authorities. Of these items the combine, through its connection with the Soviet Union (and the Russians' refusal to allow imports from other sources), had a monopoly of supply. Sovrompetrol also sought to augment the rights and output of its constituent companies by purchasing private rights on oil lands, and attempted to operate with other companies both through the A.I.P.R.[2] and bilaterally. In January 1946 approaches were made to Steaua Română to drill jointly exploitation wells on Sovrompetrol's lands, with Sovrompetrol providing the casing and sharing the output. Later, in the spring of that year, Stupin broached the idea of Sovrompetrol's entering the Distribuția as a shareholder, with representation on the board.

Sovrompetrol became what Kontinentale Oel had been designed to be in the event of a German victory. Both organizations derived their standing in Romania from comprehensive agreements of economic reconstruction at a time when their respective states were the only source of supply for materials and equipment. The founding of both companies explicitly replaced Western interests, and both issued from a policy of securing post-war sources of supply.[3]

[1] It appears that the Romanian Government committed itself initially to direct partnership with the Russian Government but withdrew: see 201.

[2] The total representation comprised the representatives of the individual companies plus the spokesman for Sovrompetrol, who invariably commanded the former's public support.

[3] Sovrompetrol was one of several such organizations representing the Soviet interest in Romania. Of the others, two impinged directly on the oil industry,

The establishment of Soviet control over the industry and the creation of a Russian undertaking showed an all but complete disregard of the established interests of Russia's allies and became a factor in the general breakdown of the wartime alliance between them.[1] Friction arose out of the day-to-day supervision of Allied property by Directorate officers, particularly over demands for information, much of which was according to the British Mission "of such a purely local nature that it could never be of service for the development of the oil industry in the Soviet Union": the Russian A.C.C. officers, for example, quite explicitly demanded "details of a patented refinery process which they stated was for the ex-German and now Soviet-controlled Petrol Block refinery". The assiduous gathering of information by Russian personnel became of more obvious relevance to Romanian conditions after the setting up of Sovrompetrol.[2]

Difficulties of a more fundamental nature derived from mutually exclusive views of the nature of German assets. These had been assigned to the Soviet Union by inter-Allied arrangements, but no agreement could be reached as to the definition of "a German asset" or what constituted a proper transfer.[3] The Russian attitude was that any property taken over by the Germans had *ipso facto* become German and on these grounds fell to the Soviet Union. The Governments of the United Kingdom and the United States dissented from this view: they stressed the notion of ultimate beneficial interest as essential to determine the real, as distinct from the apparent, national character of industrial property.

Sovrombank and Sovromtransport. Sovrombank was set up in Aug. 1945 with a capital of 1,200 million lei, of which the Russians contributed 474 million (representing the entire share capital of the Banca Comercială Română, acquired by the Germans from the French and Belgian owners during the war, and the Soc. Bancară Română), while the Romanian State contributed most of the capital of the Banca Agricolă and 30 per cent of the capital of the Banca de Credit, amounting to 120 million lei. Other Romanian banks subscribed in cash to the new institution, which had a larger capital than any other bank in Romania. The Russians also acquired as war booty various small direct participations in six other Romanian banks. Sovromtransport, with a capital of 3,000 million lei, was also set up in August 1945 to control sea and river transport, harbours, and other installations.

[1] On the general aspects of this breakdown see W. H. McNeill, *America, Britain and Russia: their Co-operation and Conflict 1941–1946*; Byrnes, Books II and III; M. F. Herz, *passim.*

[2] 202.

[3] See F. A. Mann, "German External Assets", *British Yearbook of International Law*, 1947, pp. 239–57.

Strictly wartime transfers should be considered under the terms of the Inter-Allied Declaration 5 January 1943,[1] to which the Russians had adhered. Efforts by British and American representatives to gain Russian acceptance of their points of view were unsuccessful.

The first clash in Romania of these differing views occurred over the removal of 50,678 tons of oil equipment. Heated discussions took place in the Allied Control Commission. The British argued that removals were wholly unjustifiable and constituted looting; the Russians replied that the material, having been supplied by Germany to Romania for use by the oil companies who had supplied oil to the German armed forces, should properly be regarded as war trophies under Article 7 of the Armistice Agreement; alternatively that, even if it was not booty, it was being put to a much better use for the war effort by the Russians than it could ever be by the Romanians. There being no local agreement, the matter was referred to Moscow.[2] The removals were specifically raised at Yalta but not discussed through lack of time:[3] in consequence they became part of the agenda for the Potsdam Conference.[4] In the meantime, as a result of the representations, part of what had been removed was returned, though not reassembled.

The question was eventually discussed by the Foreign Ministers at Potsdam on 22 July and 1 August 1945[5] and was referred to two "bilateral commissions of experts",[6] both of which met in Bucharest. The U.S.–U.S.S.R. commission in its deliberations never got beyond the procedural stage and was terminated on 12 June 1947.[7] The U.K.–U.S.S.R. commission, by the end of October 1945, got as far as the acceptance by the Russians of documentary evidence but achieved no more. The head of the oil section of the British Mission reported the attitude of the Russians to the examination as follows:

In fact, on 11th November the members of the Russian delegation "disappeared" and were absent for some 7 weeks. After this period we

[1] Cmd. 6418, Miscellaneous no. 1, 1943.

[2] See Briefing-Book Paper "Principal Roumanian Problems" in the section "I, Pre-Conference Documents", in *FRUS—The Conferences at Malta and Yalta, 1945*, p. 247.

[3] See ibid., pp. 893 and 964.

[4] See *FRUS—The Conference of Berlin 1945*, vol. i, pp. 420–1.

[5] See ibid., vol. ii, pp. 234–7 and 547–8: for U.S. case, see pp. 741–2; U.K. case, pp. 143–4; Soviet Union case, pp. 745–6.

[6] *Protocol of the Proceedings*, XIII.

[7] See *The Department of State Bulletin*, vol. xvii, no. 422 (3 Aug. 1947).

accidentally ran into the only member of the old delegation who in the meantime had returned to Bucharest. We were calmly asked whether we were interested in having further meetings. At the same time we were informed that the composition of the Russian delegation had been changed.[1]

The attempt to solve the problem was abandoned, and companies sustaining losses were left with claims under the Romanian Peace Treaty.[2]

This cavalier treatment of Allied interest was also evident in the setting up of Sovrompetrol. From the beginning it was maintained that part of the Russian contribution would be the Colombia and Concordia companies, both of which were put under Russian control in the autumn of 1944 on the grounds of their being German property. Control was turned into management in both companies, with the consent of the Romanian Government. In the Colombia the Russians demanded from the management a provisional certificate covering the German-held shares, and used it to convene a shareholders' meeting at which the existing directors were dismissed and replaced by Russian nominees. Concordia passed into Russian management in similar fashion: on 14 May 1945 a new board was appointed, consisting of supporters of the F.N.D., with a Russian observer: the latter was replaced on 28 November 1945, when Ciopashvili and Stupin of Sovrompetrol were co-opted as members. The transfer of the relevant share-ownership in each company was ordered by the Romanian Decree Law 573 of 18 May 1945 on the grounds of the shares having belonged to "persons or legal bodies of German nationality". The transfers were disputed by the French and Belgian Governments. In August 1945 the Colombia sale was declared illegal by the French Government, and in August 1946 the directors of Petrofina were, after prolonged hearings, found guilty of having "collaborated" with the wartime occupying authorities by selling Concordia shares to the Germans. By reason of this decision the Belgian Government confiscated the shares.[3]

Apart from giving rise to a dispute between governments over share-ownership, the Russian action in setting up Sovrompetrol represented exceptional treatment for one specific Allied interest,

[1] 203.
[2] See below, p. 297.
[3] See *Journal des tribunaux* (14 July 1946), Brussels, p. 386.

although, under the Armistice Agreement, the Romanian Government had bound itself to treat all uniformly.

The creation of a Russian enclave in the oil economy tidied up the industry, organizationally, since the new Russian interest comprised a number of small-scale units whose existence had depended on an officially guaranteed sphere of activity rather than on market considerations. There remained, however, other "Allied" concerns which, in 1945, still represented 61 per cent of crude-oil production and whose operations were essential to fulfilling Romania's external obligations and to providing for domestic consumption. At this time all companies were faced with the problems of repairing war damage, restoring output, and trying to renew their capital equipment. For the Western-owned "foreign-capital" companies, this implied the earning of foreign exchange in order to finance imports of essential materials. They were thus vitally interested in the resumption of their pre-war trading connections, particularly with the United States and Britain, and in re-engaging foreign technical and managerial staff.

Both aims, after March 1945, ran counter to Romanian state policy. The surplus remaining for export, after Armistice commitments had been met, went either to the Soviet Union, or to states falling under Russian control,[1] while the Russian authorities in Romania held up or vetoed attempts at sales to unfavoured destinations, like Turkey. The Control Commission delayed granting entry permits to foreign staff.[2] In post-war Romania the position of the "foreign-capital" companies became no stronger than their likely utility in any given supply situation: in the absence of a general resolution of problems with the Soviet Union, intervention by the governments of the shareholders in these concerns was likely to be ineffective. Furthermore, in their role as Romanian and capitalist undertakings, these companies had become involved in a revolution set in train by the Russians on their appearance in Romania, in which the setting-up of the Groza Government marked an important, ultimately crucial, stage.

After the *coup d'état*, the German organization of the oil industry collapsed.[3] To ensure compliance with Romania's anti-Axis policy

[1] See below, pp. 292–5.

[2] None, in fact, was given until after the setting up of the Groza Government.

[3] Hillgruber, p. 191. Neubacher was away from the scene in Belgrade

while the war continued, the Rădescu Government on 6 October 1944 appointed administrators to companies "in which on 23 August 1944 were functioning one or more administrators [who were] subjects of a state with which Romania was at war after that date or who had taken an attitude notably favourable to such a state or had acted prejudicially to national interests, political or economic".[1] At the same time the mandates of such officials were revoked. The administrators were specifically empowered to convene general assemblies of shareholders within six months, so that the companies, particularly those with foreign shareholdings, could put their affairs in order after the administrative upheavals of the previous four years. The administrators were appointed from within the industry, and in most cases from within the firms over which they presided, in the Government's interest. This not only served the needs of the moment but pointed to the State's adopting a policy towards the industry based on their mutual advantage rather than on preconceived political dogma.[2] The advent of the Groza Government made such ideas illusory: Government policy for the industry aimed at its complete suppression as an independent organization; legislative, financial, and, where necessary, diplomatic means were employed to this end.

At this stage the apparatus of the State as regards its policy towards industry was extended through the organization of labour. In September 1944 *Confederația Generală a Muncii* (General Confederation of Labour) was resurrected as an exclusively Communist organization with headquarters in Bucharest and branches throughout the country extending down to factory committees—the different unions being interlinked at subsidiary centres, such as, for example, Ploești and Câmpina, in local syndicates. The function of this Communist organization—in the absence of a sympathetic government—was to make the revolution;

Clodius was handed over to the Russians and was subsequently deported, and von Killinger committed suicide: the Axis personnel in the various companies fled westward.

[1] Decree Law No. 478: see *MPR*, no. IX–X (1944), pp. 42–6; no. XI–XII, p. 96: this Law was modified by the Groza Government, 29 May 1945, in Decree Law No. 405, whereby administrators could be appointed "to any enterprise when the activity of its management is harmful to the political or economic interests of the State". See also below, p. 303 n. 6.

[2] 204.

this, at the level of the factory or action committees, involved intimidating all opposition among workers or management and building up positive, ideologically articulate, support for use in organized spontaneous demonstrations and other political activities.

In the oil industry the relationship between wages and prices,[1] both state-controlled, provided a basis for an attack on the position of management: actions on salaries and Economat[2] supply were used to press demands and conditions on purging "Fascist legionary elements"[3] and the hiring and dismissal of labour. Since the pressing of these demands was accompanied by intimidation and physical violence, with the co-operation or connivance of police and Red Army personnel, managements tended to comply. Under the Groza Government, it was noted, the general attitude of labour was not fundamentally quieter, but discipline improved.[4] Henceforward Government and union policy increasingly responded to the same initiatives. Organized labour therefore came to represent an extension of the area of state authority over the oil industry and was an entirely new factor in their relationship.

The A.I.P.R. responded to industrial unrest by suggesting discussion of outstanding questions by the industry as a whole. This principle was accepted by the unions, who produced a collective labour contract on which discussions took place during 1945. The draft contract set out categories of work and various steps and grades of remuneration, and attempted to delimit union rights in staff and administrative questions. The A.I.P.R. negotiating committee tried and failed to establish a single figure for wages in which all welfare benefits would be included: the

[1] Increases in wages were promulgated by law on 27 Sept. 1944, but were inadequate to meet the increase in prices, particularly of food, which followed the breakdown of internal communications and Red Army requisitioning. The A.I.P.R. gained Government consent to pay higher wages than those stipulated by law from 1 Nov., but this merely changed the basis of industrial unrest. 205.

[2] Company stores set up during the war to ensure purchase and distribution of food and clothing to employees, at official prices.

[3] Workers in the oilfields, to which access by Bucharest management was barred by the Russian authorities, carried out their own purges without waiting for promised legislation. In practice, the definition of "Fascist" and "Legionnaire" was made by the purgers, and their selection was as arbitrary as their methods. The law was passed in 1945 and subsequently official joint worker–management purge committees were instituted within companies.

[4] 206.

unions supported the idea of a basic wage-rate to which various premiums were added. The contract was signed in January 1946 and was to operate for six months, but it proved difficult to implement even the "technical" clauses. The crucial problems of management–labour rights and industrial discipline were dealt with under Articles 82, 108, and 109, and Annex I. These set up regular meetings of authorized representatives of both employer and employee and defined the respective spheres of competence of management and unions. This system was complemented by the organization of disciplinary committees to judge and sentence any employee "guilty of an offence during working hours or on the company's property": the idea behind punishment being not to "harm an employee but to raise him to a higher civic level". The offences to be referred to the committee included not only traditional and identifiable misdemeanors such as pilfering, drunkenness, or absence without leave, but also "sabotage" and "defamation of the trade union movement", the definitions of which were more sweeping.[1] The existence of these offences and the vagueness with which they were defined assisted the Communists in their assertion of labour leadership.

[1] "6. As sabotage will be considered any act which obstructs the running of the company's activity or the normal functioning of the machinery, or any act which is committed in bad faith and leads to a reduction of output, whether individually or collectively.

"7. There shall be punished for attempts to defame the Trade Union movement any employee who brings an unfounded accusation against a Trade Union organization or any of its leaders with the clear intention of compromising the Trade Union movement.

"There shall likewise be punishable anyone who in any way seeks or endeavours to destroy Trade Union unity by acts which are clearly proved, or who obstructs any employee from joining the Trade Union.

"Complicity in all the foregoing offences shall be punishable to the same degree as the main offender if the offence was committed together or if the gain was divided between them. Anyone who does not disclose an offence shall be liable to punishment of a grade lower than that of the offender.

"The Disciplinary Committee will be advised of any offences by the body which has discovered it.

"An investigation will be carried out by the offender's immediate superior, by a controlling organization, or by a committee of inquiry consisting of a Trade Union delegate and a representative of the Management, nominated by the Disciplinary Committee.

"The Committee will act with impartiality, may take account of attenuating circumstances, and propose penalties with the object of re-education, and at the same time with the scope of strengthening the discipline in the company and of eliminating from the ranks of the workers those incorrigible elements who by their actions disturb the harmony and output. . . ."

The industry under the Armistice

The economic consequences of the Armistice were severe. At the time of the *coup d'état* the harvest was satisfactory and much of the rationing system had been abandoned: industrial output was disrupted by shortages of raw materials, due in part to air attacks in the spring of 1944 which also disorganized distribution —leading to a pattern of localized gluts and shortages. Inflationary pressures were building up, but were not unmanageable.[1]

The advent of the Russians precipitated the collapse of the already strained economy. The pursuit of the German forces in Transylvania and the requisitioning of draught animals and transport completed the disorganization of distribution. Local shortages of fuel and raw materials led to a contraction of industrial output, despite relatively little damage to plant. Imports ceased, whilst exports on reparations account began promptly. Finance was disrupted through the barring of access to Switzerland (which destroyed the basis for foreign exchange)[2] and the necessity imposed on the National Bank to exchange all roubles presented to it against lei—which encouraged wild speculation. Moreover, the expected effects of military operations were aggravated by the policy of the Russian occupying forces: the official requisitioning policy was supplemented by the efforts of individual soldiers or units.[3] The Red Army in Romania had no quartermasters' branch, and, as Otto Stern noted, the Russian troops "entered Romania as conquerers, not as allies as the majority of the Romanians—once their country had changed sides—tried to make themselves and others believe; it [the army] had come all the way from Stalingrad through its own devastated country into a land of

[1] The bank-note circulation increased by 40 milliards in 1943 and 47 milliards between Jan. and June 1944.

[2] The £ and U.S. $ were not quoted officially after Dec. 1941; quotations for Swiss francs were reintroduced by the National Bank in Apr. 1946 at lei 3,693 = Sw. fr. 1 for non-commercial transactions, and lei 5,500 = Sw. fr. 1 for approved sales of francs to individuals. To these rates various *ad hoc* premiums were added.

[3] The legal aspects of these confiscations were dealt with under the Peace Treaty of 10 Feb. 1947, which, ignoring Article I of the Armistice Agreement, stated that Romania "having concluded on 12 Sept. 1944 an armistice with the Governments of the U.S.S.R., the U.K., and the U.S.A., acting in the interest of all the United Nations, took an active part in the war against Germany". This meant that seizures prior to 12 Sept. were regarded as having taken place during "war operations" and therefore did not enter into the calculation of reparations.

plenty".[1] Requisitioning extended not only to goods but also to labour: the Netherlands Government representative commented:

this is being undertaken on a fairly large scale whereby the Russian authorities are not confining themselves to prisoners of war and enemy civilian internees but are laying their hands on the so-called Saxon and Schwabian sections of the population, that is Romanian subjects of "ethnically" German origin—the minorities. Men and women between the ages of 18 and 45 (or 35) are being transported to Russia. This has caused great consternation not only in Transylvania, where villages and towns are being emptied, but in other parts of the country as well. . . . Whoever bears a German name (except the Jews) is seized even if the family was naturalized generations ago without belonging to the minorities.[2]

A serious economic situation became more intractable, since recurrent political crises disrupted the administration of the economy; political issues pre-empted discussion, and the Government lacked the power of enforcement. More generally, the Romanians' contempt for and fear of the Russian occupiers engendered an acute feeling of insecurity—one consequence of which was a run on available goods and speculation in currency. These conditions brought about a steep and rapid rise in prices, which the Government sought to counter by cutting down available purchasing power through tax increases and the issue of a domestic loan to which the inducement to subscribe was the opportunity to convert some lei into gold.[3] This eased but did not stem the inflation which was the financial consequence of economic disruption. From September 1944 to the end of February 1945 the cost of armistice obligations was computed at 1,235,962 million lei, of which 1,148,460 million was covered by advances from the National Bank.[4] The note circulation doubled. Under the Groza regime there was no fundamental improvement, since fulfilling the Armistice as required by the Russians and maintaining the national economy intact were incompatible.[5]

The changes brought about by the Armistice are demonstrated by statistics in use at the time (see Table 5). Although the figures differ in showing the extent of the disruption, they concur in

[1] 207. [2] 208. [3] *MPR*, no. I–II (1945), p. 42. [4] 209.

[5] See speech of Tătăreanu (Governor, Romanian National Bank), 17 Feb. 1946: 210; also National Bank report for 1945, given to shareholders on 17 Feb. 1946: *MPR*, no. III–IV (1946), pp. 99–102.

showing a *relatively* stable situation prior to September 1944[1] which got out of hand after that date. The situation became steadier in August 1946, when the official reparations obligations were cut and Russian over-lifting of oil in replacement of other goods diminished. The economic context of oil-industry operations became, under the Armistice, one of unstable finance, with an excessive but increasingly disaffected labour force and a hostile political authority.

TABLE 5

	Astra Română Cost of living	Official* Cost of living	Retail prices
		1933 = 100	
1940	100	..	..
1941	133	..	..
1942	215	..	..
1943	301	..	..
1944	Aug. 570	Aug. 6,89·2	Aug. 1,052·6
	Sept. 628	Sept. 8,34·6	Sept. 1,266·2
	Oct. 731	Oct. 8,99·0	Oct. 1,386·4
	Nov. 929	Nov. 1,055·8	Nov. 1,665·2
	Dec. 1,080	Dec. 1,423·0	Dec. 2,429·2
1945	Jan. 1,198	Jan. 1,992·6	Jan. 3,453·8
	Feb. 1,389	Feb. 2,482·0	Feb. 4,365·4
	Mar. 1,555	Mar. 2,712·2	Mar. 4,928·7
	Apr. 1,860	Apr. 3,247·0	Apr. 6,197·5
	..	May 3,550·8	May 6,808·4
	..	June 3,708·1	June 7,049·1
	..	July 4,120·9	July 7,633·3

* Central Statistical Institute, quoted in *MPR*, IX–X (1945), p. 323.

(*a*) *Organization*

During 1944 the main effort within the industry was the repair of refining plants to maximize output and, particularly, cracking capacity for gasoline.[2] The achievement of this end was hampered not only by the detailed and often contradictory Russian supervision of all operations but also by frequent breakdowns of overloaded plant and the occurrence of fires: the mechanical fire-fighting equipment having been removed by the Russians on the grounds of its German origin.[3] Breakdowns followed on the

[1] The cost of living in Bucharest even decreased during the first quarter of 1944, largely through a reduction in food prices consequent on the bumper harvest of 1943: see *MPR*, IX–X (1945), p. 323.

[2] 1945 statistics of cracking, *MPR*, no. I/II (1946), p. 7; 1946 statistics, *MPR*, I–II (1947), pp. 34–5; 1947 statistics, *MPR*, no. I–II (1948), pp. 29–30.

[3] 211.

compulsory neglect of repair and maintenance schedules, but tended to be attributed by Russian officials to "sabotage" and were investigated as such. No attention was paid to economical refining practice: plants had to accept the crude oil made available by the authorities, irrespective of type and source.[1]

(*b*) *Production and trade*

Questions of production, and, in particular, drilling, only began to be considered in December 1944, when the Oil Directorate convened a series of meetings of company representatives to discuss the drilling programme. The German withdrawal had left the oilfields virtually intact. Nevertheless the Romanian Geological Institute estimated a decline in output over 1945 of 12–15 per cent: the companies argued that on the basis of existing stocks of materials, i.e. without the restoration of Russian removals, it would only be possible to produce 4·2 million tons in 1945, not 4·75 million as required by the Russians: this argument was rejected by the Directorate representatives, who demanded new proposals by the end of the month on the basis that drilling programmes originally submitted by the companies would be completed in the first six or seven months of the year. The companies were also asked to support their new proposals by detailed information relating to well-yields, redrilling and reopening of old wells, necessary materials, and so on.[2] The final Russian demand was for 4,530,000 tons, involving 118,000 metres of exploitation drilling and 22,800 metres of exploration drilling. The final output for the year was 4,680,000 tons, and 160,000 metres were drilled.[3] This pattern of official demand, refusal, and bargaining over the programme, which was then more or less complied with, followed in the other armistice years, but the results showed declining output in spite of increased metreages drilled, viz:

1946	4,257,000 tons	135,000 metres[4]
1947	3,796,351 tons[5]	164,000 metres[6]
1948 (Jan.–June)	1,999,742 tons	137,750 metres[7]

As in refining, the achievement of results depended increasingly

[1] 212.
[2] 213.
[3] *MPR*, no. III–IV (1946), p. 119.
[4] *MPR*, no. IX–X (1947), p. 283.
[5] *MPR*, no. I–II (1948), p. 22.
[6] *MPR*, no. XI–XII (1947), p. 343.
[7] See 214.

on the disregard of economic or technically reasonable practice, and the more general threat of "sabotage" held over individuals and companies. A particular restrictive factor was the absence of drilling materials and the wearing out of existing equipment. Furthermore, such essential general items as tyres, gaskets, and cement were scarce. Technical difficulties were aggravated by labour indiscipline, so that actual operations were slowed down.

In spite of indications that new drilling materials would be obtained from Britain and the United States, no imports were made from these sources until April 1948, when Dresser Industries of Cleveland delivered equipment comprising, among other things, one drilling rig, compressors, pumps, blow-out preventers, and seismic gear.[1] In the meantime an agreement concluded on 26 April 1946 between the Ministry of Mines and Petroleum and Czech manufacturers (as part of the Commercial Treaty of 9 November 1945)[2] provided for the delivery of 58,268 tons of oil against a stipulated range of tubulars (5,400 tons) and other equipment. Deliveries on both sides lagged behind schedule. The oil had to be received in Czechoslovakia before the materials would be released, and, in the absence of a freight-transit arrangement through Hungary, the Danube route had to be used: lack of tank-cars and barges from Sovromtransport combined with unexpectedly low draughts in the river to make movements from Giurgiu intermittent. On the Czech side, supply was interrupted by diversion of orders to hard-currency destinations and by the attempt to insist on the supply of crude oil—as distinct from products—by Romania. By December 1946 only 1,600 out of the total 6,200 tons of materials had arrived in Romania.[3] The only other external source of supply was the Soviet Union, which contributed materials of either Soviet or Romanian manufacture—the latter out of its reparation deliveries. These were the first charge on the output of the Malaxa works, whose operations were further restricted

[1] See below, p. 316.

[2] Text: *MPR*, no. XI–XII (1945), pp. 341–2; text of agreement on renewal: *MPR*, no. VII–VIII (1947), p. 201.

[3] Since Czechoslovakia's oil requirements were supplied by the Russians from Austria and Hungary and by U.N.R.R.A., Romania, which had only oil to offer, was in a weak bargaining position. See 215; also interventions by A.I.P.R. cited in the report of its *Activities*: *MPR* 10-11-12 (1946): particularly "Transportul materialelor comandate in Cehoslovacia", p. 28; "Amplasamentul nouilor sonde şi materialele necesare pentru forajele viitoare", p. 31; and "Dificultăţi în aducerea materialului tubular comandat", p. 37.

by the lack of fuel and raw materials. Most of the Russian-supplied material went to Sovrompetrol.[1]

The technical operations of the industry were also affected by the situation in transport. On their arrival in Romania the Russians assumed control over all rail facilities and port installations, and where necessary altered them to serve the reparations programme. This demanded the dismantling of two of the pipelines between Bucharest and Giurgiu and their relaying as a spur of the Ploeşti–Constanţa line to Reni on the river Prut: the main railway lines through Moldavia to Ploeşti were altered to broad gauge. The alterations appeared to the British Mission to have been "carried out by our Allies rather hurriedly, without any detailed and well-thought-out plan", hence, as they were not synchronized with changes in refinery marshalling yards and loading stations, the immediate effect was delay and waste.[2] The situation improved during 1945. The Russians also controlled tank-car allocation—nominally in consort with a Romanian Government Commission for Transport Co-ordination, which included representatives of C.F.R. In practice Oil Directorate authority was exercised to secure the delivery of oil as reparations as first call on available tank-cars—which themselves constituted a source of war trophies and reparations.[3]

The political reorientation of Romania meant that its oil resources were employed only in resuscitating the Soviet Union and in strengthening Romania's commercial links with its other political associates. In the immediate post-armistice period the export programme set by the Soviet authorities provided only for military use (at roughly the same level as the German offtake) and reparations. The latter comprised not only totals stipulated in the Armistice Agreement but also quantities of oil taken by the Russians to compensate for shortfalls in other reparations goods. The Russians stipulated that 1938 prices should rule for the whole reparations period: negotiations as to quantities arrived at a total of 1·7 million tons a year throughout the six years and established the relevant products and their specifications.[4] Nevertheless the

[1] See quantities tabulated in *MPR*, no. V–VI (1947), p. 157.

[2] 216. From the Romanian standpoint, the changes virtually cut off areas north of Adjud (from which point there was a single-track, broad-gauge line to Iaşi and Cernăuţi) from the rest of the country.

[3] 217.

[4] See Convention on Article 11 deliveries: *MPR*, no. I–II (1945), p. 13; and 218.

fulfilment of reparations programmes was attended by disputes arising from the fact that Russian control officers in refineries were responsible for supervising both deliveries to the front (Article 10)[1] and reparations (Article 11)—and subsequently exports under the Commercial Agreement—and for deciding the account relevant to each particular loading.[2] It was a matter of importance to the Romanian Government, since its supply obligation under Article 10 was unlimited in quantity or time, while Article 11 imposed specific demands for a limited period. The differences in computation were considerable: in the first half of 1945 Article 10 deliveries totalled 97,000 tons according to the Romanians and 342,000 tons according to the Russians.

Similar argument took place over Article 10 itself, deliveries under which reached their maximum (72,763 tons) in April 1945. By the end of the year they had declined to about 10,000–15,000 tons a month. The Romanians maintained that Article 10 deliveries should cease as from 9 May 1945, the end of hostilities: the Russians argued that such deliveries should be supplied all the time Soviet troops were actually on the scene, and added that supplies at the end of 1945 were for the needs of the troops permanently stationed in Romania. The Romanians tried to solve both disputes by gaining the Russians' agreement to the notion that all Article 10 deliveries after 9 May 1945 should be credited against Article 11, namely reparations deliveries, but they were unsuccessful.

Additional disputes were occasioned by the shortcomings in delivery practice. The oil companies delivered f.o.b. refinery; the Russians took delivery at the frontier, or at the Danube shipping port—the differences between amounts dispatched and those received were considerable. In May 1945 the companies persuaded the Romanian Government to set up a separate organization—the Office for the Delivery of Petroleum Products to the U.S.S.R.[3] (O.L.P.)—to handle the oil for the Russians from the point where the companies handed it over at their refineries to the point at which the Russians took delivery. Because the Government lacked suitable personnel, the O.L.P. was staffed by employees seconded

[1] Possibly, not all Article 10 deliveries were lost to Romania: Tantareanu (Chairman of the (Romanian State) Committee for Internal Market Allocations) voiced the suspicion that the Russians participated, through Romanian agents, in the black-market trade in fuel oil, which was in short supply for industrial purposes: 219.

[2] 220.

[3] *MPR*, no. III–IV (1945), pp. 69, 174: see also 221.

by the oil companies, but the office, whose mandate was signed in December 1945, was and remained, until it was wound up three years later,[1] a Romanian Government authority. By definition O.L.P. handled only Article 11 deliveries—supplies to the army being made through the Ministry of Mines and Petroleum. Nevertheless the decision as to which account any particular batch of oil had to be credited remained with the Directorate officers either in the refinery or at the point of export. They did not always agree, so that changes of plan were frequent, with much complication of invoicing and accounting. The Russian practice of delaying the announcement of ships' arrivals till the last moment did not assist O.L.P. in maintaining loading schedules and quality requirements, and caused considerable transport and handling losses.

After the cessation of hostilities in Europe the quantity available for export was directed entirely to Russia and to those countries falling into the Soviet sphere of influence.[2] Additional supplies to the Soviet Union were arranged under the Commercial Agreement of May 1945[3] providing for an exchange of goods, in which Romania undertook to supply *inter alia*, 267,000 tons of petroleum products during the contractual year. This amount included 100,000 tons of motor gasoline and 45,000 tons of gas oil. The execution of these deliveries was vested in O.L.P., which concluded for this object a convention of sale with Soyuznefteexport on 12 July 1945.[4]

'Satellite' trade began with the Romanian-Bulgarian Commercial Treaty of January 1945. It was conducted on a "compensation" basis of oil against goods, such as Polish coal (for the C.F.R.) or Czech oil equipment. To the extent that these markets were situated in the Danube area, this meant the re-establishment of pre-war links, but of Western countries only Switzerland reappeared as a trading partner in this category. Switzerland, for the Romanian Government, was a source of hard currency without the political implications presented by Britain or the United States. A trade and payments agreement was concluded in June 1946,[5] and

[1] 1 Jan. 1948: see *MPR*, no. XI–XII (1947), p. 332.

[2] For a discussion of general trade policies, see M. Dewar, *Soviet Trade with Eastern Europe 1945–1949*; and N. Spulber, *The Economics of Communist Eastern Europe*, Chap. 11.

[3] Concluded at the same time as the Agreement on Economic Collaboration.

[4] Text, see 222: on operation of contract, see 223; and report on O.L.P., 224.

[5] Text *MPR* 10-11-12, (1946), pp. 385–6.

Creditul Minier (Sovrompetrol) participated with Swiss interests in founding, on 13 August 1946, a subsidiary in Zürich to handle imports of Romanian and Russian oil.[1] The resumption of trade with the U.K., proposed by the British Mission in January 1945 and eventually approved by the Russians in October of that year, was hampered by lack of production and the problem of payment. The British Treasury furnished an additional complication by refusing to exempt any proceeds from being impounded under the "trading with the enemy" regulations.[2] Romanian oil imports into France were arranged under an agreement of 3 July 1946 for the exchange of goods to a total value of fr. 400 million,[3] of which the quota for oil was restricted to fr. 25 million. After the war Romanian oil virtually disappeared from its traditional markets in Western Europe and the east Mediterranean, whose demands were met from Venezuela and the Middle East. In the latter area wartime exploitation had strengthened the lead over Romanian production established before the war.[4]

Supplies to the domestic market reflected the existence of the commitments already discussed: the initial programmes set out by the Soviet authorities allowed for deliveries within Romania at approximately the same rate as during the Antonescu regime before bombing disrupted refining and distribution. These estimates, in practice, were subject to two factors—the shortage of transport and the extra liftings of the Russians themselves, either to make up deficits of other goods on reparations account or under the Commercial Agreement of May 1945. Consequently actual deliveries to the domestic market, particularly of fuels, lagged considerably. This caused cuts in transport and power,

[1] 225.

[2] Romanian exports to the U.K. from the Armistice to the nationalization comprised 42,000 tons motor gasoline, exported by the Russians from "reparations" oil to Russian Oil Products U.K., and 700 tons naphthenic acid residue, exported by Astra to Shell. Contracts had been signed also for 30,000 tons motor gasoline and 22,000 tons white spirit—deliveries being made in June and July 1948.

[3] Text: 226.

[4]

% World Production	Romania	Middle East
1938	2·3	6·03
1946	1·1	9·23

Source: *Petroleum Almanac 1946.*

which during 1945 and the winter of 1946 progressively retarded industrial output.[1] The prices of all products were set by the Romanian authorities, and the difference between the permitted price levels and operating costs, together with the dilatory payment practices of the State for goods received, created widespread indebtedness and forced the oil companies to apply to the banks for short-term loans at necessarily high interest rates. The companies, separately and, through the A.I.P.R. collectively, intervened with the Ministry of Finance for debt collection and retroactive price increases.[2]

State policy was determined by the fact that the Government "bought" on average 66 per cent of the total oil production: this was disposed of under the Armistice Agreement at prices which were only about one-third of post-war world market prices. At the same time, the cost of items of equipment needed for production and refining had risen anything from one hundred- to four hundred-fold. This situation combined with the delays in payment and the general inflation of the currency to cause the oil companies heavy losses. Sovrompetrol remained an exception, since their losses were offset by special arrangements allowing remittances in hard currency. These factors made it impossible for the "Western" oil companies, as a whole, to observe their commitments to suppliers of materials, to their staff and workers, and, particularly, in regard to supplies to Economate. The Government issued certificates compelling the purchase of specific items which were only to be had on the black market, to which the oil companies were driven in order to comply with the State's requirements. They thus left themselves open to prosecution under the regulations.[3]

[1] 227.

[2] For the complications of obtaining payment for deliveries on Romanian Government account, see 228.

[3] 229.

CHAPTER XI

The Communist Regime

THE new pattern of relationships for the Romanian oil industry could only be upset by external action, i.e. by the institution of a government more responsive to Western opinion. Allied differences in this respect were emphasized by the prompt recognition given to the Groza Government after the Potsdam Conference by the Soviet Union on 8 August 1945 and the refusal to do so by the U.K. and U.S. Governments on the grounds that the regime was unrepresentative. During the rest of 1945 and 1946 Anglo-U.S. policy attempted to improve the situation and protect the interests of American and British nationals by terminating the armistice regime—and therefore the Russian occupation—through the conclusion of a peace treaty and by promoting elections in Romania.

The Romanian Peace Treaty of 1947

At Potsdam the conclusion of peace treaties with all the ex-German satellites was approved in principle, and their preparation referred to the Council of Foreign Ministers. The Council, at their meeting in Moscow in December 1945, decided on a general conference of states which had waged war against the former satellites to consider draft treaties which the Council would present, at which "a hearing would be given to representatives of Italy, Romania, Bulgaria, Hungary, and Finland. Each would be permitted to discuss and present his views on the peace settlement with his nation." On Romania the U.K. and U.S. representatives pressed for the reorganization of the Groza Government and agreed with the Soviet Union proposals on how this was to be done, viz. that the Government should admit a "suitable representative" of each party, the National Peasants and the Liberals, and proceed to "free and unfettered" elections as soon as possible. The Government was to assure freedom of speech, press, association, and religion, and a commission of the participants at

Moscow[1] was to supervise the carrying out of these terms in discussion with the King and the Groza Government. The first condition was complied with by the admission of Hațieganu and Romniceanu: the required assurances were given, and, accordingly, the Government was recognized by the United States and Britain on 4 February 1946. The two representatives, however, remained observers of Cabinet proceedings. Shortly afterwards press censorship was resumed and the elections were repeatedly postponed on various pretexts and against protests from the U.S. and U.K. Governments until November 1946. The Romanian delegates (Tătărescu, Gheorghiu-Dej, Maurer) to the Peace Conference, which opened in Paris on 29 July 1946, were from the ineffectually amended Groza Government.

The settlement[2] attempted to provide for problems arising not only from the upheavals of war years but also from the preceding Carolist period. In the draft treaty the problems of the oil industry were considered mainly under Part V (Reparations and Restitution) and Part VI (Economic Clauses), in which were reaffirmed the provisions of the Armistice Agreement, the Inter-Allied Declaration of 5 January 1943, and the Potsdam Protocol. The U.K. delegation proposed an "Annex on Petroleum"—designed to be No. 4 of "Special Provisions Relating to Certain Kinds of Property"—with the explicit intention of bringing the outstanding problems of foreign-owned companies into the terms of the final treaty. The draft comprised, broadly, compensation for losses

[1] i.e. Vyshinsky, Harriman (U.S. Ambassador to Russia), and Clerk Kerr (British Ambassador to Russia).

[2] This analysis of the treaty is concerned only with the clauses on the oil industry and other clauses necessary to explain them or which subsequently were invoked as relevant. For the background and Paris discussions, see S. Lowery, "The Peace Settlement in Eastern Europe", in *The Realignment of Europe*, ed. A. and V. M. Toynbee, pp. 332–51; W. H. McNeil, *America, Britain and Russia: their Co-operation and Conflict 1941–1946*; J. F. Byrnes (U.S. Secretary of State), "Report on the Paris Peace Conference", 18 Oct. 1946, in *A Decade of American Foreign Policy, 1941–1949*, p. 88, and also in his memoirs *Speaking Frankly*, particularly Chap. 8. For a detailed textual analysis of the treaty as a whole, see E. C. Ciurea, *Le Traité de paix avec la Roumanie du 10 février 1947*: for the views of exiled Romanians, see article by G. Gafencu in *New York Times* (8 Aug. 1946); *Roumania at the Peace Conference* (memorandum submitted to conference 6 Jan. 1946); and *Observations concerning the Draft Peace Treaty with Roumania*, Paris, Oct. 1946—these last two being publications of an *ad hoc* group of the historic parties led by Tilea, Cretzianu, and Gafencu. See also A. Martin, "Private Property Rights and Interests in the Paris Peace Treaties", *British Yearbook of International Law*, 1947, pp. 273–300.

sustained by United Nations nationals during the war, alterations to mining legislation, maintenance of acquired rights, and freedom of entry for personnel. Both the United States and the Soviet delegations dissented from this annex on the grounds that it was in general covered by the proposed article (24) on the restoration of United Nations' property—the United States on the condition that the provisions on compensation eventually arrived at were adequate: the Soviet delegation merely saw "no reason to include any special provisions relating to petroleum". In the course of negotiations the U.K. delegation also attempted to obtain an amendment to the proposed Article 22 (repeating part of Article 11 of the Armistice Agreement and the related Annex) on reparations, in order to enforce payment by the Romanian Government of fair prices for the deliveries made on its behalf by the oil companies to the Russians. These manœuvres failed—as did also an attempt by the U.K. delegation to have a special "Annex on Shipping" included, and a combined U.S.–U.K. proposal for reinternationalizing the Danube. The latter was left over for discussion by a conference of interested states to be convened within six months of the coming into force of the peace treaty.

The issues were fought out in the Balkan Economic Commission of the conference, where the basic political factor was the Soviet Union's determination not to allow its special position in Romania to be upset by an international conference.[1] This entailed their supporting the Romanian delegation in their resistance to proposals designed to give equality of treatment to all Allied nationals, and, notwithstanding a number of defeats in committee, to defer certain major issues for further discussion by the Council of Foreign Ministers, who were responsible for completion of the text.[2] Allied oil interests necessarily entered into the discussions, particularly in regard to the proposals for repeal of the 1942 Mining Law (which the Soviet Union denounced as an infringement of Romanian sovereignty), for compensation for damage to property, and for the price differential in reparations deliveries. These occasioned acrimonious exchanges over the oil companies' role as a source of fuels for the invasion of the Soviet Union, and

[1] See, e.g., Molotov's speech: *New York Times* (27 Aug. 1946) and *Soviet News* (29 Aug. 1946).

[2] See "Report by the U.S. State Department on the Third Meeting of the Council of Foreign Ministers", in *A Decade of American Foreign Policy 1941–1949*, p. 92.

profits on wartime trading. The Russian charges on the first topic were not such as could be rebutted by argument or appeal to fact,[1] and the accusations in respect of the second rested on selected figures from balance sheets.[2]

The treaty was signed in Paris on 10 February 1947[3]—the oil industry being essentially concerned with Part V, comprising Articles 22 and 23 on reparations and restitution, Part VI, particularly Article 24, on the restoration of legal rights and interests and compensation for United Nations' nationals, and Article 33 on "fair prices". These articles represented the final form of the draft "Annex on Petroleum". Article 26 confirmed Russia's right to German assets and bound Romania to effect their transfer. In Part VII, comprising Article 36, the Danube was declared to be "free and open for the nationals' vessels of commerce and grades of all States": this was the basis for the special conference, referred to above.

The peace treaty with Italy, negotiated at the same time, was also in one respect relevant to the Romanian scene, since it envisaged (Article 74. A2 (*b*))[4] the cession of Italian assets on account of reparations to a value of U.S. $100 million. These assets included the companies "Prahova", "Xenia", and "Petrol Latina", which eventually became subject to transfer under an agreement concluded in Moscow on 11 December 1948.

The aims of the Western governments, in negotiating the peace treaties, of promoting Romania's independence by terminating the armistice regime and ensuring equality of access in economic relations were not fulfilled, since they rested entirely on the extent to which the Romanian Government could be made to fulfil its obligations. Events soon made it apparent that the Romanian Government remained beyond compulsion or inducement from the West. The elections of November 1946[5] demonstrated the invulnerable position of Communism within the country, so that the Soviet occupation forces could be dispensed with.[6] A phase

[1] See report of speech by Vyshinsky in *Soviet News* (2 Oct. 1946), and article by V. Spiru, "Profit and Loss in Roumanian Oil", ibid. (10 Oct. 1946).

[2] 230.

[3] The text of the treaty for signature was published, together with the other satellite treaties, in Cmd. 7022, Miscellaneous no. 1 (1947): a commentary was also published, Cmd. 7026, Miscellaneous no. 2 (1947): the treaty as signed appeared as Cmd. 7686, Treaty Series, no. 55 (1948): see also 231.

[4] Cmd. 7022, p. 25.

[5] See below.

[6] Selected personnel merely changed out of uniform and remained as "specialists": see below, p. 313.

of Soviet external support of Romania came to an end in the Danube Conference of July and August 1948, at which, under Vyshinsky's leadership, Western states were again excluded from the political and technical direction of the Danube.[1] Action on the "oil" clauses was confined to protest at Romanian non-compliance with them. For the United States and the United Kingdom Governments, the effect of the treaty was merely to define a responsible party with whom to negotiate on questions raised by the Romanian Government's treatment of their interests.

The "Communization" of Romanian state authority

The peace treaty officially restored full sovereignty to Romania: the often postponed elections were held by the Groza Government on 19 November 1946. A violent campaign,[2] even by Romanian standards, resulted in a sweeping victory for the Communist-led Bloc of Democratic Parties (B.P.D.) according to official figures. The Cabinet was reconstituted, with Gheorghiu-Dej becoming Minister of National Economy. The two nominees of the opposition—the vestige of Anglo-American responsibility for the working of the Armistice—resigned. As the National Peasants and the Brătianu Liberals decided to boycott the parliament, there remained only the small group of Tătărescu Liberals within the B.P.D. and the Petrescu Independent Social Democrats as a focus of resistance in the Chamber.

The election and the signing of the peace treaty[3] in February

[1] See 232; statement by U.S. Ambassador, 16 Aug. 1948, printed in *A Decade of American Foreign Policy 1941–49*, pp. 804–7; and U.S. protest on the establishment of the Danubian Commission, ibid., pp. 807–8. Vyshinsky's speeches putting the Soviet case are published in *Soviet News* (3, 5, 6, 9, 10, 16, and 20 Aug. 1948).

[2] On the elections see Markham; Prost, pp. 185–9. Astra Română, e.g., suffered directly from the election in that a large number of personnel and motor vehicles were forcibly mobilized and requisitioned in the interests of the Democratic Party's Bloc. During the period of the election, 165 men were mobilized into the Gendarmerie, and some 350 man-days were lost by personnel being taken away by the Syndicates to act as election stewards, etc.: 10 cars and 36 lorries were requisitioned for various Party purposes. See 233.

[3] The Romanian Parliament ratified the treaty on 23 Aug. 1947. In the debate in the Chamber speakers from all B.P.D. parties, including Tătărescu, dilated on its harshness, particularly regarding the economic clauses. One member, Th. Iordăchescu (left-wing Social Democrat) singled out as particularly obnoxious "the attempt by our reactionary circles . . . to impose on Romania

1947 were followed by the undisguised emergence of the Communists as the driving force in Romanian politics and by their destruction of actual or potential sources of opposition, both in the country and in Parliament. In July, Maniu and his colleagues were arrested, their party dissolved by decree, and its property impounded. The leaders were brought to trial before a military court in October, on charges of "anti-state activity", a term which related not only to the immediate past of the National Peasants but to their entire history, on the grounds that they had from the beginning been an "infamous conspiracy" with "no other aim than to turn Romania into a semi-colony of international trusts, into the puppet of imperialist circles with the old role of anti-Soviet provocation and a springboard for attack on the Soviet Union".[1] The Party's attitude to the "foreign-capital" sector of the oil industry entered into the general indictment. The charges were elaborated in detail in *Viața Sindicală* on the eve of the trial.[2]

The 1929 Mining Law delivered our oil to British, Dutch, and American oil companies, for its provisions gave to these companies the possibility of exploiting our oil as they pleased. The only aim of Maniu's supporters was to enrich oil companies beyond all limits to the detriment of wage-earners and consumers. Firstly foreign companies were given preference in the granting of concessions. Secondly they were given such facilities in the granting of areas that within a short time all certain or probable oil deposits fell into the hands of the American Standard Oil Company of New Jersey, the British Royal Dutch Company, etc. One of the clauses of the Law permitted the State to grant concessions for its own deposits: in this way, the National Peasants sold to foreigners even the private property of the State against considerable commissions. Another clause, for which we are still paying heavily, provided that oil companies should be able to export crude oil. As a result of that paragraph foreign oil companies began to plunder our oil deposits, regardless of their future, and serious unemployment was caused in the oil area.

international measures for assuring Romania's effective political independence, the integration of the economy in the international economy to impede relations with the U.S.S.R., the international control of reparations, and the international control and guarantee of navigation on the Danube. These patriots wish the country to be subjected to an international regime similar to that of the time of the Turks." He gave the International Danube Commission as an example of hindrance by foreign powers.

[1] *Scînteia* (13 Sept. 1947).

[2] 28 Sept. 1947.

These sentiments could have been expressed by one of the more doctrinaire followers of Vintila Brătianu, but in the context of 1947 they served to condemn the National Peasants not only on domestic grounds but also as regards their foreign policy, since the oil companies were held to represent the "Anglo-American imperialist circles",[1] which according to Gheorgiu-Dej, were trying to "prevent Romania's democratization and maintain us in our former position as vassals". The full implications of this aspect of the trial did not appear for another month:[2] Maniu and Mihalache were sentenced to life imprisonment with solitary confinement.[3] The National Peasant Party disintegrated. The admissions of the defence were also used to bring about the downfall of Tătărescu—through the alleged complicity of the Ministry of Foreign Affairs—and the resignation or expulsion of his supporters from their bureaux.[4] For the time being the Brătianu[5] Liberal group was allowed to continue to exist but the Banca Românească was taken over by the State (November 1947).[6]

[1] The analyses of the position of the "foreign-capital" companies published in *Scînteia* and the other journals of the members of the B.P.D., at this time and in succeeding months, were unduly *simpliste*. The available evidence shows that, while the shareholders and individual members of Bucharest managements did not approve of, or protested against, the policies of the Romanian Government, the companies themselves and their operations were under effective Romanian legal and political control. There could be no possibility of their organizing a counter-revolution, even if their shareholders' governments (whose position in post-war Romania was, at best, weak and insecure) had entered on such a policy, and could have brought Astra Română and Romăno-Americană to consider operating it.

[2] See below, p. 315 n. 1.

[3] They died in Sighet prison in 1953: see A. Berkowitz, "The Prisoners of Sighet", *Europeast Bulletin*, no. 6 (June 1962).

[4] The Foreign Affairs Committee passed a vote of no confidence in Tătărescu on the grounds that "the acts of treason disclosed at the [Maniu] trial were committed with the active aid of the services of the Ministry of Foreign Affairs and could not have taken place without the knowledge of the leaders of that Ministry". This was followed by a virtually unanimous vote of censure in the Chamber of Deputies. Tătărescu resigned on 5 Nov. and Ana Pauker became Foreign Minister; Vasile Luca replaced Alexandrini as Minister of Finance.

[5] Brătianu was later arrested and died in Sighet prison in Aug. 1950: see Berkowitz, op. cit.

[6] See *Monitor Oficial*, no. 246 (24 Oct. 1947): the take-over was based on Article I of the Law No. 478 of Oct. 1944 for the temporary control of joint-stock companies (see above, p. 284 n. 1). The directors were charged with having "led the bank along the line of constant political and economic support for the former dictatorship of Antonescu, as shown by the reports to the General Assembly submitted [by them] in the years 1940, 1941, 1942, and

The Social Democrats were overcome by being merged into a "United Workers' Party" (February 1948), while Petrescu's independent organization was intimidated into dissolution and its leader imprisoned (May 1948). Also during this period the judiciary was reformed in conformity with Marxist theories of law. This was the work of Pătrăscanu. Existing lawyers' associations were abolished and replaced by new institutions to which admission had to be sought by lawyers who had hitherto derived their main income from independent practice and who had shown no "anti-democratic attitude" in their professional or public life: failure to apply or to survive commissions to scrutinize applications spelt the end of professional practice. At the same time the rights of defence lawyers were curtailed, and People's Assessors were created to give critical co-operation to the existing judiciary, "to prevent professional judges from giving unjust verdicts under the cover of legal tricks":[1] in cases of dispute their verdict over-ruled that of the professional judge.

Outside the sphere of party politics, the regime sought, during 1947, to eradicate those sections of society whom it regarded as undesirable through reform of the currency and trade union action. The army was put under Bodnaraş, as Minister of Defence, in place of General Lascar, who in Russia had helped to raise and then commanded a division of Romanian prisoners of war. On assuming office, Bodnaraş began to reorganize the higher echelons, to ensure their sympathy with the regime. The remaining possible focus of anti-Communist action—the monarchy—was overcome by the forced abdication of King Michael on 30 December 1947; the same day the monarchy was replaced by the Romanian People's Republic.[2] *Scînteia* indicated political and constitutional motives, namely that the monarchy, as the second exponent of interests hostile to the people—the Liberals being the first—was

1943, thus actively promoting Hitlerist economic and political penetration in Romania". Their mandates were cancelled and functions vested in a sole administrator, Gheorghe Radulescu of the National Bank.

[1] See article by Pătrăscanu in *Scînteia*, 4 Dec. 1947; amendment to procedure code, published 5 Dec. 1947; Law for the reorganization of the bar, published 19 Jan. 1948. The Assessors were elected by simple majority on an open vote. The electoral body was, for rural magistrates courts, the local inhabitants of the area of the court's jurisdiction, and, for town courts and all higher courts, the members of the area trade union. Civil servants, merchants, and anyone who employed paid labour were not eligible to vote.

[2] See *MPR*, no. XI–XII (1947), p. 301.

used as an instrument by reaction at home and abroad: it could no longer be allowed to hold authority higher than that of Parliament, elected by the people, and the Government, the result of the will of the people.[1] On 13 April 1948 the new regime established a new constitution which declared Romania to be "a People's State . . . born from the struggle waged by the Romanian people, headed by the working classes, against Fascism, reaction, and imperialism".[2] The State which had been founded seventy years before, and which has been one pole of the discussion in this study, ceased to exist.

The liquidation of non-Communist authority in Romanian society was accomplished in the eighteen months between November 1946 and April 1948. It had important repercussions in the sphere of foreign and economic policy. The procedure adopted for the peace treaty—discussion in two stages (July–September and December 1946), formal signature (February 1947), and entry into force (15 September 1947)—allowed time for the Russians from outside and for the Communists within Romania to consolidate their interests during a period of increasing conflict between the Soviet Union and its wartime allies.

The effects of the Communist programme on the oil industry

On 12 December 1946 the Soviet Department of Foreign Trade made a *démarche* in respect of the Soviet interest in Sovrompetrol,[3] setting out a number of grievances and demanding redress in the form of world (not Romanian state) prices for the products deriving from the oil royalties contributed by the Romanian Government to Sovrompetrol (U.S. $ 3·5 million), a further U.S. $ 2·5 million for loss of profits due to the Romanian Government's low-price policy, a guaranteed 15 per cent net profit, and an unstated amount for amortization on the grounds that the Romanian Government "used all Sovrompetrol's production" and thus denied the organization the opportunity of making its own

[1] Article by Serin Toma, 1 Jan. 1948. On the same day *Frontul Plugarilor* traced the King's "opposition to democratic achievements" in detail: see also survey by Gheorghiu-Dej on the "Consolidation of the Romanian regime as a People's Democracy" in the Cominform journal *For a Lasting Peace, For a People's Democracy* (10 Jan. 1948).

[2] See *BFSP*, vol. 152, 1948, part IV, pp. 769–86.

[3] And Sovromlemn, the joint company for the timber industry.

reserves. At the same juncture, Sovrompetrol pressed its claims for favourable treatment in the award of exploration rights on isolated structures and mining districts. All these demands were conceded in principle by the Romanian Government,[1] and agreement in detail was reached during February 1947 in intergovernmental discussions in Moscow.[2]

The agenda also included the arrangements for permanent economic collaboration between the two states;[3] these took the form of a Permanent Commission, consisting of three Russian and three Romanian delegates, charged with assisting the normal functioning of the joint companies and the development of their activities, co-ordinating measures to remove difficulties, and examining questions arising out of the activities and interests of the respective parties in the joint companies, as well as other problems arising from the Convention of Economic Collaboration of 8 May 1945. The Commission was to meet not less than four times a year. A Treaty of Commerce and Navigation (with an annex concerning the Soviet Trade Mission in Romania) and a Trade and Payments Agreement—the first of a series annually negotiated—were concluded on the same day.[4] These arrangements signified a shift in emphasis from collaboration to integration, and were followed by the founding of a number of joint "Sovrom" companies, covering all aspects of Romanian production, and services such as transport and insurance. In the course of the year they were supplemented by arrangements on entailed questions such as direct rail communications, and by a programme of training Romanian specialists in Soviet techniques. Finally, on 4 February 1948, the integrated economic relationship was formally complemented by a Treaty of Friendship, Collaboration, and Mutual Assistance.[5]

In this situation the British and American Governments could

[1] 234.

[2] See Protocol (i.e. to the Convention of 8 May 1945) on Problems of Economic Collaboration between Romania and the U.S.S.R., 20 Feb. 1947, Annexes II, III, and IV: 235.

[3] Protocol, Annex I.

[4] Texts: Treaty of Commerce and Navigation; *MPR*, no. III–IV (1947), pp. 61–2; Annex not printed but see references and discussion in article by S. L. S., "The Soviet Commercial Mission in Roumania", *American Perspectives* (Jan. 1948), Foundation for Foreign Affairs, Washington D.C. Trade and Payments Agreement; not published, but see Gheorghiu-Dej's statement summarized *Scînteia* (7 Mar. 1947); and 236.

[5] Text: *BFSP*, vol. 152, 1948, pp. 801–3.

do nothing but protest against infringements of the peace treaty or other international instruments, and take such external action as blocking Romania's membership of the United Nations. The protests were regularly rebutted by reference to Romania's sovereignty, but outright hostility was muzzled until after the failure of the meeting of the Council of Foreign Ministers in November 1947. The revelations at Maniu's trial, that two American officers had discussed with the National Peasants the possibility of organizing resistance (the precise status and implications of these conversations are far from clear), and that Maniu had sought American help in setting up a "free" Romanian government outside the country, identified the United States as the enemy *par excellence*, though in a context of an "Anglo-American imperialism" which allegedly aimed at keeping Romania in a state of vassalage. Since France, immediately after the war, lacked the resources to re-establish her traditional position in Romanian life and was involved in the "Big Three" reorganization of Europe only to a limited extent, the Franco-Romanian connection lapsed. Equally Germany had ceased to be a competitor for political influence in the Balkans. By 1948, therefore, Romania achieved a diametrically opposite international position to that obtaining ten years before.

While changes in internal and foreign politics were taking place, Romania's economic situation approached the point of collapse. The recovery of agriculture was retarded through the application of the land reform of 22 March 1945,[1] and through drought in the summers of 1945 and 1946,[2] causing not only a shortage of cereals but also the slaughter of livestock. A particularly severe winter in 1946–7 paralysed collection and transport and completed the necessary conditions for food shortages everywhere and famine in Moldavia.[3] Industrial production recovered slightly during 1946,[4] but the diversion of both agricultural and industrial output to Russia undermined the reconstruction of the Romanian economy. The post-Armistice inflation, already noted,[5] increased. Roberts cites[6] an official cost-of-living index, showing increases over 1938 (= 100) as follows: 1945—3,678·7; December 1946—46,140;

[1] See Roberts, pp. 292–9. [2] 237.

[3] 238: famine conditions were exacerbated by an epidemic of typhus (March–April). [4] See Roberts, p. 316. [5] See above, p. 289.

[6] See Roberts, p. 317.

and July 1947—525,688. Another indication is provided by the A.I.P.R., which worked out coefficients, based on selected costs of labour and materials, by which to multiply the known value of the leu in 1939 in order to arrive at a basis for inter-company transactions after the war, when the leu was tied to no external official currency yardstick and was falling in value.[1] Translated into U.S. dollars, on the basis of the last (March 1940) National Bank quotation (U.S. $ = lei 211·66), the coefficients yielded the following rates:

	Lei	
July 1945	6,673·64	= 1 U.S. $
Jan. 1946	20,651·67	
July 1946	33,179·82	
Jan. 1947	57,395·84	
July 1947	858,463·33	

These rates were, of course, based on a sample of goods and services essentially concerned with the oil industry, but serve to demonstrate the inflation of the currency, particularly in the first six months of 1947.[2] The rapid deterioration at this time reflects not only the physical factors mentioned above but also the beginnings of demoralization as the Communists overtly asserted their leadership of the Government and, supported by the Red Army, were clearly revealed as the only effective authority in Romania.

A decisive change in the relationship of industry with the State was signified by the installation of Gheorghiu-Dej as Minister of National Economy on 2 December 1946, and the assumption by his ministry[3] of control over the whole Romanian economy, with authority over the related departments, e.g. the National Bank, the Ministry of Mines and Petroleum, and the Ministry of Finance. In his acceptance speech[4] Gheorghiu-Dej ascribed Romania's economic difficulties to the economic and financial consequence of a criminal war, two years successive droughts, and economic sabotage practised by elements hostile to the democratic regime. Recovery was already under way, thanks to the reconstruction of communications and the support by the Soviet Union. Future policy would be based on the B.P.D.

[1] 239.

[2] For the effect on wages and salaries see 240.

[3] The Ministry of National Economy was reorganized in Feb. 1947.

[4] *MPR*, no. 10-11-12 (1946), pp. 369–70.

programme, which his ministry would carry out. The immediate task was to find out exactly the real possibilities of the economy, its difficulties, and its objectives: this would enable work to be carried out within the framework of a realistic plan.

Nationalization

The immediate evidence of realistic planning[1] was the nationalization of the Romanian National Bank and various restrictive measures on gold and share transfers, enacted on 21 December 1946.[2] New scales of wages and prices were also forecast, with revision of the law to permit the reallocation of labour, and the investigation of costs and the prices of Economate supply. These were problems arising from the practice of the immediate past. Innovations were proposed in organization by the creation of *oficii industriale* (industrial corporations). Their purpose, as appeared from the draft law published in January 1947, was to organize all Romanian industry under the Ministry of National Economy through a series of *oficii* (bureaux), each representing a given industry and established by compulsory enrolment of individual firms. The *oficii* were to be set up as limited-liability companies, the members of which were to subscribe their entire share capital. Constituent meetings were to be held by companies concerned, at which subscriptions of capital would be decided on, and boards of directors elected. The boards, elected for one year, were to be composed of representatives of the constituent companies, government departments most closely connected with the activity of the corporation, organized labour, and nominees of the Ministry of National Economy. They would control the activity of the member firms—for which they would be answerable to the Ministry of National Economy, which was entitled to revoke the mandates and decisions of the board of directors, and to order at any time control of, or an inquiry into, the affairs of the *oficii* or of any firm within its sphere of activity. The scope of the *oficii* comprised the "stimulation, control, and rationalization of production, circulation, and distribution of products" and "commercial

[1] State planning to develop the country's economic power, ensure the welfare of the people, and safeguard national independence was subsequently written into the Constitution of 13 Apr. 1948 (Article 15): *BFSP*, vol. 152, 1948, part III, p. 771.

[2] 241.

functions carried out in the name of and on behalf of the industry member companies". The precise implication of these two attributes was enumerated in detail and covered all the functions and activities of any industrial concern. The *oficii* were to be non-profit-making, but expenses were to be financed by fixed commission on sales and on the value of imports by the component companies, which, however, having subscribed all the share capital, were responsible for any losses. To enable the *oficii* to discharge these tasks, the member companies were bound to supply the necessary technical information: the return of false or forged information, along with failure or refusal to comply with, or obstruction of, the directives of the *oficii*, carried severe penalties.

The draft terms amounted to nationalization without the concomitant payment of compensation: to the extent that, if applied, they affected foreign ownership by United Nations' nationals, they also contravened Article 24 of the Peace Treaty. The result was uncompromising opposition. The General Federation of Romanian Industries (U.G.I.R.), to which the scheme was referred for consideration, attacked the scheme as contravening both the Constitution and the Commercial Code, and proposed instead that *oficii*, representing the free collaboration of those directly interested, should be set up for each industry, where there was some real need of them, by special and separate laws—the State reserving only the right of control. The functions of such *oficii* would be to collect statistics, to plan and control production, prices, and sales, to supervise the application of economic legislation, to allocate quotas of raw materials, and to exchange information. These, the U.G.I.R. pointed out, were to some degree already done by the existing professional organizations.[1] The A.I.P.R.'s reaction was in the same vein; the Association made technical criticisms but stressed the more general view that the draft law, if applied to the oil industry, would deprive the member companies of the power to finance their own activity and market the resultant products, and leave them only with the responsibility for manufacturing programmes set by an outside authority, which by law would command the requisite technical installations and raw materials.[2]

The opposition of interested parties in industry was paralleled

[1] 242.

[2] Sovrompetrol abstained from comment.

by that inside the Government:[1] the measure was denounced by Tătărescu and was withdrawn. This is noteworthy as the last occasion under the *ancien régime* when public opposition brought about a reversal of Communist plans. It did not last long. On 15 May a new draft was published, in substantially the same terms as before.[2] Since, in the meantime, the designated responsible department, the Ministry of Industry and Commerce, had been reorganized[3] and purged of its non-Communist officials, there could be no doubt that the new draft was intended to put industry on a Communist basis. Sponsored by Gheorghiu-Dej, the Bill became law on 6 June 1947,[4] over protests by the U.G.I.R. and the U.S. Government.[5]

The Law gave the State all that was necessary, short of formal nationalization, to control systematically the activity of any designated industry. The Government's determination was emphasized by Maurer in replying to the U.G.I.R.: the Law had to work; if industry was co-operative it would be applied with understanding, but if there were any resistance the powers conferred would be used to the full.[6]

While the dispute over the *oficii* was in progress, the principle of state ownership was inscribed in the Romanian Constitution in respect of "oil and mineral deposits . . . still in the hands of private persons" (Article 6). Further it provided for expropriation "for reasons of public utility" (Article 10)—particularly with regard to "means of production or bank or insurance companies owned by private persons or bodies corporate" (Article 11). These terms were not so much formal constitutional rights as an implied political programme. A fortnight after the Constitution became law, all private mining royalties were abrogated, and in June 1947 the Romanian Geological Institute was empowered to make a complete inventory of subsoil resources on the basis of data obligatorily registered by the oil companies.[7]

Political upheaval and the fear engendered by the "Communization" of personnel in government did nothing to allay the rapidly deteriorating economic situation. In the memorandum[8] circulated

[1] 243. [2] 244. [3] 245.

[4] 246. Text: *MPR*, no. V–VI (1947), pp. 120–3.

[5] Text of U.S. protest: *New York Herald Tribune* (14 June 1947). Romanian reply: *New York Times* (22 June 1947).

[6] 247. [7] *MPR*, no. VII–VIII (1947), pp. 233–5.

[8] 248.

in May 1947 (which represented his last independent political initiative) Tătărescu called for retrenchment and currency stabilization, but identified as fundamental to the reconstruction of the economy the restoration of confidence through the ending of arbitrary actions and abuses. As a counter the Communist members of the Government produced their own programme,[1] in which the same general remedies of economy, increases in production, and currency reform were stressed without Tătărescu's political preconditions. This policy was passed by the Council of Ministers on 14 June[2] and implemented by a ministerial commission, set up on 15 July, headed by Gheorghiu-Dej,[3] with the widest powers to decide on specific measures and enforce them. The ministerial commission, which worked through various sub-commissions and a corps of "economic inspectors",[4] became the executive authority for the Government's economic policy.

The immediate inflationary problem[5] was dealt with by a drastic stabilization of the leu—finally announced on 12 August and completed three days later—at 20,000 "old"-lei to one "new" leu. Limits on permitted holdings of new lei were imposed according to the social class of the holder; at the same time all gold and foreign currencies and credit balances abroad had to be declared to the National Bank, and the gold medallion, issued as an anti-inflationary measure by the Rădescu Government, ceased to be legal tender.[6]

The general Communist programme was applied to the oil industry through an *ad hoc* sub-commission headed by Razvan Penescu,[7] a former oil adviser to Marshal Antonescu. On 27 August[8] Penescu convened discussions on drilling programmes and the rationalization of refining—from which it emerged that "the Ministry of Mines and Petroleum is superseded by the Sub-Commission . . . in all save the duty of following up the

[1] 249.

[2] See extracts from Gheorghiu-Dej's report: *MPR*, no. V–VI (1947), pp. 115–16.

[3] Law No. 248: *Monitor Oficial*, no. 159.

[4] 250.

[5] In May the Government took advantage of the currency depreciation to pay off all consolidated internal public debt at par or on the terms originally stipulated: 251.

[6] On the stabilization operation, see 252; Prost, pp. 209–11.

[7] On personnel and background, see 253.

[8] 254.

Sub-Commissions decisions". For the first time the day-to-day technical operations of the industry were subjected to detailed Communist control. This was complementary to that exercised by Russian officers of the Oil Directorate, who acknowledged the entry into force of the peace treaty (15 September 1947) by changing out of uniform into mufti.[1]

The ministerial commission also took steps to achieve a more rational allocation of manpower. Industry as a whole suffered from excess staff, recruited from farm labour which had migrated to the towns since 1944.[2] The cutting of costs of production and the rehabilitation of agriculture demanded a reversal of this trend. Accordingly, on 12 July, the Council of Ministers published a decision on the reduction of staff by means of "compression". This involved the sending by managements of a schedule of staff, together with a nominal roll of employees, to a local commission established *ad hoc* and composed of representatives of various ministries. Personnel declared surplus by the commission were to receive as compensation one month's salary if they were transferred to another job, and two months' if no further employment could be found.[3]

As with currency reform, the economic need to get rid of redundant personnel was made to serve a political purpose. In practice the commissions stipulated not only the numbers but also specific individuals to be removed from any particular undertaking. Prominent among the latter were anti-Communist workmen and officials: since the first compressions took place immediately prior to the August elections to the factory committees, they helped to eliminate opposition in the trade unions and reduce uncompressed operatives to a more reasonable frame of

[1] General Susaikov and his staff left Bucharest on 29 Sept. The U.K. and U.S. missions withdrew and were replaced by diplomats: the R.A.F. Transport Command air service from Vienna to Bucharest ceased without replacement.

[2]

	Salaried staff	Workmen	Total	Monthly salaries '000 lei
Dec. 1944	8,455	37,370	45,825	633,688
,, 1945	8,985	34,953	43,938	2,511,502
,, 1946	10,725	39,479	50,204	7,225,540
May 1947	12,499*	44,887*	57,386*	96,056,000*

Source: *MPR*, no. V–VI (1947), p. 157. * = approximately.

[3] These terms were less than those provided in the current law on labour contracts.

mind. Compressions were drastic throughout the oil industry[1] and evoked protests from individual managements and the U.G.I.R., in which the oil companies were represented. The A.I.P.R. could not register a corporate protest, through the refusal of Sovrompetrol, in whose organization compression was applied strictly to eliminate redundancy.

The main "foreign-capital" companies, Astra Română, Steaua Română, Româno-Americană, and Unirea, presented a joint petition on 22 August 1947 to the three main departments concerned (Mines and Petroleum, Commerce and Industry, and Labour) and to the President of the Council of Ministers, accepting compression on grounds of labour efficiency and reduction in costs, but questioning the legality of the way in which it was carried out, particularly in respect of the application of Article 2, which explicitly reserved the decision as to the number and identity of those to be compressed to the companies themselves. This petition was followed by a second on 27 August 1947, addressed to the same departments and also to the Penescu commission, setting out the whole effect of compression on the oil industry, and showing that, of a total of 3,058 employees declared surplus by the companies, the commissions had accepted only 2,080, to which they had added, however, 2,027 on their own initiative and without previous consultation with the management of the companies affected. The petitioners urged that this removal was particularly detrimental in view of the need to raise output, to keep up with the paper-work necessarily entailed by increased government regulation of the industry, and to work out the effect of such measures as monetary stabilization on its position.

In consequence of these petitions, the Government set up a committee of inquiry to investigate the situation in the industry, but by this time employment was in a state of chaos and the workers demoralized by fear of unemployment[2] and victimization, particularly as it appeared that appeals against compression which were upheld by the government commissions would not necessarily be accepted by local factory committees.[3] Lacking redress from

[1] The commissions included representatives of the Siguranţa, with dossiers of undesirable characters.

[2] Which automatically involved, e.g., surrender of food and clothing ration cards and loss of housing where provided by the undertaking concerned.

[3] Compression, in its political aspects, was heralded by purges by factory committees in refineries and oilfields beginning on 19 June: 255.

Romanian authority, Astra, Steaua, and Unirea on 21 October 1947 jointly addressed a memorandum to the British Legation asking for diplomatic intervention on the grounds of the violation by the Romanian Government of Article 24 (1) of the peace treaty.

In its oil aspects the application of the Communist programme "for economic rehabilitation and monetary stabilization" related particularly to the foreign—i.e. "Western"—capital sector of the industry. Sovrompetrol, reinforced by the privileges granted in February 1947, was beyond the Romanian Government's reach. From October 1947 the attack on the "Western" oil companies, which had been carried on within the industry, gathered force. The Union of Oil Workers' Syndicates, meeting at Ploeşti, called on the Government to take measures against the companies with Anglo-American capital, which were sabotaging production. This charge was repeated in a systematic campaign in *România Libera* and *Viaţa Sindicală*, augmented by accusations of their having assisted the Fascist war effort, of their refusing to collaborate with Sovrompetrol, and of fraud and espionage. Culprits were identified. In November *România Libera* featured a gallery of managers "each with his own crime sheet" and demanded "a close inquiry which should lead to the establishment of the guilt, and the mopping up of all scoundrels and saboteurs in the managements of Româno-Americană and Astra Română".[1] In the absence of any sanctions which could be applied, the "foreign-capital" companies were forced to continue their defensive policy.

In November it was rumoured that the State would try to control all the managements by direct supervision.[2] Rumour became reality on 15 December 1947, when, by order of the ministerial commission, "supervisory administrators"[3] were appointed to Astra Română and Unirea alleging their failure to fulfil drilling and production plans laid down by the Penescu sub-commission. Two days later these moves evoked diplomatic protests[4] on the grounds of the invalidity of the reasons adduced by the Romanian Government. The motives behind the appointments were political rather than legal, viz. the elimination of foreign influence.[5] Other means discussed to this end were the

[1] The press attacks began before the Maniu trial but were resumed more comprehensively in Oct.

[2] 256.

[3] On the powers of administrators, see 257.

[4] 258.

[5] 259.

cancellation of "acquired rights" (by bringing forward the terminal date from 1973), the retraction of state lands and seizure of private lands from the companies concerned, the restriction of credit to accelerate bankruptcy, and measures for the compulsory liquidation of companies not fulfilling drilling programmes. In the short term it suddenly appeared (23 January 1948) that the "Anglo-American" companies would be combined with others into an *oficiu industriale* for the oil industry. This plan was not proceeded with, apparently by reason of objections from Sovrompetrol. In the meantime the general and technical managers of Astra Română disappeared, and the British majority shareholders, supported by the British Government, formally declared on 6 March that they regarded the supervisory administrators' activity as detrimental to their interests and incompatible with the corporate function of the company: accordingly the latter was in a state of "forcible dissolution". The effect of this declaration was to sever all contractual relationships with the shareholders, who became entitled to exercise all the company's rights, including those to external assets. Foreign nationals in the management automatically ceased to function, and were permitted to leave the country. The Romanian board members also took the view that the declaration ended their tenure as shareholders' representatives. On 10 March the ministerial commission reorganized the management by appointing a "sole administrator" to take over from the supervisory administrator, who then became general manager.

Although in the campaign against "the Anglo-American oil trusts" Româno-Americană was singled out as particularly culpable, no administrator was appointed to that company at the time of the appointments to the "British" concerns, since the Romanian Government was in negotiation with an American firm for the supply of oil equipment.[1] Representatives of Dresser Industries of Cleveland arrived in Bucharest in December 1947, and, with the agreement of the U.S. State Department, negotiated an arrangement to supply materials during 1948 to a value of U.S. $ 10 million with three years' credit and against a gold guarantee lodged in Switzerland. The Swiss bankers refused to accept gold without the establishment of Romania's title to it, and the payment terms were therefore re-negotiated, viz. with payment to be made in U.S. dollars over the period 4 February 1948 to

[1] 260.

24 January 1950. The Romanian Government's oil-export policy thereupon became concerned with maximizing dollar earnings. In the meantime the U.S. Office of International Trade published licensing regulations governing trade with various destinations, including Romania; but before these came into effect on 1 March the first shipment left the United States.[1] The remaining deliveries were embargoed.

The deal occasioned a public dispute in the United States. Dresser Industries protested their loyalty to Government policies and disclosed that the original suggestion for the deal came from the Russians (with whom Dresser concluded a contract for gas equipment in 1946) and was negotiated without reference to the official Romanian Purchasing Mission in the United States: so the matter may have been essentially an item in United States-*Soviet* relations. The episode contains many unexplained problems, but its ironical outcome was that the United States became the first supplier of modern equipment to the Romanian oil industry[2] on the eve of its nationalization and the consequent disappearance of the American investment.

The immediate Romanian reaction to the embargo was the appointment of a supervisory administrator to Româno-Americană, on 18 April.[3] Administrators were successively appointed to other "Western" oil interests—Distribuția (23 April 1948), Steaua Română (30 April 1948), and Socony Vacuum (12 May 1948). As the powers of these officials were virtually unlimited, and as they were all party activists, the authority of foreign shareholders was eclipsed, and the companies could no longer represent a "Western" interest. It remained, however, to administer the *coup de grâce* to the private sector.

On 9 and 10 June,[4] the Central Committee of the Romanian Workers' Party held a plenary meeting at which Gheorghiu-Dej reported on the problems of nationalizing Romanian industry. In consequence the Committee decided to propose immediate

[1] Delivery had been contracted, f.a.s. U.S. port.

[2] Dresser Industries two years later settled their outstanding commitment on the advance payment made by the Romanian Government by supplying machinery to the Argentine state petroleum company Y.P.F. to cover shipments of hides and tannin to Romania.

[3] The company intervened directly with the Minister of Mines and Petroleum (22 Apr. 1948) and with Petre Groza (5 May 1948), but without effect.

[4] *Roumanian News* (12 June 1948).

nationalization to the Romanian Government. On the morning of 11 June all head offices were sealed by the police. A Bill based on Gheorghiu-Dej's report, which had not been previously announced on the Assembly's agenda, was introduced in the morning and passed unanimously after three hours' discussion. Its terms appeared in the afternoon newspapers and were broadcast over the radio in the same evening. The new managers arrived at their offices in the late afternoon of 11 June after the passage of the Law,[1] which was published the next day, 12 June, in the *Monitor Oficial*, no. 133 bis. The United Press correspondent reported:[2]

> Extraordinary precautions were taken to prevent leakage of news from either the Cabinet or the Parliament session before the final passage of the Bill. The gates of Parliament were closed and guarded after the Deputies had entered. . . . Newspapers temporarily were restricted in their movements and foreign correspondents were especially watched. Some were held incommunicado for half an hour or more before being permitted to ask questions of Parliament authorities.

The Law provided that twenty-five petroleum and natural-gas companies, as listed, were to be among those enterprises taken over by the State (Chapter I, Article 1, para 7). The relevant list (No. IV), which was one of a number in which individual enterprises were enumerated, and which was deemed to be an integral part of the Law (Chapter I, Article 1, para 77), included all the Western oil concerns. The shares of these companies passed "free of any charge into the property of the State as common property of the people and were to be administered by the Ministry of Finance". The enterprises themselves, as distinct from the share capital, were to be run by the appropriate ministries—for oil companies, the Ministry of Commerce and Industry. All subsidiaries and installations permanently serving a nationalized enterprise, even though such installations came under different ownership, were declared nationalized, as were organizations possessing several units of which only one or more fell under the terms of the Nationalization Law (Chapter I, Articles 2 and 3). Excepted from these provisions, under Chapter I, Article 5, were "enterprises or that part of their capital which belongs to a state which is a member of the United Nations which acquired such properties as a consequence of the implementation of the peace treaty, or through payment of some of the obligations resulting from a state of war".

[1] 261.   [2] *New York Herald Tribune* (13 June 1948).

This description applied only to Sovrompetrol, which remained outside the scope of the act.

Chapters II and III dealt respectively with the effects and procedure of nationalization. Chapter IV outlined the terms of compensation, establishing those qualified and those disqualified (Articles 11 and 15), and creating (Articles 11 and 12) a "Fund for Nationalized Industry" under the control of the Ministry of Finance as the means whereby the State would discharge its obligation to compensate owners and shareholders of nationalized enterprises by issuing bonds to be redeemed from the net profits of their operations.

Chapter V accorded to the State the sole right to establish new enterprises where existing ones had been nationalized (Article 16), though it could make exceptions to this by granting concessions on the proposal of the appropriate ministry (Article 17). The remaining chapters set out the penalties attached to "those who hinder or try to hinder" the process of nationalization (Chapter VI) and the various arrangements for implementing the Act (Chapter VII).

According to Gheorghiu-Dej,[1] nationalization was designed to secure a political and an economic objective—the removal of "economic arms from the hands of those from whom we have taken political weapons" and to enable a planned national economy to be instituted. However, the circumstances of nationalization suggest a number of motives and present a number of problems which cannot, on present evidence, be conclusively resolved. Nationalization, implying the elimination of private ownership, was and is an essential doctrine of Communism, and the example set by the Soviet Union in 1918 was followed in the other states of Eastern Europe where Communist regimes had been installed after 1944. Its application to Romania, once the Communists were in power, was therefore likely. Nevertheless it remains curious that eighteen months elapsed between that time and the nationalization of Romanian industry. This could have been expected, for example, to have been put through at the time of the nationalization of the National Bank, or, as an emergency measure to deal with the economic crisis, early in 1947. Part of the explanation for delay may be the necessity to "mop up" the opposition in Romania, or to await the entry into force of the peace treaties

[1] Speech in support of Nationalization Bill: *Roumanian News*, 129a (12 June 1948).

(September 1947) in order to minimize foreign complications. The difficulty with the former argument is that nationalization would have served as an instrument to overcome the opposition equally as well as the *oficii industriale* to which the Government resorted. The latter implies a regard for formality conspicuously absent from the regime's general dealings with Western governments. Furthermore, during 1947 Gheorghiu-Dej, as Minister of National Economy, was fully in charge, with the widest possible powers; the managements of the various oil companies, Romanian and other, were sufficiently under control both through government-imposed administrators and because of the general atmosphere of hostility and suspicion generated, for example, by trade union action; the companies were virtually deprived of independence in their operations through government control of exports, and their financial assets were cut down through the application of the monetary reform. The existence of the monarchy presented no obstacle, since the foundation of Sovrompetrol, for example, had demonstrated its legal and administrative irrelevance. The factors adduced by Gheorghiu-Dej for nationalization had already existed for some time before June 1948. The domestic opponents of Communism had been disarmed, and no private interest could impede the setting up of a planned economy.

It appears more likely that the timing of nationalization depended on external factors. Steanu[1] supposes that the Romanian Communists' original policy was to allow private interests to carry the burdens, while the State took over control and direction: nationalization took place in response to *Russian* pressure, since the Russians were concerned at the chaos into which the industry was falling and the consequences for production and the deliveries to the Soviet Union under one form or another. On this reading, nationalization was imposed in order to establish one identifiable authority for the Russians to prod over oil matters and to remove "frictional losses" in the existing machinery which made it insufficiently responsive to Russian demands. Steanu gives no evidence for his view, but it seems plausible, the more so as Gheorghiu-Dej and Maurer were even in 1948 by no means pro-Russian, and may well have improvised the *oficii industriale* as a—temporary—means of squaring their Marxist presuppositions with Romania's advantage, as they saw it, in regard to the

[1] *Journal of Central European Affairs* (Jan.–Apr. 1957), pp. 51–3.

Soviet Union. The haste of the formal act of nationalization on 10 and 11 June would then be explained as their reaction to the collapse of this position.

Alternatively, nationalization represented the liquidation of Western interests, and its timing was a matter both of ensuring the necessary constitutionality (April 1948) under conditions of totally unfettered sovereignty (i.e. since September 1947), and of not alienating a major source of supply for the oil industry until it could safely be dispensed with (December 1947–March 1948). The anti-Western motive was mentioned by Gheorghiu-Dej in the Chamber, and repeated by spokesmen then and subsequently. It is supported by the provision for compensation made in the Nationalization Law and its entailed legislation. Under Articles 11 to 15, a Nationalized Industry Fund was to be created in Bucharest to issue bonds as compensation to dispossessed owners. It was subsequently decided (15 September 1948) that the bonds could be redeemed only out of a quota determined by the Ministry of Finance from the net profit of the nationalized corporations, and they were nominal, not transferable, bore no interest, and could not be pledged. The amount of compensation was, under the Nationalization Law, to be assessed by a commission of three representatives of the Ministry of Justice, against whose decisions there was no appeal. The commission was empowered to reduce its assessments by taking into account liabilities existing before nationalization, as well as deductions for "mismanagement". These provisions referred, of course, to all former owners. Their relevance to the former *foreign* owners was demonstrated by the actions of the Romanian authorities, who immediately after the take-over investigated the managements, inviting denunciations from employees on grounds of sabotage. These led in some instances to trials for espionage or for aiding Anglo-American imperialism.[1] Meanwhile members of managements who were foreign nationals were refused exit visas. One object of these activities was the building up of counter-claims against the former shareholders to set off against the latter's claims for loss of property, rights, and interests.[2] Whatever the predominant motive, the

[1] The A.I.P.R. was dissolved on 17 June, on the grounds that it had acted as an information and espionage agency for the benefit of foreign interests. Its leading officials were tried in 1952.

[2] A counter-claim against Romanofir, the J. & P. Coats subsidiary, was afterwards submitted at a value slightly less than that of the company's assets.

nationalization of the Romanian oil industry was sudden and rigorous. There is no evidence to suggest that purely economic circumstances, other than those of supply advanced by Steanu, had anything to do with it.

The nationalized sector was reorganized under the terms of a decree law of 8 July[1] creating industrial centres as the executive agencies to control the activity of state industrial concerns. In consequence the assets of the former privately owned oil companies were merged in the "petroleum centres" "Muntenia" (12 July)[2] and "Moldova" (27 July),[3] based on the main geographical areas of oil activity. The natural-gas industry was also combined into a "centre", "Gaz Metan". The assets of the former Distribuţia, Unirea, Aragaz, and Prahova[4] organizations were absorbed into "Competrol", the state inland-distribution monopoly for oil products and gas (31 July).[5] The management of these concerns was entrusted to Communist Party nominees—"Muntenia" the largest being put in the charge of a former tram-driver who was Gheorghiu-Dej's brother-in-law. Technical personnel were drafted from the former organizations.

Thus the privately owned foreign-capital sector of the oil industry, whose existence had been essential to the industry's growth, and which at the same time had created a particularly intractable problem for Romanian nationalism, was eliminated—but only in circumstances which got rid of private capitalist enterprise altogether. Sovrompetrol remained to represent a foreign-owned and managed organization within the oil industry. It was, however, from the Romanian point of view, the interest of an allied government whose external support was vital to the disappearance of the traditional foreign-capital sector. The governments of the foreign shareholders whose property was nationalized protested on 7 September[6] on the grounds of violation of the peace treaty (Article 31 (i) and 31 (i) (c)), absence of provision for appeal on compensation, and non-compliance with the requirements of international law regarding compensation. They therefore reserved the rights of their shareholders, demanding either restitution or prompt, adequate, and effective compensation.

[1] *Monitor Oficial*, no. 156 (9 July 1948). [2] Ibid., no. 158 (12 July 1948).
[3] Ibid., no. 171, First part A (27 July 1948).
[4] On Prahova, see above, p. 300.
[5] *Monitor Oficial*, no. 175, First part A (31 July 1948); 262.
[6] Text of U.S. protest: State Department Press Release, 14 Sept. 1948; 263.

Conclusions

The foreign-capital sector of the Romanian oil industry just outlasted the parties which had for so long sought to bring it under Romanian ownership.

Nationalist policy had two main aspects—the creation of large-scale firms as the basis of a national industry, and, as a corollary, the subordination of foreign capital to the position it occupied at the time the programme was enacted (1920–4). The positive element was successful: the companies were brought into existence and established connections in the European oil trade, even though financially their performance, after spectacular beginnings, can hardly have fulfilled their founders' expectations. It can be argued that in the aftermath of the First World War the attempt to build up a "national" industry involved a misdirection of effort. Romania was in great need of capital, and available sums could better have been used on a range of internal projects: foreign capital could have been encouraged into the oil industry on terms dictated by the current notions about a world oil shortage and therefore favourable to Romania. Control could have been exerted, as from the beginning, through the traditional state instruments—legislation and taxation. Such a course might have been more prudent, but from the nationalist standpoint was not enough—partly because the exploitation of oil resources came to signify the nation's advance, and partly because the proceeds were intended to strengthen the hold of the Liberal Party on the Romanian economy. The same political forces which created the "national" concerns also prevented the disappearance of the multiplicity of small businesses. In this sector technical advances and market forces might otherwise have enforced some degree of rationalization.

Between the two World Wars the most important formative influence on the development of the Romanian oil industry was political. After the First World War any Romanian policy was bound to be nationalist, but it is crucial to any assessment of the situation that the effective political party, enjoying a heyday of virtually uninterrupted power, was led by the Brătianus, whose xenophobic tendencies had been stimulated by the experience of the Peace Conference. They were not alone among Romanian political leaders in being ignorant of the technical and economic complexities of the oil business, but they held power and therefore dictated the context of oil-development. Their policies were designed to serve political ends and prevailed over economic and technical solutions to the industry's problems or distorted their application. It was only after the break-up of the Liberal Party machine in the 1930s that any possibility of an agreed solution to current problems in the industry as a whole, particularly on its relations with the State, began to emerge. This was promoted by the financial failure of the "Liberal" companies and the fear of exhaustion of oil resources. At this stage, however, the political factor remained important for reasons of foreign policy. Until 1938 official decisions on Romania's oil represented what the Romanian State wished to do about that particular raw material: after that date Romania had to share these decisions with other powers, Nazi Germany and Soviet Russia, who saw Romanian oil as a component in their respective *Grossräume*.

Because political necessities, whatever they may have entailed, overrode other considerations, the nationalist programme was achieved. This turned on a number of contingencies, both internal and external. *Politica de stat în industria petrolului* could easily have remained, like so much else in this study, a course of action which was discussed but not implemented. The specific opportunity to translate it into practice arose from a combination of two circumstances—that its authors and sponsors were the "pro-Allied" party, and that they came to power when substantial German oil assets were available to provide a basis for a "national" industry. In 1911 even Vintila Brătianu could hardly have foreseen that within eight years one of the largest and best-equipped units of the industry would fall into his lap, with his right to retain it supported by foreign governments. That the restrictive element in the programme succeeded as far as it did was due to the respect for sovereignty shown by the British and U.S. Governments and to the narrowness of the questions at issue. Also Romania was in practice marginal to their oil policies. The French Government never with any consistency succeeded in underpinning the general cultural and ideological relations with Romania by economic arrangements and was experimenting with an oil policy of its own, based on Iraq. On pre-war experience, more forceful intervention could have been expected from the German Government, but there were no German shareholders whose rights needed protecting. The return later of German influence rested on inter-governmental agreement, not on investment.

These were all matters of politics: the understanding of technical and operational factors was, with very few exceptions, not within the grasp of Romanian politicians, who thought of the country's oil as a source of wealth rather than as a part of a world oil industry. This limitation brought about the failure of the restrictive aspects of their programme. It was clearly demonstrated that the international position of Romanian oil as a whole depended on the technical and commercial resources of Standard Oil and Royal Dutch/Shell, whose affiliates in Romania were also the best equipped to overcome the fundamental problems of exploration and production. The demonstration ultimately compelled a revision of nationalist policies and the consequent surrender of the privileged position of the national concerns. Further development was frustrated by war and revolution. The Communist take-over put an end to nationalist oil policies, as considered in this study, although Romanian publicists continue to stress anti-foreign (because anti-Western) themes in their writings.[1] Like their "bourgeois-nationalist" predecessors, they also ignore or underrate the technical and operational considerations governing the investment and behaviour of foreign capital in Romanian oil.

[1] See, e.g., Gh. Ravaş, *Din Istoria Petrolului Romînesc.*

Note on Romanian Export Statistics

THE statistics of Romanian exports as recorded in the customs returns or by the Ministry of Finance and reproduced in the *Moniteur* do not always correspond with those of import countries (or with each other). Apart from minor discrepancies due to leakage and transport losses, divergences between exports from Romania and imports received commonly arise for three main reasons.

(*a*) Variations in customs definitions—so that what left Romania as "gas oil" might appear in the destination as "fuel oil". As specifications became more exacting, this sort of discrepancy declined.

(*b*) The destination shown in the bill of lading (on which official returns were compiled) was not always the ultimate destination, either through the rerouting of the tanker at sea or through shipment to storage for disposal elsewhere. Thus, for example, the destinations Rotterdam, Antwerp, or Vienna, on Romanian showing, would be equivalent to the Netherlands, Belgium, and Austria, though the ultimate destination would often be Germany or France. Similarly a ship directed to "Land's End for orders" appeared in Romanian statistics as an export to the U.K., though cargoes were customarily discharged elsewhere in north-west Europe.

(*c*) Shipments of fuel oil for use as bunker fuel were taken into bond and therefore not recorded in the returns of the country shown as importer in the Romanian statistics. Thus, for example, Spain consistently appeared as the importer of large quantities of fuel oil which in fact were stored at Tenerife for the international bunker trade.

Furthermore, in the 1930s the effective delivery of products shipped from Romania often awaited the solution of difficulties in the inter-state clearing arrangements, pending which the products would be shown as a Romanian export but not as an import into the country of destination.

Glossary of Technical Terms

THE following definitions relate to the text: they are based on *A Glossary of Petroleum Terms* (fourth edition) published by the Institute of Petroleum, to which the reader is referred for a more comprehensive list.

Alkylation. Process in which olefins are combined with either iso-paraffins or aromatics in the presence of a catalyst. The process may be used in the production of high octane gasoline.

Anti-knock value. A measure of the resistance of a fuel to those conditions which cause explosive combustion or detonation (commonly known as "knocking") of the fuel–air mixture in a spark-ignition engine. See Octane number.

Aromatics. (Aromatic hydrocarbons.) Group of hydrocarbons related to benzene, having relatively high specific gravities and good solvent properties: some of them have valuable anti-knock characteristics. They are called "aromatics" because many of them have sweet or aromatic odours. Typical examples are benzene, toluene, xylene.

Aviation gasoline. A specially blended gasoline suitable for use in aircraft engines, characterized by high anti-knock value, high stability, high overall volatility, and low freezing-point.

Base. Term relating to the chemical nature of crude petroleum. Crude oil may be of paraffinic, asphaltic, or mixed base, according to the presence of paraffin wax, bitumen, or both paraffin wax and bitumen in the residue after distillation. Naphthenic base is approximately synonymous with asphaltic base.

Bean. Removable nipple inserted, for example, in the flow line of a well to control oil or gas flow. Sometimes called a choke.

Benzene. Pure aromatic hydrocarbon of ring structure and characteristic odour, which boils at 80 °C. Occurs in small proportions in certain crude oils (but is usually obtained from by-products of the carbonization of coal). See Aromatics and Benzole.

Benzine. Obsolescent term for gasoline. It should not be confused with Benzene.

Benzole. Commercial aromatic fraction from coal tar, consisting mainly of benzene and small quantities of other aromatics.

Bitumen. A viscous liquid, or a solid, consisting essentially of hydrocarbons and their derivatives, which is soluble in carbon disulphide: it is substantially non-volatile and softens gradually when heated. It is black or brown in colour and possesses water-proofing and adhesive properties. It is obtained by refinery processes from petroleum and is also found as a natural deposit or as a component of naturally occurring asphalt, in which it is associated with mineral matter.

Blending. Intimate mixing of the various components in the preparation of a product of specified properties.

Butanes. (Normal butane, Isobutane.) Two gaseous saturated hydrocarbons of different structure containing four carbon atoms. Used for heating purposes as constituents of gasoline to improve volatility and anti-knock value and as raw materials in certain synthetic processes.

Casing. Steel lining of a well used to prevent caving of the sides of a well, to exclude unwanted fluids, and to provide means for the control of well pressures and oil and gas production.

Cetane number. A measure of the ignition quality of a diesel fuel, indicating the relative ease with which the fuel will ignite when injected into the combustion chamber of a compression-ignition engine.

Core. Cylindrical rock sample taken by means of an annular cutter.

Cracking. Process whereby large oil molecules are broken into smaller molecules. This may be accomplished by the use of heat or catalysts.

Dewaxing. Process for removing wax from a petroleum product (usually lubricating oil) in order to improve its flow characteristics at low temperatures.

Diesel fuel. Oil used as fuel in diesel and other compression-ignition engines.

Distillate. Any product obtained by condensing the vapours distilled from petroleum or its products.

Distillation range (boiling range). The range of temperature usually determined at atmospheric pressure and by means of standard apparatus, over which boiling or distillation of a liquid proceeds. Only a pure substance has one definite boiling or distillation temperature at a given pressure; mixtures of substances (such as petroleum distillates) exhibit a distillation range.

Drilling mud. Fluid, commonly consisting of clay suspended in water, circulated through the drill pipe and drilling bit in the rotary method of drilling.

Drilling pipe. Steel pipe used for carrying and rotating the drilling tools in a well and for permitting the circulation of the drilling mud.

Edeleanu process. Process whereby the selective solvent action of liquid sulphur dioxide is used to reduce the sulphur and aromatics content of a petroleum distillate.

Extract. Substance, or substances, separated from petroleum products by the selective action of a solvent in solvent-extraction processes other than dewaxing. Extracts are generally aromatic in character. See Raffinate.

Feedstock. Material to be processed, introduced into a plant.

Fraction. An individual product obtained during the distillation of petroleum or its products.

Fractional distillation (fractionation). Distillation process in which the distillate is collected as a number of separate fractions each having a different boiling-range.

Fuel oil. Heavy distillates, residues, or blends of these, used as fuel for the production of heat or power.

Gas oil. A petroleum distillate having a viscosity and distillation range intermediate between those of kerosine and light lubricating oil. Distillation range normally falls within 200–340 °C. Used as a fuel for high-speed diesel engines, as a burner fuel in heating installations, for carburetting (enriching) water gas, and as a wash oil in the extraction of benzole.

Gasoline. Refined petroleum distillate, normally boiling within the limits of 30 to 220 °C., suitable for use as a fuel in spark-ignition internal-combustion engines.

Gum. The dark-coloured residue which may remain, in small quantities, when all the volatile constituents of certain types of motor fuels have been evaporated under the prescribed conditions.

Hydrogenation. (*a*) Chemical addition of hydrogen to unsaturated compounds, as in the preparation of octanes from the corresponding unsaturated hydrocarbons produced by polymerization. (*b*) Treatment of oil or coal with hydrogen at high temperatures and pressures, whereby cracking and other reactions occur, in addition to true hydrogenation.

Illuminating oil. Oil suitable for use in wick-fed or mantle type lamps.

Inhibitor. Substance added in small proportion to a petroleum product to prevent or retard undesirable changes in the quality of the product, or in the condition of the equipment in which the product is used. The changes which are prevented or retarded are usually oxidation, e.g. gum formation in stored gasolines, colour change in lubricating oils, corrosion in turbines.

Initial boiling point (IBP). The temperature indicated on the distillation thermometer at which the first drop of distillate falls from the condenser during a standard laboratory distillation.

Kerosine. Refined petroleum distillate intermediate in volatility between gasoline and gas oil. Its distillation range normally falls within the limits of 150 to 300 °C. Its main uses are for lighting and heating, and as a fuel for certain types of internal-combustion engines.

Lubricating oil. A petroleum oil, with or without additives, suitable for lubricating purposes.

Mercaptans. Foul-smelling organic sulphur compounds which occur naturally in petroleum.

Naphthenes. Saturated hydrocarbon compounds in which the molecule contains at least one closed ring of carbon atoms.

Naphthenic acids. Organic acids, principally of monocyclic structure, which occur naturally in crude petroleum.

Natural gas (casinghead gas). Gas issuing from the earth under pressure in certain localities, often in association with crude petroleum. Natural gas is usually classified as "wet" or "dry" according as the proportions of gasoline constituents in it are large or small.

Natural gasoline (casinghead gasoline). A low-boiling liquid petroleum product extracted from natural gas. In its "wild" or unstabilized condition, it contains fairly high proportions of propane and butanes. The removal of the propane yields a stabilized gasoline suitable for blending with other gasoline.

Octane number. A measure of the anti-knock quality of a gasoline.

Petroleum. Mineral oil, normally a liquid mixture consisting essentially of many different hydrocarbons, occurring naturally and having a wide range of colours from yellow to black and characteristic odours. It is the raw material from which gasoline, kerosine, lubricating oil, fuel oil, paraffin wax, bitumen, and other products are obtained. In modern technical usage the term includes gaseous and solid as well as liquid hydrocarbons.

Raffinate. Refined product resulting from a solvent extraction process. See Extract.

Tetraethyl lead. Volatile lead compound which, when added in small proportions to gasoline, increases the octane number. For this purpose tetraethyl is used in the form of ethyl fluid.

Toluole. Commercial aromatic fraction from coal tar boiling at approximately 110 °C. and consisting essentially of toluene together with minor quantities of benzene and zylene.

Topped crude. Crude oil from which some of the lighter constituents have been removed by distillation.

Topping plant. Distillation equipment employed for the removal of the more volatile fractions of an oil.

Viscosity. That property of a liquid by virtue of which it offers resistance to motion or flow. Viscosity decreases as temperature rises.

Viscosity index. Arbitrary scale used to show the magnitude of viscosity changes in lubricating oils with changes in temperature.

Volatility. The degree to which a substance tends to vaporize or evaporate.

Wildcat. Well drilled in search of a new oil or gas accumulation.

Bibliography

This list indicates published sources used and consulted in preparing this study, and comprises:

(*a*) Official and semi-official documents: reports, bulletins, and compilations.

(*b*) Journals.

(*c*) Articles and books.

For unpublished material, see note above, p. xvi.

(*a*) OFFICIAL AND SEMI-OFFICIAL DOCUMENTS

ROMANIA

Bulletin de la Banque Nationale de Roumanie, Bucharest.

Central Statistical Office: *Progresele Economei a/c Românei 1866–1906*.

Concession des terrains pétrolifères de l'Etat. Exposé des motifs, loi, règlement, contrat type, tableau des lots, Bucharest, 1906.

Ministry of Agriculture and Domains: *Rumänien 1886–1906*.

Ministry of Foreign Affairs: *Documents diplomatiques: Événements de la péninsule balkanique—Action de la Roumanie 1912/13*, Bucharest, 1913.

Ministry of Industry and Commerce: *Correspondance économique roumaine*.

—— *Roumania before the Peace Conference*, Paris, 1919.

—— *Statistica Minieră a Românei*.

—— *The planned development of the Rumanian People's Republic*, 1950.

Moniteur du pétrole roumain, Bucharest, 1900–48.

Observations concerning the draft Peace Treaty with Roumania, Paris, 1946.

Roumania at the Peace Conference, Paris, 1946.

AUBOIN, R., *Rapport annual sur la situation financière et monétaire de la Roumanie*, Bucharest, 1935.

BRĂTIANU, I., *Activitatea Corpurilor Legiuitoare şi a Guvernului, 1922–6*, Bucharest, 1926.

BRĂTIANU, V., *La Situation financière de la Roumanie*, Bucharest, 1923.

IORGA, N., *Correspondance diplomatique roumaine sous le roi Charles I (1866–1880)*, Paris, 1923.

MADGEARU, V., *La Roumanie à la Conférence de Stresa: memorandum présenté par le gouvernement roumain*, Geneva, 1923.

PETRESCU, G., STURDZA, D. C., and COLESCU, C., *Actes et documents relatifs à l'histoire de la régénération de la Roumanie*, Vartic, Bucharest, 1888–91.

ROSETTI, D. R., *Lois, règlements, décrets concernant les mines et les pétroles*, Bucharest, 1900.

AUSTRIA-HUNGARY

Allgemeines Berggesetz für das Kaiserthum Oesterreich, Vienna, 1854.

Sammlungen der kriegswirtschaftlichen Vereinbarungen mit dem verbündeten Ausland, Vienna, 1918.

BELGIUM

Journal des tribunaux (14 July 1946).

FRANCE

Foreign Office: *Documents diplomatiques français.*

Ministry of National Economy: *Études et conjonctures.*

GERMANY

Die Anglo-Rumänische Zerstörung der rumänischen Erdölindustrie, Berlin, 1918.

"Die rumänische Volkswirtschaft", in *Handbuch, herausgegeben von der Militärverwaltung in Rumänien*, Berlin, 1917.

Die Ursachen des deutschen Zusammenbruchs, Berlin, 1925–8.

Documents on German foreign policy 1918–1945, Series D, H.M.S.O., London.

Dokumente der deutschen Politik und Geschichte, ed. Hohfeld, Berlin, 1951–5.

Nazi conspiracy and aggression, 1946–8.

Nazi–Soviet relations, Washington, 1948.

ITALY

Documenti diplomatici Italiani, Settima Serie, 1922–35. Rome, from 1953.

U.K.

B.I.O.S. Report No. 1016: "The war structure of the German oil industry 1934–45", in *Oil fields investigation*, Part IV, Section I.

British and Foreign State Papers.

Conditions of an armistice with Roumania, Cmd. 6585, 1945.

Department of Overseas Trade: *Economic conditions in Roumania* (Reports issued periodically, 1919–37).

Documents on British Foreign Policy 1919–1939.

Foreign Office: A. P. Bennet, *Roumania: report on the petroleum industry of Roumania*, Misc. Series, no. 23, 1896.

—— Reports in the annual series *Trade and commerce in Roumania*, 1891–1914.

—— *Roumania*: Peace Handbooks, no. 23.

H.M. Petroleum Department: *Report of Anglo-French commission to Roumania*, 23 September 1920.

H.M.S.O.: *Conditions in occupied territories 1942.*

Inter-allied declaration against acts of dispossession, Cmd. 6418, 1943.

Ministry of Fuel and Power: *Report on the petroleum and synthetic oil industry of Germany*, 1947.

U.S.A.

Department of Commerce: *Roumania: an economic handbook*, Special Agents Series, No. 222, Bureau of Foreign and Domestic Commerce, Washington D.C., 1924.

Department of State: *Foreign relations of the United States* (annual series).

Bulletin (bi-monthly).

Also special series:

(i) *The conferences at Malta and Yalta 1945*, 1955.
(ii) *The conference of Berlin 1945*, 2 vols., 1960.
(iii) *The conferences at Cairo and Tehran 1943*, 1961.
(iv) *A decade of American foreign policy 1941–1949.*

—— *The Roumanian petroleum industry in 1921.*

Federal Oil Conservation Board: *Reports*, Washington D.C., 1926 onwards.

U.S. Military Government in Germany: *Dresdner and Deutsche Banks*, 1947.

U.S. Tariff Commission: *Foreign trade and exchange controls in Germany*, 1942.

U.S.S.R.

Soviet documents on foreign policy, ed. J. Degras, London.

Soviet treaty series, ed. L. Shapiro, Washington D.C., 1950 and 1955.

LEAGUE OF NATIONS

Commercial policy in the interwar period, 1942.

Enquiry into clearing agreements, 1935.

Études relatives au problème des rapprochements économiques européens, Series I and II, 1932.

Le Commerce international des pays danubiens après la guerre, May 1937.

Memoir sur le commerce entre la Roumanie et les états danubiens, April 1937.

Memorandum on commercial banks, 1931–2, 1934.

Report on Danube navigation, 1925.

Report on general transport situation in 1921, 1922.

Report on Stresa Conference, 1932.

(*b*) JOURNALS

Le Courrier des pétroles, Paris.

Oel und Kohle, Berlin.

Petroleum, Berlin/Vienna.

Petroleum Press Service, London.

The Petroleum Industrial and Technical Review, London.

The Petroleum Times, London.

(*c*) ARTICLES AND BOOKS

AISINMAN, S., "Bustenari petroleum", *The Petroleum Review* (26 October 1907).

AITKEN, H. G. J. (ed.), *The State and economic growth*, New York, 1959.

ALMOND, N., and LUTZ, R. H., *The Treaty of Trianon*, California, 1935.

AMERICAN BAR ASSOCIATION, *The legal history of conservation of oil and gas*, New York, 1938.

ANNINOS, E. G., *Der Wirtschaftliche Einfluss Deutschlands auf die Petroleum Industrie Rumäniens und ihre Bedeutung für die internationale Wirtschaft*, Darmstadt, 1927.

ANTIPA, A., *L'Occupation ennemie de la Roumanie et ses conséquences sociales*, Paris, 1929.

ANTONUCCI, A., *La Liquidation financière de la guerre et la reconstruction en Europe centrale*, Paris, 1928.

ARMSTRONG, H. F., "Danubia: relief or ruin?", *Foreign Affairs* (July 1932).

ARNDT, H. W., *Economic lessons of the 1930's*, London, 1944.

BAHR, H. "Neue Erdoel Politik im kontinental-europäischen Raum", *Wirtschaftsdienst*, no. 29 (19 July 1940).

BAILEY, E. M., "The oil shales of the Lothians", in *The geological survey for 1927*, D.S.I.R. of Scotland, Edinburgh.

BARBUL, G., *Memorial Antonescu—le troisième homme de l'Axe*, Paris, 1950.

BASCH, A., *The Danube Basin and the German economic sphere*, London, 1944.

BASILESCU, N., *La Roumanie dans la guerre et la paix*, Paris, 1919.

BEATON, K., "Dr. Gesner's kerosene, the start of American oil refining", *The Business History Review* (March 1955).

BEEBY THOMPSON, A., *The oilfields of Russia*, London, 1900.

BÉRANGER, H., *La Pétrole et la France*, Paris, 1919.

—— *La Politique du pétrole*, Paris, 1920.

BERINDEY, A., *La Situation économique et financière de la Roumanie sous l'occupation allemande*, Paris, 1921.

BERKOWITZ, A., "The prisoners of Sighet", *Europeast Bulletin* (June 1962).

BERNARD, F., *Les Finances de la France, de L'Angleterre et de la Roumanie, 1914–24*, Paris, 1925.

BIHEL, A., *Le Pétrole et l'état*, Paris, 1938.

BIRKENFELD, W., *Der synthetische Treibstoff 1933–1945*, Goettingen, 1964.

BLACK, C. E. (ed.), *Readings on contemporary Eastern Europe*, New York, 1952.

BLAISDELL, D. C., *European financial control in the Ottoman Empire*, New York, 1929.

BONESCU, G., "Roumania and foreign investments", *International Communications Review*, VI (1930).

BONNELL, A. T., *German control over international economic relations—1930–40*, Illinois, 1940.

BRACKEL, O. VON, *Rumäniens Staatskredit in deutscher Beleuchtung*, Munich, 1902.

—— and LEIS, J., *Der dreissigjährige Petroleumkrieg*, Berlin, 1903.

BRĂTIANU, V., *Petrolul şi politica de stat*, Bucharest, 1919.

—— *Rapport général la projectul de budget pe 1916–1917*, Bucharest, 1917.

—— *La Situation financière de la Roumanie*, Bucharest, 1923.

—— and HALACEANU, C., *Politica de stat în industria petrolului*, Bucharest, 1911.

BRUIKMANN, C., "Mihai Manoilescu und die klassische Aussenhandelstheorie", *Weltwirtschaftliches Archiv* (September 1938).

BURKS, R. V., *The dynamics of Communism in Eastern Europe*, New Jersey, 1961.

BYRNES, J. F., *Speaking frankly*, London (n.d.).

CAMERON, R. E., *France and the economic development of Europe, 1880–1914*, New Jersey, 1961.

CAMPBELL, J. C., "The influence of Western political thought in the Roumanian principalities 1821–1848", *Journal of Central European Affairs* (October 1944).

CHAMBERLAIN, J. P., *The regime of international rivers: Danube and Rhine*, New York, 1923.

CHURCHILL, W. S., *Triumph and tragedy*, London, 1954.

CIORICEANU, G., *La Dette publique de la Roumanie*, Paris, 1927.

—— *La Roumanie économique 1860–1915*, Paris, 1928.

CIOROGARU, G. J., *Aktuelle wirtschaftspolitische Fragen in Rumaenien*, Heidelberg, 1932.

CIUREA, E. C., "La Vie politique et les partis en Roumanie après la seconde guerre mondiale", *Revue française de science politique* (October/December 1957).

—— *Le Traité de paix avec la Roumanie du 10 février 1947*, Paris, 1954.

COHEN, J. G., *Le Régime des mines et du pétrole en Roumanie*, Paris, 1926.

COHN, E. J., "German enemy property", *International Law Quarterly*, no. 3 (1950).

CONRAD, G. J., *Die Wirtschaft Rumäniens von 1945 bis 1952*, Berlin, 1954.

CREANGA, G. D., *Les Finances roumaines sous le régime d'occupation et de la paix allemande*, Paris, 1919.

CRETZIANU, A., "The Soviet ultimatum to Roumania (26 June 1940)", *Journal of Central European Affairs* (January 1950).

—— "The Rumanian armistice negotiations, Cairo, 1944", ibid. (October 1951).

—— *The lost opportunity*, London, 1957.

—— (ed.), *Captive Rumania*, New York, 1956.

CZERNIN, O., *In the World War*, London, 1920.

DALLIN, A., *Odessa 1941–1944*, California, 1957.

DEAKIN, F. W., *The brutal friendship*, London, 1952.

DELZELL, C. F., "Russian power in Central Eastern Europe", in J. L. Snell (ed.), *The meaning of Yalta*, Baton Rouge, 1956.

DENNET, R., and JOHNSON, J. E., *Negotiating with the Russians*, Boston, 1951.

DEWAR, M., *Soviet trade with Eastern Europe 1945–1949*, London, 1951.

DUNSTAN, A. E., *et al.*, *The science of petroleum*, Oxford, 1953.

EARLE, E. M., *Turkey, the Great Powers and the Bagdad railway*, New York, 1923.

EAST, W. G., *The union of Moldavia and Wallachia 1859*, Oxford, 1929.

EASTERMAN, A. L., *King Carol, Hitler and Lupescu*, London, 1942.

ECONOMIST "Roumanian supplement" (20 July 1929). "*Economist*" *survey of Roumania*, 1930.

L'ÉCONOMISTE EUROPÉEN (Special Supplement), "Le Pétrole roumain et les capitaux français", Paris, 27 July 1928.

EDELEANU, J., *Le Pétrole roumain, sa composition et ses propriétés physiques et techniques*, Bucharest, 1907.

ELLIS, H. S., *Exchange control in Central Europe*, Cambridge, Mass., 1941.

ENOS, J. L., *Petroleum, progress and profits*, Cambridge, Mass., 1962.

ETHRIDGE, M., and BLACK, C. E., "Negotiating on the Balkans, 1945–47", in R. Dennet and J. E. Johnson (eds.), *Negotiating with the Russians*, Boston, 1951.

FEILCHENFELD, E. H., *The international economic law of belligerent occupation* I, Washington, 1942.

FEIS, H., *Petroleum and American policy*, Stanford, 1944.

—— *Europe, the world's banker*, New Haven, 1930.

—— *Churchill, Roosevelt, Stalin*, New Jersey, 1957.

—— *Between war and peace*, New Jersey, 1960.

FISCHER, F., *Griff nach der Weltmacht: die Kriegszielpolitik des kaiserischen Deutschlands*, Düsseldorf, 1961.

FORBES, N., *et al.*, *The Balkans: a history of Bulgaria, Serbia, Greece, Roumania, Turkey*, Oxford, 1915.

FORBES, R. J., *A short history of the art of distillation*, Leiden, 1953.

—— *More studies in early petroleum history*, Leiden, 1959.

FOSTER, N. L., and ROSTOVSKY, D. B., *The Roumanian handbook*, London, 1931.

FRANGES, O. VON, "Die Donaustaaten Südosteuropas und der deutsche Grosswirtschaftsraum", *Weltwirtschaftliches Archiv* (March 1941).

FRIEDENSBURG, F., "Die gegenwärtige und zukünftige Mineralölwirtschaft von Süd-Ost Europa", *Oel und Kohle*, no. 40 (22 October 1940).

—— *Das Erdöl im Weltkrieg*, Stuttgart, 1939.

FUNDATIANU, C. H., *L'Industrie du pétrole en Roumanie*, Paris, 1912.

GAFENCU, G., *The last days of Europe*, London, 1948.

—— *Prelude to the Russian campaign*, London, 1945.

GELBER, N. M., "The intervention of German Jews at the Berlin Congress 1878", in *Year Book V*, Leo Baeck Institute, London, 1960.

GEORGESCU, A., *Le Régime juridique et financière des capitaux étrangers en Roumanie*, Paris, 1932.

GEORGIAN, I., *Bedeutung der rumänischen Ölgewinnung für den Weltmarkt*, Tübingen, 1922.

GERARD, A., "Eastern Europe: vassal or free", *Foreign Affairs* (April 1938).

GERRETSON, F. C., *History of the Royal Dutch*, The Hague, 1955.

GIBB, G. S., and KNOWLTON, E. H., *The resurgent years 1911–1927*, New York, 1956.

GIDDENS, P. H., *The birth of the oil industry*, New York, 1938.

GOLYER, E. de, "Historical notes on the development of the technique of prospecting for petroleum", in *The Science of Petroleum*, vol. i, Oxford, 1938.

GRAF, O., *Die Industriepolitik Alt-Rumäniens und die Grundlagen der Industrialisierung Gross-Rumäniens*, Leipzig, 1926.

GRÄTZ, G., and SCHULLER, R., *The economic policy of Austria-Hungary during the War in its external relations*, New Haven/London, 1928.

GUTTERIDGE, J., "Expropriation and nationalisation in Hungary, Bulgaria and Roumania", *International Comparative Law Quarterly*, no. 1 (1952).

HAASE, F., *Die Erdöl-Interessen der Deutschen Bank und der Direktion der Disconto-Gesellschaft in Rumänien*, Berlin, 1922.

HAJNAL, H., *The Danube: its historical, political and economic importance*, The Hague, 1920.

HALACEANU, C., *Pacea de la Bucareşti şi chestiunea petrolului*, Bucharest, 1919.

HAMILTON, D. C., *Competition in oil*, Cambridge, Mass., 1958.

HARTSHORN, J. E., *Oil companies and governments*, edn. 2, London, 1967.

HEATHCOTE, D., "Brătianu and the new Roumania", *Fortnightly Review*, cxxii (1927), 808–16.

HENRIQUES, R., *Marcus Samuel*, London, 1960.

HERTZ, F., *The economic problem of the Danubian States*, London, 1947.

HERZ, M. F., *Beginnings of the Cold War*, Bloomington, 1966.

HESSE, K., *Kriegführung und Kriegswirtschaft im Feindland*, Hamburg, 1936.

HIDY, R. W., and M. E., *Pioneering in big business*, New York, 1955.

HILLGRUBER, A., *Hitler, König Karol und Marshall Antonescu*, Wiesbaden, 1957.

—— *Hitlers Strategie, Politik und Kriegführung*, Frankfurt-a-M, 1965.

HODSON, H. V., *Slump and recovery*, London, 1938.

HOISESCU, C., *The revision of the economic legislation of the Liberal Party*, (Report to National Peasant Party Congress), Bucharest, 1928.

HOOG, E. VAN DER, *Betrachtungen über die Rentabilität des Ausfuhrhandels mit rumänischen Petroleum-Produkten*, Hilversum, 1935.

HOPPER, S., "The War for Eastern Europe", *Foreign Affairs* (October 1941).

HORNBECK, S. K., "The struggle for petroleum", *Annals of the American Academy of Political and Social Science* (March 1924).

HULL, CORDELL, *The memoirs of Cordell Hull*, London, 1948.

IBBEKEN, R., *Das aussenpolitische Problem: Staat und Wirtschaft in der deutschen Reichspolitik, 1880–1914*, Schleswig, 1928.

IONESCU, G., *Communism in Roumania 1944–1962*, London, 1964.

IONESCU, T., *Les Questions roumaines du temps présent*, Paris, 1921.

IORGA, N., *Histoire des relations entre la France et les Roumains*, Paris, 1918.

—— *Comment la Roumanie s'est détachée de la Triplice*, Bucharest, 1933.

ISE, J., *The United States oil policy*, New Haven, 1926.

JELAVICH, C. and B. (eds.), *The Balkans in transition*, California, 1963.

—— "The Danubian principalities and Bulgaria under Russian protectorship", in *Jahrbücher für die Geschichte Osteuropas*, October 1961.

JEUNE, A. L., *La Politique française du pétrole*, Paris, 1936.

JOHNSON, D. W., "The conquest of Roumania", *Geographical Review* (1917).

JONES, J. M., "Claims on behalf of Nationals who are shareholders in foreign companies", in *The British Year Book of International Law 1949*, London.

JORDAN, C. N., *The Romanian methane gas industry*, New York, 1955.

—— *The Romanian oil industry*, New York, 1955.

KASTRIS, D., *Les Capitaux étrangers dans la finance roumaine*, Paris, 1921.

KERTESZ, W. D. (ed.), *The fate of East Central Europe*, Indiana, 1953.

KEWLEY, J., and JACKSON, J. S., "The burning of mineral oil in wick-fed lamps", *The Journal of the Institution of Petroleum Technologists*, vol. 13, no. 62 (June 1927).

KISSEL, H., *Die Katastrophe in Rumänien*, Darmstadt, 1964.

KNAPTON, E. J., "The duel for Central Europe: some aspects of French diplomacy, 1938–1939", *Journal of Central European Affairs*, vol. ii, no. 1 (April 1942).

KREJCI-GRAF, K., *Die rumänischen Erdöllagerstätten*, Stuttgart, 1929.

KRÖGER, H., "Rumäniens Bedeutung für die europäische Oelversorgung", in *Der Wirtschaftsring*, 13. Jahrgang, Berlin, 1940.

LANDER, W., *Le Pétrole en Pologne*, Caen, 1934.

LAUFS, L., "Das Donau-Versorgungsgebiet", *Oel und Kohle*, no. 40 (20 October 1940).

LAVES, W. H. C., "German governmental influence on foreign investments 1871–1915", *Political Science Quarterly* (December 1928).

LEAGUE OF NATIONS, *Chronique des événements politiques et économiques dans le bassin Danubien 1918–1936: Roumanie*, Paris, 1938.

LEHMAN, J., *Down river: a Danubian study*, London, 1939.

LOGIO, G. C., *Rumania: its history, politics and economics*, Manchester, 1932.

LUDENDORFF, E., *The General Staff and its problems*, London (n.d.).

—— *My war memoirs*, London (n.d.).

LUKACS, J. A., *The Great Powers and Eastern Europe*, New York, 1953.

MACARTNEY, C. A., *October fifteenth*, edn. 2, Edinburgh, 1961.

—— and PALMER, A. W., *Independent Eastern Europe*, London, 1962.

MACHRAY, R., *The Little Entente*, London, 1929.

MACKINTOSH, J. M., *Strategy and tactics of Soviet foreign policy*, Oxford, 1962.

MADGEARU, V., *Evoluţia economiei româneşti după războiul mondial*, Bucharest, 1940.

—— *Roumania's new economic policy*, London, 1930.

—— *Mémoire sur le contrôle des changes en Roumanie*, Paris, 1939.

MAMATEY, B. S., *The United States and East Central Europe*, New Jersey, 1957.

MANN, F. A., "German external assets", in *The British Yearbook of International Law 1947*, London.

MANN, F. K., *Kriegswirtschaft in Rumänien*, Bucharest, 1918.

MANOILESCU, M., 'La Politique d'industrialisation dans la Roumanie', *Annales Économiques*, no. 1, Paris, 1937.

—— "Arbeitsproduktivität und Aussenhandel", *Weltwirtschaftliches Archiv* (July 1936).

MANOLIOU, F. E., *La Reconstruction économique et financière de la Roumanie et les partis politiques*, Paris, 1931.

MARKHAM, R. H., *Rumania under the Soviet yoke*, Boston, 1949.

MARTIN, A., "Private property rights and interests in the Paris peace treaties", in *The British Yearbook of International Law 1947*, London.

MCKENNA, J. C., *Diplomatic protest in foreign policy*, Chicago, 1962.

MEDLICOTT, W. A., *The economic blockade*, London, 1952, 1954/5.

MERGL, G., "Der Weg zur Kapitulation 1940–44", *Osteuropa* (October 1952).

MEYER, H. C., *"Mitteleuropa" in German thought and action 1815–1945*, The Hague, 1955.

MIDDLEMASS, R. K., *The master builders*, London, 1963.

MILENKOVIC, J., *La Question du pétrole et des carburants de remplacement en France*, Paris, 1936.

MILLER, W., and OSBORN, H. G., "History and development of some important phases of petroleum refining in the United States", in *The Science of Petroleum*, Oxford, 1938.

MITRANY, D., "German peaceful penetration of Roumania", *Quarterly Review* (October 1916).

—— "The problem of South East Europe", *Round Table*, no. 93 (1933).

—— "The U.S.S.R. and South Eastern Europe, permanent factors", *International Affairs*, vol. 20, no. 3 (1944).

—— *Introduction to "Economic development in South East Europe"*, Political and Economic Planning, London, 1945.

—— "The new Roumanian Constitution", *Journal of Comparative Legislation and International Law*, 3rd Series (February 1924).

—— *The land and the peasant in Roumania*, London, 1930.

—— *The effect of the War on South-Eastern Europe*, London, 1936.

—— *Roumania: her history and politics*, London, 1915.

MOMTCHILOFF, N., *Ten years of controlled trade in South-Eastern Europe* (National Institute of Economic and Social Research. Occasional Papers, No. 6), London, 1944.

MOSELEY, P. E., "Is Bessarabia next?", *Foreign Affairs* (April 1940).

——"Transylvania partitioned", ibid. (October 1940).

—— "Hitler and South-Eastern Europe", *The Yale Review* (Winter 1939).

NANO, F., "The first Soviet double cross", *Journal of Central European Affairs* (October 1952).

NEUBACHER, H., *Sonderauftrag Südost*, Goettingen, 1957.

NEVINS, A., *Rockefeller*, New York, 1953.

NOVO, J. de, "The Movement for an aggressive American oil policy abroad, 1918–1920", *American Historical Review*, vol. xvi, no. 4 (July 1956).

OBBEKE, K., "Bayern und die rumänische Petroleumindustrie", in *Festschrift für den III. Internationalen Petroleum Kongress*, Berlin, 1907.

OPIE, R. (ed.), *The search for peace settlements*, Brookings Institution, Washington D.C., 1951.

PASVOLSKY, L., *Economic nationalism in the Danubian States*, New York, 1928.

PĂTRĂSCANU, L., *Sous trois dictateurs*, Paris, 1949.

PATRICIU, V., *Problèmes fondamentaux pour l'avenir de notre industrie du pétrole*, Bucharest, 1937.

PAVEL, P., *Why Roumania failed*, London (n.d.).

PIZANTY, M., *Le Pétrole en Roumanie*, Bucharest, 1933.

—— *Le Situation de la Roumanie dans le commerce mondial du pétrole*, Bucharest, 1937.

POHL, G., *Die rumänische Petroleumindustrie während des Krieges 1914–1918*, Berlin, 1926.

POPP, D. C., *La Condition juridique des sociétés étrangères en Roumanie*, Paris, 1929.

PROST, H., "La Roumanie et la seconde guerre mondiale", *Revue d'histoire de la deuxième guerre mondiale* (April 1952).

—— *Destin de la Roumanie*, Paris, 1954.

QUINDRY, S. E., *Bonds and bondholders*, Chicago, 1934.

RAVAS, GH., *Din Istoria Petrolului Romînesc*, Bucharest, 1955.

RAZMIRIȚA, N., *Essai d'économie roumaine moderne 1831–1931*, Paris, 1932.

—— *Les Problèmes actuels de l'industrie petrolière roumaine*, Bucharest, 1929.

REDWOOD, Sir B., "The Russian petroleum industry", *Journal of the Society of Chemical Industry* (28 February 1885).

—— *A treatise on petroleum*, London, 1922.

RICHARD, A. de, *Loi sur les mines*, Bucharest, 1900.

RIEBER, A. J., "Russian diplomacy and Roumania", in H. L. Roberts (ed.), *Russian Diplomacy and Eastern Europe*, New York, 1963.

RITTER, K., "Germany's experience with clearing agreements", *Foreign Affairs* (October 1935).

ROBERTS, H. L., *Rumania: the political problems of an agrarian state*, New Haven, 1951.

ROMMENHOELLER, C. G., *La Grande Roumanie, sa structure économique, sociale, financière, politique et particulièrement, ses richesses*, The Hague, 1926.

ROUCEK, J., *Contemporary Roumania and her problems*, Stanford, 1932.

Roumanian Consolidated Oilfields Ltd., *An account of the destruction of its stocks, wells plants, machinery, etc.*, Claim for Compensation, London, 1917

Royal Institute of International Affairs, *International river and canal transport*, 1944.

—— *The problem of international investments*, 1937.

—— *The Balkan States*, 1936.

—— *South-East Europe*, 1939.

—— *Survey of international affairs*, 1926; 1936; 1938; 1939–46.

—— *The world in March 1939*.

—— *The eve of war 1939*.

—— *The initial triumph of the Axis*.

—— *Hitler's Europe*.

—— *America, Britain and Russia*.

—— *The war and the neutrals*.

—— *The realignment of Europe*.

—— *Four power control in Germany and Austria*.

SACHANEN, A. N., "Hydrocarbons in petroleum", in *The Science of Petroleum*, vol. v, pt. I, Oxford, 1950.

ST. JOHN, R., *Foreign correspondent*, London, 1960.

SCARTZAN, TH. N., *La Production et la consommation du pétrole et ses dérivés dans la Roumanie*, Paris, 1907.

SCHMIDT, P., *Statist auf diplomatischer Bühne*, Bonn, 1949.

SCHWARZ, P., "Die Kapitalsinvestionen in der rumänischen Petroleumindustrie in ihrer geschichtlichen Entwicklung", in *Festschrift für den III. Internationalen Petroleumkongress*, Berlin, 1907.

SCHWARZ, R., *Pétroleum Vademecum*, Berlin, 1933.

SERDARU, V. S., *Le Pétrole Roumain: aperçu historique, économique, politique et législatif, chiffres et interprétations 1825–1920*, Paris, 1921.

SETON-WATSON, H., *Nationalism and Communism*, London, 1964.

SETON-WATSON, R. W., *A history of the Roumanians*, Cambridge, 1934.

SHARP, S. L., *Nationalisation of key industries in Eastern Europe*, New York, 1946.

—— "The Soviet commercial mission in Roumania", *American Perspectives* (January 1948).

SIEBERT, F., *Italiens Weg in den zweiten Weltkrieg*, Frankfurt-a-M, 1962.

SKOGSTAD, A., *Petroleum industry of Germany during the war*, California, 1950.

SMITH, C. J., *Russian struggle for power, 1914–1917*, New York, 1956.

SMITH, H. A., "The Danube", *Yearbook of world affairs, 1950*, London, 1950.

SOBOTKA, F., "Der Anteil Oesterreich-Ungarns an der Militaerverwaltung in Rumaenien 1917–1918", in *Die Militaerverwaltung in den von den Oesterreichisch-Ungarischen Truppen besetzten Gebieten*, Vienna, 1928.

SPECTOR, S. D., *Rumania at the Paris Peace Conference*, New York, 1962.

SPIESS, G., *Die rumaenische Erdoelindustrie und ihre Bedeutung in der Weltwirtschaft*, Bucharest, 1918.

SPULBER, N., *The economics of Communist Eastern Europe*, London, 1957.

STAUSS, E. G. VON, *Bemerkungen zur Petroleumfrage: Denkschrift zum Erdoelabkommen im Bukarester Friedensvertrage*, Berlin, 1918.

STEANU, P. B., "The nationalisation of Rumanian industry", *Journal of Central European Affairs* (January–April 1951).

STURDZA, D. A., *Europa, Russland und Rumaenien*, Berlin, 1915.

—— *La Question du pétrole en Roumanie*, Berlin, 1906.

STURDZA, M., *Avec l'armée roumaine 1915–1918*, Paris, 1918.

SZILASI, B., *Die rumaenische Petroleum-Industrie*, Vienna, 1913.

TANASESCU, I., and TACIT, V., *Exploatarea Petrolului în România*, Bucharest, 1907.

TAUBINGER, L. M. VON, "Die sowjetisch-rumänischen Gesellschaften", *Osteuropa* (April 1956).

TEMPERLEY, H. W., *History of the Peace Conference*, London, 1921.

THÜMEN, K. H. VON, *Die deutsche Mineralölwirtschaft*, Hamburg, 1956.

TILEA, V., *et al.*, *Observations concerning the draft Peace Treaty with Roumania*, Paris, October 1946.

—— *Roumania at the Peace Conference*, Paris, August 1946.

TORREY, G. E., "Rumania and the belligerents 1914–1916", *The Journal of Contemporary History*, vol. 1, no. 3 (July 1966).

TREUE, W., "Das Dritte Reich und die Westmächte auf dem Balkan", *Vierteljahreshefte für Zeitgeschichte*, no. 1, 1953.

TRUMAN, H. S., *Year of decisions*, London, 1955.

—— *Years of trial and hope*, London, 1956.

VILBURG, G., *Die Rumaenische Etappe des Weltkriegs wie ich ihn sah*, Munich, 1930.

WALDECK, R. G., *Athene Palace, Bucarest*, London, 1943.

WANDYCZ, P. S., *France and her Eastern allies 1919–1925*, Minneapolis, 1962.

WARLIMONT, W., *Im Hauptquartier der deutschen Wehrmacht 1939–1945*, Frankfurt, 1962.

WEBER, E., "The men of the Archangel", *Journal of Contemporary History*, vol. 1, no. 1 (1966).

WEBSTER, C., and FRANKLAND, N., *The strategic air offensive against Germany 1939–45*, London, 1961.

WEINBERG, G. L., *Germany and the Soviet Union 1939–1941*, Leiden, 1954.

WEIZSAECKER, E. VON, *Erinnerungen*, Munich, 1950.

WILLIAMSON, H. E., and DAUM, A. R., *The American petroleum industry: the age of illumination 1859–1899*, Evanston, 1959.

WILLIAMSON, H. E., *et al.*, *The American petroleum industry: the age of energy 1899–1959*, Evanston, 1963.

WIRTH, F., *Deutsche Arbeit und deutsches Kapital in der rumänischen Erdölindustrie*, Erkelenz, 1927.

WISKEMANN, E., "The *Drang nach Osten* continues", *Foreign Affairs* (July 1939).

—— *Undeclared war*, London 1939.

WOLBUT, G. S., *American pipelines*, Oklahoma, 1952.

WOLF, J. B., *The diplomatic history of the Bagdad railroad*, Columbia, Ma., 1936.

WOLFERS, A., *Britain and France between two wars*, New York, 1940.

WOLFF, R. L., *The Balkans in our time*, Cambridge, Mass., 1956.

ZAGAROFF, S. D., *The agricultural economy of the Danube countries 1934–1945*, Stanford, 1955.

ZALESKI, E., *Les Courants commerciaux de l'Europe danubienne*, Paris, 1952.

Le développement de l'industrie du pétrole en Roumanie et la participation de la Direction der Disconto-Gesellschaft et S. Bleichröder à ce développement, Bucharest, 1907.

Petroleum development and technology (Transactions of the American Institute of Mining and Metallurgical Engineers), New York, 1933.

Index

PRINTED IN GREAT BRITAIN
AT THE UNIVERSITY PRESS, OXFORD
BY VIVIAN RIDLER
PRINTER TO THE UNIVERSITY